Texts in Applied Mathematics 15

Springer
New York
Berlin
Heidelberg
Hong Kong
London
Milan
Paris
Tokyo

Texts in Applied Mathematics

(continued after index)

Susanne C. Brenner   L. Ridgway Scott

The Mathematical Theory of Finite Element Methods

Second Edition

With 41 Illustrations

 Springer

Susanne C. Brenner
Department of Mathematics
University of South Carolina
Columbia, SC 29208
USA
brenner@math.sc.edu

L. Ridgway Scott
University of Chicago
Chicago, IL 60637
USA
ridg@cs.uchicago.edu

Series Editors

J.E. Marsden
Control and Dynamical Systems, 107-81
California Institute of Technology
Pasadena, CA 91125
USA

L. Sirovich
Division of Applied Mathematics
Brown University
Providence, RI 02912
USA

M. Golubitsky
Department of Mathematics
University of Houston
Houston, TX 77204-3476
USA

S.S. Antman
Department of Mathematics
and
IPST
University of Maryland
College Park, MD 20742-4015
USA

Cover illustration: Flow of a Newtonian fluid at Reynolds number 58, computed with the techniques developed in Chapters 12 and 13.

Mathematics Subject Classification (2000): 65N30, 65-01, 46N40, 65M60, 73V05

Library of Congress Cataloging-in-Publication Data
Brenner, Susanne C.
    The mathematical theory of finite element methods / Susanne C. Brenner, L. Ridgway
Scott.—2nd ed.
        p. ; cm. – (Texts in applied mathematics ; 15)
    Includes bibliographical references and index.
    ISBN 0-387-95451-1 (hc : alk. paper)
        1. Boundary value problems—Numerical solutions.   2. Finite element
method—Mathematics.   I. Scott, L. Ridgway.   II. Title.   III. Series.
    QA379 .B74 2002
    515 .35—dc21   2002019726

Printed on acid-free paper.

Printed in the United States of America.   (MVY)

9 8 7 6 5 4 3 2   SPIN 10983232

Springer-Verlag is a part of *Springer Science+Business Media*

springeronline.com

Series Preface

Mathematics is playing an ever more important role in the physical and biological sciences, provoking a blurring of boundaries between scientific disciplines and a resurgence of interest in the modern as well as the classical techniques of applied mathematics. This renewal of interest, both in research and teaching, has led to the establishment of the series Texts in Applied Mathematics (TAM).

The development of new courses is a natural consequence of a high level of excitement on the research frontier as newer techniques, such as numerical and symbolic computer systems, dynamical systems, and chaos, mix with and reinforce the traditional methods of applied mathematics. Thus, the purpose of this textbook series is to meet the current and future needs of these advances and to encourage the teaching of new courses.

TAM will publish textbooks suitable for use in advanced undergraduate and beginning graduate courses, and will complement the Applied Mathematical Sciences (AMS) series, which will focus on advanced textbooks and research-level monographs.

Pasadena, California	J.E. Marsden
Providence, Rhode Island	L. Sirovich
Houston, Texas	M. Golubitsky
College Park, Maryland	S.S. Antman

Preface to the Second Edition

This edition contains two new chapters. The first one is on the additive Schwarz theory with applications to multilevel and domain decomposition preconditioners, and the second one is an introduction to *a posteriori* error estimators and adaptivity. We have also included a new section on an example of a one-dimensional adaptive mesh, a new section on the discrete Sobolev inequality and new exercises throughout. The list of references has also been expanded and updated.

We take this opportunity to extend thanks to everyone who provided comments and suggestions about this book over the years, and to the National Science Foundation for support. We also wish to thank Achi Dosanjh and the production staff at Springer-Verlag for their patience and care.

Columbia, SC   Susanne C. Brenner
Chicago, IL   L. Ridgway Scott
20/02/2002

Preface to the First Edition

This book develops the basic mathematical theory of the finite element method, the most widely used technique for engineering design and analysis. One purpose of this book is to formalize basic tools that are commonly used by researchers in the field but never published. It is intended primarily for mathematics graduate students and mathematically sophisticated engineers and scientists.

The book has been the basis for graduate-level courses at The University of Michigan, Penn State University and the University of Houston. The prerequisite is only a course in real variables, and even this has not been necessary for well-prepared engineers and scientists in many cases. The book can be used for a course that provides an introduction to basic functional analysis, approximation theory and numerical analysis, while building upon and applying basic techniques of real variable theory.

Chapters 0 through 5 form the essential material for a course. Chapter 0 provides a microcosm of what is to follow, developed in the one-dimensional case. Chapters 1 through 4 provide the basic theory, and Chapter 5 develops basic applications of this theory. From this point, courses can bifurcate in various directions. Chapter 6 provides an introduction to efficient iterative solvers for the linear systems of finite element equations. While essential from a practical point of view (our reason for placing it in a prominent position), this could be skipped, as it is not essential for further chapters. Similarly, Chapter 7, which derives error estimates in the maximum norm and shows how such estimates can be applied to nonlinear problems, can be skipped as desired.

Chapter 8, however, has an essential role in the following chapters. But one could cover only the first and third sections of this chapter and then go on to Chapter 9 in order to see an example of the more complex systems of differential equations that are the norm in applications. Chapter 10 depends to some extent on Chapter 9, and Chapter 11 is essentially a continuation of Chapter 10. Chapter 12 presents Banach space interpolation techniques with applications to convergence results for finite element methods. This is an independent topic at a somewhat more advanced level.

To be more precise, we describe three possible course paths that can be chosen. In all cases, the first step is to follow Chapters 0 through 5. Someone interested to present some of the "hard estimates" of the subject could then choose from Chapters 6 through 8, and 12. On the other hand, someone interested more in physical applications could select from Sect. 8.1, Sect. 8.3 and Chapters 9 through 11. Someone interested primarily in algorithmic efficiency and code development issues could follow Chapters 6, 8, 10 and 11.

The omissions from the book are so numerous that is hard to begin to list them. We attempt to list the most glaring omissions for which there are excellent books available to provide material.

We avoid time-dependent problems almost completely, partly because of the existence of the book of (Thomée 1984). Our extensive development of different types of elements and the corresponding approximation theory is complementary to Thomée's approach. Similarly, our development of physical applications is limited primarily to linear systems in continuum mechanics. More substantial physical applications can be found in the book by (Johnson 1987).

Very little is said here about adaptivity. This active research area is addressed in various conference proceedings (cf. Babuška, Chandra & Flaherty 1983 and Babuška, Zienkiewicz, Gago & de A. Oliveira 1986).

We emphasize the variety of discretizations (that is, different "elements") that can be used, and we present them (whenever possible) as families depending on a parameter (usually the degree of approximation). Thus, a spirit of "high-order" approximations is developed, although we do not consider increasing the degree of approximation (as is done in the so-called P-method and spectral element method) as the means of obtaining convergence. Rather, we focus on mesh subdivision as the convergence parameter. The recent book by (Szabo & Babuška 1991) may be consulted for alternatives in this direction.

Although we provide a brief introduction to mixed methods, the importance of this subject is not appropriately reflected here. However, the recent book by (Brezzi & Fortin 1991) can be consulted for a thorough treatment of the subject.

We draw extensively on the book of (Ciarlet 1978), both following many of its ideas and using it as a reference for further development of various subjects. This book has recently been updated in (Ciarlet & Lions 1991), which also contains an excellent survey of mixed methods. Moreover, the Handbook series to which the latter reference belongs can be expected to provide valuable reference material in the future.

We take this opportunity to thank the many people who have helped at various stages, and in many different ways, in the preparation of this book. The many students who struggled through early drafts of the book made invaluable contributions. Many readers of the preliminary versions will find their specific suggestions incorporated.

This book was processed by the authors using the TEX macro package from Springer-Verlag.

Contents

Chapter 0

Basic Concepts

The finite element method provides a formalism for generating discrete (finite) algorithms for approximating the solutions of differential equations. It should be thought of as a black box into which one puts the differential equation (boundary value problem) and out of which pops an algorithm for approximating the corresponding solutions. Such a task could conceivably be done automatically by a computer, but it necessitates an amount of mathematical skill that today still requires human involvement. The purpose of this book is to help people become adept at working the magic of this black box. The book does *not* focus on how to turn the resulting algorithms into computer codes, although this is at present also a complicated task. The latter is, however, a more well-defined task than the former and thus potentially more amenable to automation.

In this chapter, we present a microcosm of a large fraction of the book, restricted to one-dimensional problems. We leave many loose ends, most of which will be tied up in the theory of Sobolev spaces to be presented in the subsequent chapter. These loose ends should provide motivation and guidance for the study of those spaces.

0.1 Weak Formulation of Boundary Value Problems

Consider the two-point boundary value problem

(0.1.1)
$$-\frac{d^2u}{dx^2} = f \text{ in } (0,1)$$
$$u(0) = 0, \qquad u'(1) = 0.$$

If u is the solution and v is any (sufficiently regular) function such that $v(0) = 0$, then integration by parts yields

(0.1.2)
$$(f, v) := \int_0^1 f(x)v(x)dx = \int_0^1 -u''(x)v(x)dx$$
$$= \int_0^1 u'(x)v'(x)dx =: a(u, v).$$

Let us define (formally, for the moment, since the notion of derivative to be used has not been made precise)

$$V = \{v \in L^2(0,1): \quad a(v,v) < \infty \text{ and } v(0) = 0\}.$$

Then we can say that the solution u to (0.1.1) is characterized by

(0.1.3) $u \in V$ such that $a(u,v) = (f,v)$ $\forall v \in V,$

which is called the *variational* or *weak* formulation of (0.1.1).

The relationship (0.1.3) is called "variational" because the function v is allowed to vary arbitrarily. It may seem somewhat unusual at first; later we will see that it has a natural interpretation in the setting of *Hilbert spaces*. (A Hilbert space is a vector space whose topology is defined using an inner-product.) One example of a Hilbert space is $L^2(0,1)$ with inner-product $(\cdot, \cdot)$. Although it is by no means obvious, we will also see that the space V may be viewed as a Hilbert space with inner-product $a(\cdot, \cdot)$, which was defined in (0.1.2).

One critical question we have not yet dealt with is *what sort of derivative is to be used* in the definition of the bilinear form $a(\cdot, \cdot)$. Should this be the classical derivative

$$u'(x) = \lim_{h \to 0} \frac{u(x+h) - u(x)}{h} \, ?$$

Or should the "almost everywhere" definition valid for functions of bounded variation (BV) be used? We leave this point hanging for the moment and hope this sort of question motivates you to study the following chapter on Sobolev spaces. Of course, the central issue is that (0.1.3) still embodies the original problem (0.1.1). The following theorem verifies this under some simplifying assumptions.

(0.1.4) Theorem. *Suppose* $f \in C^0([0,1])$ *and* $u \in C^2([0,1])$ *satisfy* (0.1.3). *Then* u *solves* (0.1.1).

Proof. Let $v \in V \cap C^1([0,1])$. Then integration by parts gives

(0.1.5) $(f,v) = a(u,v) = \displaystyle\int_0^1 (-u'')v\,dx + u'(1)v(1).$

Thus, $(f - (-u''), v) = 0$ for all $v \in V \cap C^1([0,1])$ such that $v(1) = 0$. Let $w = f + u'' \in C^0([0,1])$. If $w \not\equiv 0$, then $w(x)$ is of one sign in some interval $[x_0, x_1] \subset [0,1]$, with $x_0 < x_1$ (continuity). Choose $v(x) = (x - x_0)^2(x - x_1)^2$ in $[x_0, x_1]$ and $v \equiv 0$ outside $[x_0, x_1]$. But then $(w, v) \neq 0$, which is a contradiction. Thus, $-u'' = f$. Now apply (0.1.5) with $v(x) = x$ to find $u'(1) = 0$. Of course, $u \in V$ implies $u(0) = 0$, so u solves (0.1.1).   □

(0.1.6) *Remark.* The boundary condition $u(0) = 0$ is called *essential* as it appears in the variational formulation explicitly, i.e., in the definition of V.

This type of boundary condition also frequently goes by the proper name "Dirichlet." The boundary condition $u'(1) = 0$ is called *natural* because it is incorporated implicitly. This type of boundary condition is often referred to by the name "Neumann." We summarize the different kinds of boundary conditions encountered so far, together with their various names in the following table:

Table 0.1. Naming conventions for two types of boundary conditions

Boundary Condition	Variational Name	Proper Name
$u(x) = 0$	essential	Dirichlet
$u'(x) = 0$	natural	Neumann

The assumptions $f \in C^0([0,1])$ and $u \in C^2([0,1])$ in the theorem allow (0.1.1) to be interpreted in the usual sense. However, we will see other ways in which to interpret (0.1.1), and indeed the theorem says that the formulation (0.1.3) is a way to interpret it that is valid with much less restrictive assumptions on f. For this reason, (0.1.3) is also called a *weak* formulation of (0.1.1).

0.2 Ritz-Galerkin Approximation

Let $S \subset V$ be any (finite dimensional) subspace. Let us consider (0.1.3) with V replaced by S, namely

(0.2.1) $u_S \in S$ such that $a(u_S, v) = (f, v)$ $\forall v \in S$.

It is remarkable that a discrete scheme for approximating (0.1.1) can be defined so easily. This is only one powerful aspect of the Ritz-Galerkin method. However, we first must see that (0.2.1) does indeed *define* an object. In the process we will indicate how (0.2.1) represents a (square, finite) system of equations for u_S. These will be done in the following theorem and its proof.

(0.2.2) Theorem. *Given $f \in L^2(0,1)$, (0.2.1) has a unique solution.*

Proof. Let us write (0.2.1) in terms of a basis $\{\phi_i : 1 \le i \le n\}$ of S. Let $u_S = \sum_{j=1}^n U_j \phi_j$; let $K_{ij} = a(\phi_j, \phi_i), F_i = (f, \phi_i)$ for $i, j = 1, ..., n$. Set $\mathbf{U} = (U_j), \mathbf{K} = (K_{ij})$ and $\mathbf{F} = (F_i)$. Then (0.2.1) is equivalent to solving the (square) matrix equation

(0.2.3) $\mathbf{KU} = \mathbf{F}$.

For a square system such as (0.2.3) we know that uniqueness is equivalent to existence, as this is a *finite dimensional* system. Nonuniqueness would imply that there is a nonzero $\mathbf{V}$ such that $\mathbf{KV} = \mathbf{0}$. Write $v = \sum V_j \phi_j$ and note that the equivalence of (0.2.1) and (0.2.3) implies that $a(v, \phi_j) = 0$ for all j. Multiplying this by V_j and summing over j yields $0 = a(v, v) = \int_0^1 (v')^2(x) \, dx$, from which we conclude that $v' \equiv 0$. Thus, v is constant, and, since $v \in S \subset V$ implies $v(0) = 0$, we must have $v \equiv 0$. Since $\{\phi_i : 1 \leq i \leq n\}$ is a basis of S, this means that $\mathbf{V} = \mathbf{0}$. Thus, the solution to (0.2.3) must be unique (and hence must exist). Therefore, the solution u_S to (0.2.1) must also exist and be unique. $\square$

(0.2.4) *Remark.* Two subtle points are hidden in the "proof" of Theorem (0.2.2). Why is it that "thus v is constant"? And, moreover, why does $v \in V$ really imply $v(0) = 0$ (even though it is in the definition, i.e., why does the definition make sense)? The first question should worry those familiar with the Cantor function whose derivative is zero almost everywhere, but is certainly not constant (it also vanishes at the left of the interval in typical constructions). Thus, something about our definition of V must rule out such functions as members. V is an example of a *Sobolev* space, and we will see that such problems do not occur in these spaces. It is clear that functions such as the Cantor function should be ruled out (in a systematic way) as candidate solutions for differential equations since it would be a nontrivial solution to the o.d.e. $u' = 0$ with initial condition $u(0) = 0$.

(0.2.5) *Remark.* The matrix $\mathbf{K}$ is often referred to as the *stiffness* matrix, a name coming from corresponding matrices in the context of structural problems. It is clearly symmetric, since the *energy* inner-product $a(\cdot, \cdot)$ is symmetric. It is also *positive definite*, since

$$\sum_{i,j=1}^n k_{ij} v_i v_j = a(v, v) \quad \text{where} \quad v = \sum_{j=1}^n v_j \phi_j.$$

Clearly, $a(v, v) \geq 0$ for all (v_j) and $a(v, v) = 0$ was already "shown" to imply $v \equiv 0$ in the proof of Theorem 0.2.3.

0.3 Error Estimates

Let us begin by observing the fundamental *orthogonality* relation between u and u_S. Subtracting (0.2.1) from (0.1.3) implies

(0.3.1) $a(u - u_S, w) = 0 \quad \forall w \in S.$

Equation (0.3.1) and its subsequent variations are the key to the success of all Ritz-Galerkin/finite-element methods. Now define

$$\|v\|_E = \sqrt{a(v,v)}$$

for all $v \in V$, the energy *norm*. A critical relationship between the energy norm and inner-product is Schwarz' inequality:

$$(0.3.2) \qquad |a(v,w)| \le \|v\|_E \|w\|_E \qquad \forall v, w \in V.$$

This inequality is a cornerstone of Hilbert space theory and will be discussed at length in Sect. 2.1. Then, for any $v \in S$,

$$
\begin{aligned}
\|u - u_S\|_E^2 &= a(u - u_S, u - u_S) \\
&= a(u - u_S, u - v) + a(u - u_S, v - u_S) \\
&= a(u - u_S, u - v) \qquad \text{(from 0.3.1 with } w = v - u_S) \\
&\le \|u - u_S\|_E \|u - v\|_E \qquad \text{(from 0.3.2)}.
\end{aligned}
$$

If $\|u - u_S\|_E \ne 0$, we can divide by it to obtain $\|u - u_S\|_E \le \|u - v\|_E$, for any $v \in S$. If $\|u - u_S\|_E = 0$, this inequality is trivial. Taking the infimum over $v \in S$ yields

$$\|u - u_S\|_E \le \inf\{\|u - v\|_E : v \in S\}.$$

Since $u_S \in S$, we have

$$\inf\{\|u - v\|_E : v \in S\} \le \|u - u_S\|_E.$$

Therefore,

$$\|u - u_S\|_E = \inf\{\|u - v\|_E : v \in S\}.$$

Moreover, there is an element (u_S) for which the infimum is attained, and we indicate this by replacing "infimum" with "minimum." Thus, we have proved the following.

(0.3.3) Theorem. $\|u - u_S\|_E = \min\{\|u - v\|_E : v \in S\}.$

This is the basic error estimate for the Ritz-Galerkin method, and it says that the error is optimal in the energy norm. We will use this later to derive more concrete estimates for the error based on constructing approximations to u in S for particular choices of S. Now we consider the error in another norm.

Define $\|v\| = (v,v)^{\frac{1}{2}} = (\int_0^1 v(x)^2 dx)^{\frac{1}{2}}$, the $L^2(0,1)$-norm. We wish to consider the size of the error $u - u_S$ in this norm. You might guess that the $L^2(0,1)$-norm is weaker than the energy norm, as the latter is the $L^2(0,1)$-norm of the *derivative* (this is the case, on V, although it is not completely obvious and makes use of the essential boundary condition incorporated in V). Thus, the error in the $L^2(0,1)$-norm will be at least comparable with the error measured in the energy norm. In fact, we will find it is considerably smaller.

To estimate $\|u - u_S\|$, we use what is known as a "duality" argument. Let w be the solution of

$$-w'' = u - u_S \quad \text{on } [0, 1] \quad \text{with} \quad w(0) = w'(1) = 0.$$

Integrating by parts, we find

$$
\begin{aligned}
\|u - u_S\|^2 &= (u - u_S, u - u_S) \\
&= (u - u_S, -w'') \\
&= a(u - u_S, w) \quad (\text{since } (u - u_S)(0) = w'(1) = 0) \\
&= a(u - u_S, w - v) \quad\quad\quad\quad\quad (\text{from } 0.3.1)
\end{aligned}
$$

for all $v \in S$. Thus, Schwarz' inequality (0.3.2) implies that

$$
\begin{aligned}
\|u - u_S\| &\le \|u - u_S\|_E \, \|w - v\|_E \, / \|u - u_S\| \\
&= \|u - u_S\|_E \, \|w - v\|_E \, / \|w''\|.
\end{aligned}
$$

We may now take the infimum over $v \in S$ to get

$$\|u - u_S\| \le \|u - u_S\|_E \inf_{v \in S} \|w - v\|_E \, / \|w''\|.$$

Thus, we see that the L^2-norm of the error can be much smaller than the energy norm, provided that w can be approximated well by some function in S. It is reasonable to assume that we can take $v \in S$ close to w, which we formalize in the following *approximation assumption*:

$$(0.3.4) \qquad\qquad \inf_{v \in S} \|w - v\|_E \le \epsilon \|w''\|.$$

Of course, we envisage that this holds with ϵ being a small number. Applying (0.3.4) yields

$$\|u - u_S\| \le \epsilon \|u - u_S\|_E \,,$$

and applying (0.3.4) again, with w replaced by u, and using Theorem 0.3.3 gives

$$\|u - u_S\|_E \le \epsilon \|u''\|.$$

Combining these estimates, and recalling (0.1.1), yields

(0.3.5) Theorem. *Assumption* (0.3.4) *implies that*

$$\|u - u_S\| \le \epsilon \|u - u_S\|_E \le \epsilon^2 \|u''\| = \epsilon^2 \|f\|.$$

The point of course is that $\|u - u_S\|_E$ is of order ϵ whereas $\|u - u_S\|$ is of order ϵ^2. We now consider a family of spaces S for which ϵ may be made arbitrarily small.

0.4 Piecewise Polynomial Spaces – The Finite Element Method

Let $0 = x_0 < x_1 < ... < x_n = 1$ be a partition of $[0,1]$, and let S be the linear space of functions v such that

  i) $v \in C^0([0,1])$
  ii) $v|_{[x_{i-1},x_i]}$ is a linear polynomial, $i = 1, ..., n$, and
  iii) $v(0) = 0$.
We will see later that $S \subset V$. For each $i = 1, .., n$ define ϕ_i by the requirement that $\phi_i(x_j) = \delta_{ij}$ = the Kronecker delta, as shown in Fig. 0.1.

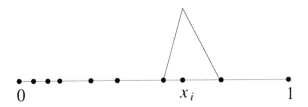

Fig. 0.1. piecewise linear basis function ϕ_i

(0.4.1) Lemma. $\{\phi_i : 1 \le i \le n\}$ *is a basis for* S.

(0.4.2) *Remark.* $\{\phi_i\}$ is called a **nodal** basis for S, and $\{v(x_i)\}$ are the **nodal values** of a function v. (The points $\{x_i\}$ are called the **nodes**.)

Proof. The set $\{\phi_i\}$ is linearly independent since $\sum_{i=1}^{n} c_i \phi_i(x_j) = 0$ implies $c_j = 0$. To see that it spans S, consider the following:

(0.4.3) Definition. *Given* $v \in C^0([0,1])$, *the* **interpolant** $v_I \in S$ *of* v *is determined by* $v_I := \sum_{i=1}^{n} v(x_i)\phi_i$.

Clearly, the set $\{\phi_i\}$ spans S if the following is true.

(0.4.4) Lemma. $v \in S \Rightarrow v = v_I$.

Proof. $v - v_I$ is linear on each $[x_{i-1}, x_i]$ and zero at the endpoints, hence must be identically zero.   □

We will now prove the following approximation theorem for the interpolant.

(0.4.5) Theorem. *Let* $h = \max_{1 \le i \le n} (x_i - x_{i-1})$. *Then*

$$\|u - u_I\|_E \le Ch\|u''\|$$

for all $u \in V$, *where* C *is independent of* h *and* u.

Proof. Recalling the definitions of the two norms, it is clearly sufficient to prove the estimate piecewise, i.e., that

$$\int_{x_{j-1}}^{x_j} (u - u_I)'\,(x)^2\,dx \le c\,(x_j - x_{j-1})^2 \int_{x_{j-1}}^{x_j} u''(x)^2\,dx$$

as the stated result follows by summing over j, with $C = \sqrt{c}$. Let $e = u - u_I$ denote the error; since u_I is a linear polynomial on the interval $[x_{j-1}, x_j]$, the above is equivalent to

$$\int_{x_{j-1}}^{x_j} e'(x)^2\,dx \le c\,(x_j - x_{j-1})^2 \int_{x_{j-1}}^{x_j} e''(x)^2\,dx.$$

Changing variables by an affine mapping of the interval $[x_{j-1}, x_j]$ to the interval $[0, 1]$, we see that this is equivalent to showing

$$\int_0^1 \tilde{e}'(\tilde{x})^2\,d\tilde{x} \le c \int_0^1 \tilde{e}''(\tilde{x})^2\,d\tilde{x},$$

where $x = x_{j-1} + \tilde{x}\,(x_j - x_{j-1})$ and

$$\tilde{e}(\tilde{x}) = e\,(x_{j-1} + \tilde{x}\,(x_j - x_{j-1}))\,.$$

Note that we have arrived at an equivalent estimate that does not involve the mesh size at all. The technique of reducing a mesh-length dependent estimate to a mesh-independent one in this way is called a *homogeneity argument* (or scaling argument) and will be used frequently in Chapter 4 and thereafter.

The verification of the latter estimate is a simple calculus exercise. Let $w = \tilde{e}$ to simplify the notation, and write x for $\tilde{x}$. Note that w vanishes at both ends of the interval (the interpolation error is zero at all nodes). By Rolle's Theorem, $w'(\xi) = 0$ for some ξ satisfying $0 < \xi < 1$. Thus,

$$w'(y) = \int_\xi^y w''(x)\,dx.$$

By Schwarz' inequality,

$$
\begin{aligned}
|w'(y)| &= \left| \int_\xi^y w''(x)\,dx \right| \\[2mm]
&= \left| \int_\xi^y 1 \cdot w''(x)\,dx \right| \\[2mm]
&\le \left| \int_\xi^y 1\,dx \right|^{1/2} \cdot \left| \int_\xi^y w''(x)^2\,dx \right|^{1/2} \\[2mm]
&= |y - \xi|^{1/2} \left| \int_\xi^y w''(x)^2\,dx \right|^{1/2} \\[2mm]
&\le |y - \xi|^{1/2} \left(\int_0^1 w''(x)^2\,dx \right)^{1/2}.
\end{aligned}
$$

(0.4.6)

Squaring and integrating with respect to y completes the verification, with

$$c = \sup_{0 < \xi < 1} \int_0^1 |y - \xi|\, dy = \frac{1}{2}. \qquad \square$$

(0.4.7) Corollary. $\|u - u_S\| + Ch\,\|u - u_S\|_E \le 2\,(Ch)^2\,\|u''\|.$

Proof. Theorem 0.4.5 implies that the approximation assumption (0.3.4) holds with $\epsilon = Ch$. $\qquad \square$

(0.4.8) *Remark.* The interpolant defines a linear operator $\mathcal{I} \colon C^0([0,1]) \to S$ where $\mathcal{I}v = v_I$. Lemma 0.4.4 says that $\mathcal{I}$ is a *projection* (i.e., $\mathcal{I}^2 = \mathcal{I}$). The estimate (0.4.6) for w' in the proof of (0.4.5) is an example of Sobolev's inequality, in which the pointwise values of a function can be estimated in terms of integrated quantities involving its derivatives. Estimates of this type will be considered at length in Chapter 1.

0.5 Relationship to Difference Methods

The stiffness matrix $\mathbf{K}$ as defined in (0.2.3), using the basis $\{\phi_i\}$ described above, can be interpreted as a difference operator. Let $h_i = x_i - x_{i-1}$. Then the matrix entries $K_{ij} = a(\phi_i, \phi_j)$ can be easily calculated to be

$$(0.5.1) \quad K_{ii} = h_i^{-1} + h_{i+1}^{-1}, K_{i,i+1} = K_{i+1,i} = -h_{i+1}^{-1} \qquad (i = 1, ..., n-1)$$

and $K_{nn} = h_n^{-1}$ with the rest of the entries of $\mathbf{K}$ being zero. Similarly, the entries of $\mathbf{F}$ can be approximated if f is sufficiently smooth:

$$(0.5.2) \qquad (f, \phi_i) = \frac{1}{2}(h_i + h_{i+1})(f(x_i) + \mathcal{O}(h))$$

where $h = \max h_i$. (This follows easily from Taylor's Theorem since the integral of ϕ_i is $(h_i + h_{i+1})/2$. Note that the error is *not* $\mathcal{O}(h^2)$ unless $1 - (h_i/h_{i+1}) = \mathcal{O}(h)$.) Thus, the $i - th$ equation of $\mathbf{KU} = \mathbf{F}$ (for $1 \le i \le n - 1$) can be written as

$$(0.5.3) \quad \frac{-2}{h_i + h_{i+1}}\left[\frac{U_{i+1} - U_i}{h_{i+1}} - \frac{U_i - U_{i-1}}{h_i}\right] = \frac{2(f, \phi_i)}{h_i + h_{i+1}} = f(x_i) + \mathcal{O}(h).$$

The difference operator on the left side of this equation can also be seen to be an $\mathcal{O}(h)$ accurate approximation to the differential operator $-d^2/dx^2$ (and *not* $\mathcal{O}(h^2)$ accurate in the usual sense unless $1 - h_i/h_{i+1} = \mathcal{O}(h)$.) For a uniform mesh, the equations reduce to the familiar difference equations

(0.5.4) $$-\frac{U_{i+1} - 2U_i + U_{i-1}}{h^2} = f(x_i) + \mathcal{O}(h^2)$$

which are well known to be second-order accurate. However, for a general
mesh (e.g., $h_i = h$ for i even and $h_i = h/2$ for i odd), we know from Corollary 0.4.7 that the answer is still second-order accurate (in $L^2(0,1)$ at least,
but it will also be proved to be so in the maximum norm in Sect. 0.7), even
though the difference equations are formally only consistent to first order.
This phenomenon has been studied in detail by Spijker (Spijker 1971), and
related work has recently been done by (Kreiss, et.al. 1986). See exercises
0.x.11 through 0.x.15 for more details.

We will take this opportunity to philosophize about some powerful characteristics of the finite element formalism for generating discrete
schemes for approximating the solutions to differential equations. Being
based on the variational formulation of boundary value problems, it is quite
systematic, handling different boundary conditions with ease; one simply re-
places infinite dimensional spaces with finite dimensional subspaces. What
results, as in (0.5.3), is the same as a finite difference equation, in keeping
with the *dictum* that different numerical methods are usually more similar
than they are distinct. However, we were able to derive very quickly the
convergence properties of the finite element method. Finally, the notation
for the discrete scheme is quite compact in the finite element formulation.
This could be utilized to make coding the algorithm much more efficient if
only the appropriate computer language and compiler were available. This
latter characteristic of the finite element method is one that has not yet
been exploited extensively, but an initial attempt has been made in the system `fec` (Bagheri, Scott & Zhang 1992). (One could also argue that finite
element practitioners have already taken advantage of this by developing
their own "languages" through extensive software libraries of their own, but
this applies equally well to the finite-difference practitioners.)

0.6 Computer Implementation of Finite Element Methods

One key to the success of the finite element method, as developed in engineering practice, was the systematic way that computer codes could be im-
plemented. One important step in this process is the *assembly* of the inner-
product $a(u,v)$ by summing its constituent parts over each sub-interval, or
element, which are computed separately. This is facilitated through the use
of a numbering scheme called the *global-to-local* index. This index, $i(e,j)$,
relates the local node number, j, on a particular element, e, to its position
in the global data structure. In our one-dimensional example with piecewise
linear functions, this index is particularly simple: the "elements" are based

on the intervals $I_e := [x_{e-1}, x_e]$ where e is an integer in the range $1, \ldots, n$ and

$$i(e, j) := e + j - 1 \quad \text{for} \quad e = 1, \ldots, n \quad \text{and} \quad j = 0, 1.$$

That is, for each element there are two nodal parameters of interest, one corresponding to the left end of the interval $(j = 0)$ and one at the right $(j = 1)$. Their relationship is represented by the mapping $i(e, j)$.

We may write the interpolant of a continuous function for the space of all piecewise linear functions (no boundary conditions imposed) via

$$(0.6.1) \qquad\qquad f_I := \sum_e \sum_{j=0}^{1} f(x_{i(e,j)}) \phi_j^e$$

where $\{\phi_j^e : j = 0, 1\}$ denotes the set of basis functions for linear functions on the single interval $I_e = [x_{e-1}, x_e]$:

$$\phi_j^e(x) = \phi_j \left((x - x_{e-1})/(x_e - x_{e-1}) \right)$$

where

$$\phi_0(x) := \begin{cases} 1 - x & x \in [0, 1] \\ 0 & \text{otherwise} \end{cases} \quad \text{and} \quad \phi_1(x) := \begin{cases} x & x \in [0, 1] \\ 0 & \text{otherwise.} \end{cases}$$

Note that we have related all of the "local" basis functions ϕ_j^e to a fixed set of basis functions on a "reference" element, $[0, 1]$, via an affine mapping of $[0, 1]$ to $[x_{e-1}, x_e]$. (By definition, the local basis functions, ϕ_j^e, are extended by zero outside the interval I_e.)

The expression (0.6.1) for the interpolant shows (cf. Lemma 0.4.4) that any piecewise linear function f (no boundary conditions imposed) can be written in the form

$$(0.6.2) \qquad\qquad f := \sum_e \sum_{j=0}^{1} f_{i(e,j)} \phi_j^e$$

where $f_i = f(x_i)$ for all i. In particular, the cardinality of the image of the index mapping $i(e, j)$ is the dimension of the space of piecewise linear functions. Note that the expression (0.6.2) represents f incorrectly at the nodal points, but this has no effect on the evaluation of multilinear forms involving integrals of f.

The bilinear forms defined in (0.1.2) can be easily evaluated (assembled) using this representation as well. For example,

$$a(v, w) = \sum_e a_e(v, w)$$

where the "local" bilinear form is defined (and evaluated) via

$$a_e(v, w) := \int_{I_e} v'w' \, dx$$

$$= (x_e - x_{e-1})^{-1} \int_0^1 \left(\Sigma_j v_{i(e,j)} \phi_j \right)' \left(\Sigma_j w_{i(e,j)} \phi_j \right)' \, dx$$

$$= (x_e - x_{e-1})^{-1} \begin{pmatrix} v_{i(e,0)} \\ v_{i(e,1)} \end{pmatrix}^t \mathbf{K} \begin{pmatrix} w_{i(e,0)} \\ w_{i(e,1)} \end{pmatrix}.$$

Here, the *local stiffness matrix*, $\mathbf{K}$, is given by

$$\mathrm{K}_{i,j} := \int_0^1 \phi'_{i-1} \phi'_{j-1} \, dx \quad \text{for} \quad i, j = 1, 2.$$

Note that we have identified the space of piecewise linear functions, v, with the vector space of values, (v_i), at the nodes. The subspace, S, of piecewise linear functions that vanish at $x = 0$, defined in Sect. 0.4, can be identified with the subspace $\{(v_i) : v_0 = 0\}$. Including v_0 in the data structure (with a value of zero) makes the assembly of bilinear forms equally easy in the presence of boundary conditions.

0.7 Local Estimates

We wish to derive estimates for the error, $u - u_S$, in the pointwise sense. As in the case for the L^2-norm, we begin by writing the error that we wish to bound in terms of the energy bilinear form applied to $u - u_S$ and some other function. In this case, this other function is the so-called Green's function for the problem (0.1.1), which in this case is simply

$$g_x(t) := \begin{cases} t & t < x \\ x & \text{otherwise} \end{cases}$$

where x is any point in $[0, 1]$. Integration by parts shows that

$$v(x) = a(v, g_x) \quad \forall v \in V$$

since g''_x is identically zero on either side of x. Therefore,

$$(u - u_S)(x) = a(u - u_S, g_x)$$
$$= a(u - u_S, g_x - v) \quad \forall v \in S.$$

One conclusion is that, if S is the space of piecewise linear functions defined on a partition $\{x_i : i = 1, \dots, n\}$ as in Sect. 0.4, then

$$(u - u_S)(x_i) = 0 \quad \forall i = 1, \dots, n$$

since $g_{x_i} \in S$ in this case. Thus, we conclude that $u_S = u_I$, and a variant of Theorem 0.4.5 yields

(0.7.1) $$\|u - u_I\|_{\max} \leq Ch^2\|u''\|_{\max}.$$

(Recall that $\|f\|_{\max} = \max_{0 \leq x \leq 1} |f(x)|$.) Combining the above estimates, we have proved the following.

(0.7.2) Theorem. *Let u_S be determined by (0.2.1) using the space of piecewise linear functions defined in Sect. 0.4. Then*

$$\|u - u_S\|_{\max} \leq Ch^2\|u''\|_{\max}.$$

Local estimates for higher-dimensional problems are much more difficult to derive, but the use of the Green's function is similar. However, the local character of the singularity of the one-dimensional Green's function disappears, and the distributed nature of the higher-dimensional Green's function requires techniques that are illustrated in the next section.

0.8 Adaptive Approximation

In many cases, the solution to a differential equation is rapidly varying only in restricted regions. For such problems, it makes sense to adapt the mesh to match the variation in the solution. The difference in approximation power between a mesh chosen to solve general problems versus one adapted to a particular one can be substantial. We present a particularly simple approximation problem here to illustrate this effect. For more complex results, see (DeVore, Howard & Micchelli 1989).

Let us consider the problem of approximating functions of one variable whose derivatives are integrable. This is an even weaker condition than what we used in section 0.3, and we wish to consider approximation in a stronger norm, the maximum norm. We consider approximation by the space S_Δ of piecewise constant functions on a partition

(0.8.1) $$\Delta = \{x_0, x_1, \ldots, x_n : 0 = x_0 < x_1 < \cdots < x_n = 1\}.$$

In this case, we will say that $\text{size}(\Delta) = n$. It is not hard to see that the best result of the form

(0.8.2) $$\inf_{v \in S_\Delta} \|u - v\|_{\max} \leq Cn^{-p} \int_0^1 |u'(x)|\, dx$$

to hold for *all* u (with a fixed mesh) is to have $p = 0$. Indeed, whatever the mesh, we can let u go from zero at x_0 to one at x_1 (and stay at one the rest of the interval). This particular u has $\int_0^1 |u'(x)|\, dx = 1$ and yet

$$\inf_{v \in S_\Delta} \|u - v\|_{\max} = \frac{1}{2}.$$

Of course, writing u as the integral of u' (cf. (0.4.6)) allows us to prove (0.8.2) with $C = 1$ and $p = 0$, simply by taking $v \equiv 0$.

On the other hand, suppose that we fix a particular u and ask that (0.8.2) hold for *some* partition Δ as in (0.8.1). That is, what if we are allowed to choose Δ based on properties of u? To be more precise, we are making the distinction between a statement that $\forall u \; \exists \Delta$ such that (0.8.2) holds versus our earlier statement that, given Δ, (0.8.2) holds $\forall u$.

To see that there is a better estimate possible with an adaptively chosen mesh, suppose that we have a u such that $\int_0^1 |u'(x)| \, dx = 1$. The function

$$(0.8.3) \qquad \phi(x) = \int_0^x |u'(t)| \, dt$$

vanishes at $x = 0$ and is a non-decreasing function. Moreover, $\phi(1) = 1$, so there must be a points x_i where $\phi(x_i) = i/n$ and such that $x_i < x_{i+1}$ for all i. If by chance we have $x_n < 1$ in this process, we set $x_n = 1$. One property of this partition is that

$$(0.8.4) \qquad \int_{x_{i-1}}^{x_i} |u'(t)| \, dt = \phi(x_i) - \phi(x_{i-1}) = \frac{1}{n}$$

for all $i = 1, \ldots, n$.

To approximate u on the interval $[x_{i-1}, x_i]$ we use the constant $c_i = u(x_{i-1})$. Then for $x \in [x_{i-1}, x_i]$

$$(0.8.3) \qquad |u(x) - c_i| = \left| \int_{x_{i-1}}^x u'(t) \, dt \right| \leq \int_{x_{i-1}}^{x_i} |u'(t)| \, dt = \frac{1}{n}$$

proving that (0.8.2) holds for all n with $p = 1$ and $C = 1$, at least when $\int_0^1 |u'(x)| \, dx = 1$. In the general case, simply divide everything in (0.8.2) by $\int_0^1 |u'(x)| \, dx$.

Again to get the quantifiers right, let us define the approximation quotient

$$(0.8.5) \qquad Q(u, \Delta) = \inf_{v \in S_\Delta} \|u - v\|_{\max} \Big/ \int_0^1 |u'(x)| \, dx$$

for a given u such that $0 < \int_0^1 |u'(x)| \, dx < \infty$ and a given partition Δ. Then the first result we proved is that

$$(0.8.6) \qquad \forall \Delta \;\; \exists u \;\; \text{such that} \;\; Q(u, \Delta) \geq \frac{1}{2}$$

and yet in the second result we constructed a Δ to prove that

$$(0.8.7) \qquad \forall u \;\; \exists \Delta \text{ with size}(\Delta) = n \;\; \text{such that} \;\; Q(u, \Delta) \leq \frac{1}{n}.$$

These results indicate what a dramatic difference in approximation power there can be in using a fixed mesh versus a mesh adapted to a particular function.

0.9 Weighted Norm Estimates

Suppose $h(x)$ is a function that measures the local mesh size near the point x. In particular, we will assume that h is a piecewise linear function satisfying

$$h(x_j) = h_j + h_{j+1}$$

where $h_j = x_j - x_{j-1}$ (and we set $h_{n+1} = h_n$ and $h_0 = h_1$). Note that for all $j = 1, \dots, n$

(0.9.1) $$h(x) \geq h_j \quad \forall x \in [x_{j-1}, x_j],$$

since this holds at each endpoint of the interval and h is linear between them.

We begin by deriving a basic estimate analogous to (0.4.5). From its proof and (0.9.1), we have

$$
\begin{aligned}
\|u - u_I\|_E^2 &= \sum_{i=1}^{n} \int_{x_{i-1}}^{x_i} (u - u_I)'(x)^2 \, dx \\
&\leq \frac{1}{2} \sum_{i=1}^{n} h_i^2 \int_{x_{i-1}}^{x_i} u''(x)^2 \, dx \\
&\leq \frac{1}{2} \sum_{i=1}^{n} \int_{x_{i-1}}^{x_i} h(x)^2 u''(x)^2 \, dx \\
&= \frac{1}{2} \|hu''\|^2.
\end{aligned}
$$

Therefore,

(0.9.2) $$\|u - u_S\|_E \leq \frac{1}{\sqrt{2}} \|hu''\|.$$

We next derive an L^2 estimate analogous to the first inequality in Theorem 0.3.5. Choosing w as was done in the proof of that result, we find

$$\|u - u_S\|^2 = a(u - u_S, w)$$

where w solves the boundary value problem (0.1.1) with $u - u_S$ as right-hand-side. For simplicity of notation, let $e := u - u_S$. Using the orthogonality relation (0.3.1) and Schwarz' inequality, we find

$$a(e, w) = a(e, w - w_I)$$

$$= \int_0^1 h(u - u_S)'(w - w_I)'/h \, dx$$

$$\leq \left(\int_0^1 (h(u - u_S)')^2 \, dx \right)^{1/2} \left(\int_0^1 ((w - w_I)'/h)^2 \, dx \right)^{1/2}.$$

From the results of Sect. 0.4 we have

$$\int_0^1 ((w - w_I)'(x)/h(x))^2 \, dx = \sum_{i=1}^n \int_{x_{i-1}}^{x_i} ((w - w_I)'(x)/h(x))^2 \, dx$$

$$\leq \sum_{i=1}^n h_i^{-2} \int_{x_{i-1}}^{x_i} (w - w_I)'(x)^2 \, dx$$

$$\leq \sum_{i=1}^n \frac{1}{2} \int_{x_{i-1}}^{x_i} w''(x)^2 \, dx.$$

Combining the previous inequalities, we have

$$(0.9.3) \qquad a(e, w) \leq \|he'\| \left(\sum_{i=1}^n \frac{1}{2} \int_{x_{i-1}}^{x_i} w''(x)^2 \, dx \right)^{1/2}.$$

Recalling that $-w'' = e$, we find

$$\|e\|^2 = a(e, w)$$

$$\leq \frac{1}{\sqrt{2}} \|he'\| \left(\sum_{i=1}^n \int_{x_{i-1}}^{x_i} e(x)^2 \, dx \right)^{1/2}$$

$$= \frac{1}{\sqrt{2}} \|he'\| \, \|e\|.$$

Dividing by $\|e\|$ and recalling that $e = u - u_S$, we have proved

$$(0.9.4) \qquad \|u - u_S\| \leq \frac{1}{\sqrt{2}} \left(\int_0^1 (h(u - u_S)')^2 \, dx \right)^{1/2}.$$

This says that the L^2 error can always be estimated in terms of a weighted integral of the squared derivative error, where the weight is given by the mesh function (0.9.1). Now we proceed to estimate the "weighted energy" norm on the right hand side of (0.9.4).

Let us write $e := u - u_S$ for simplicity. Then first observe that

$$\int_0^1 (h(u - u_S)')^2 \, dx = \|he'\|^2 = a(e, h^2 e) - \int_0^1 2hh'ee' \, dx$$

simply by expanding the expression $a(e, h^2 e)$. We will begin to make the assumption that h' is small, i.e., that the mesh does not change rapidly (for a uniform mesh, $h' \equiv 0$). This will allow us to neglect the term

$$\int_0^1 2hh'ee'\,dx$$

in comparison with the other terms in the preceding equation. To do so, we will make frequent use of the *arithmetic-geometric mean inequality*, which is nothing more than the simple observation that, for any real numbers a and b,

$$ab \le \frac{1}{2}\left(a^2 + b^2\right)$$

(just observe that $0 \le (a-b)^2 = -2ab + a^2 + b^2$). A slightly more complicated version of the inequality comes by writing

$$ab = (\epsilon a)(b/\epsilon) \le \frac{1}{2}\left((\epsilon a)^2 + (b/\epsilon)^2\right).$$

Writing δ in place of ϵ^2, we find

$$(0.9.5) \qquad\qquad ab \le \frac{\delta}{2}a^2 + \frac{1}{2\delta}b^2$$

for any $\delta > 0$.

Let $M := \|h'\|_{\max}$. Then Schwarz' inequality and the arithmetic-geometric mean inequality imply

$$\left|\int_0^1 2hh'ee'\,dx\right| \le 2M \int_0^1 |hee'|\,dx$$
$$\le 2M\|he'\|\,\|e\|$$
$$\le M\left(\|he'\|^2 + \|e\|^2\right).$$

Therefore,

$$\|he'\|^2 \le a(e, h^2 e) + M\left(\|he'\|^2 + \|e\|^2\right)$$

and hence,

$$(1-M)\|he'\|^2 \le a(e, h^2 e) + M\|e\|^2.$$

We now estimate the term $a(e, h^2 e)$. Let $w := h^2 e$. From (0.9.3) and the arithmetic-geometric mean inequality,

$$a(e, h^2 e) = a(e, w)$$
$$\le \frac{1}{\sqrt{2}}\|he'\|\left(\sum_{i=1}^n \int_{x_{i-1}}^{x_i} (w'')^2\,dx\right)^{1/2}$$
$$\le \frac{1-M}{2}\|he'\|^2 + \frac{1}{4(1-M)}\sum_{i=1}^n \int_{x_{i-1}}^{x_i} (w'')^2\,dx$$

which, combined with the previous estimate, implies that

$$\frac{1-M}{2}\|he'\|^2 \le \frac{1}{4(1-M)} \sum_{i=1}^{n} \int_{x_{i-1}}^{x_i} (w'')^2 \, dx + M\|e\|^2.$$

Expanding, we have (on each interval (x_{i-1}, x_i) separately)

$$w'' = h^2 e'' + 4hh'e' + 2\left(h'\right)^2 e$$

since $h'' \equiv 0$. Expanding again, and using the arithmetic-geometric mean inequality, we find

$$\begin{aligned}(w'')^2 \le &h^4 \left(e''\right)^2 + 16M^2 \left(he'\right)^2 + 4M^4 e^2 \\ &+ 8Mh^3|e''||e'| + 4M^2h^2|e''||e| + 16M^3h|e'||e| \\ \le &7h^4 \left(e''\right)^2 + 28M^2 \left(he'\right)^2 + 14M^4 e^2.\end{aligned}$$

Integrating, we find

$$\begin{aligned}\frac{1-M}{2}\|he'\|^2 \le &\frac{1}{4(1-M)} \sum_{i=1}^{n} \int_{x_{i-1}}^{x_i} 7h^4 \left(e''\right)^2 \, dx \\ &+ \frac{7M^2}{1-M}\|he'\|^2 + \left(M + \frac{7M^4}{2(1-M)}\right)\|e\|^2,\end{aligned}$$

which implies

$$\begin{aligned}\left(\frac{1-M}{2} - \frac{7M^2}{1-M}\right)\|he'\|^2 \le &\frac{1}{4(1-M)} \sum_{i=1}^{n} \int_{x_{i-1}}^{x_i} 7h^4 \left(e''\right)^2 \, dx \\ &+ \left(M + \frac{7M^4}{2(1-M)}\right)\|e\|^2.\end{aligned}$$

Letting $c_1 = \left(\frac{1-M}{2} - \frac{7M^2}{1-M}\right)^{-1}$ and recalling that $e'' = u''$, we have

$$\|he'\|^2 \le \frac{7c_1}{4(1-M)}\|h^2 u''\|^2 + c_1 \left(M + \frac{7M^4}{2(1-M)}\right)\|e\|^2$$

provided that

$$M < \frac{1}{1+\sqrt{14}}.$$

Combining with estimate (0.9.4), we find that

$$\left(2 - c_1 \left(M + \frac{7M^4}{2(1-M)}\right)\right)\|u - u_S\|^2 \le \frac{7c_1}{4(1-M)}\|h^2 u''\|^2.$$

Finally, we assume that M is sufficiently small so that

$$2 - c_1 \left(M + \frac{7M^4}{2(1-M)}\right) > 0$$

(observe that $c_1 \to 2$ as $M \to 0$), and we conclude that

$$(0.9.6) \qquad \|u - u_S\|^2 \le C(M)\|h^2 u''\|^2$$

where $C(M) \to 7/4$ as $M \to 0$.

We summarize the above results in the following theorem.

(0.9.7) Theorem. *Without any restrictions on the mesh, we have*

$$\|u - u_S\|_E \le \frac{1}{\sqrt{2}}\|hu''\|$$

and

$$\|u - u_S\| \le \frac{1}{\sqrt{2}}\|h(u - u_S)'\|.$$

Provided that the mesh-size variation, $M := \|h'\|_{\max}$, is sufficiently small, there is a constant, C, depending on M but otherwise independent of the mesh, such that

$$\|u - u_S\| \le C\|h^2 u''\|.$$

The condition that the derivative of h be small is easy to interpret. From its definition,

$$h'|_{(x_{i-1}, x_i)} = \frac{h_{i+1} - h_{i-1}}{h_i} = r_{i+1} - \frac{1}{r_i},$$

where r_i is the ratio of lengths of adjacent mesh intervals, $r_i = h_i/h_{i-1}$. Thus, $|h'|$ is small whenever these ratios are sufficiently close to one. However, this does not preclude strong mesh gradings, e.g., a geometrically graded mesh, $x_i = e^{\delta(i-n)}$ for δ sufficiently small.

0.x Exercises

0.x.1 Verify the expressions (0.5.1) for the "stiffness" matrix $\mathbf{K}$ for piecewise linear functions. If f is piecewise linear, i.e.,

$$f(x) = \sum_{i=1}^{n} f_i \phi_i(x)$$

determine the matrix $\mathbf{M}$ (called the "mass" matrix) such that

$$\mathbf{KU} = \mathbf{MF}.$$

0.x.2 Give weak formulations of modifications of the two-point boundary-value problem (0.1.1) where

a) the o. d. e. is $-u'' + u = f$ instead of $-u'' = f$ and/or
b) the boundary conditions are $u(0) = u(1) = 0$.

0.x.3 Explain what is wrong in both the variational setting and the classical setting for the problem

$$-u'' = f \qquad \text{with } u'(0) = u'(1) = 0.$$

That is, explain in both contexts why this problem is not well-posed.

0.x.4 Show that piecewise *quadratics* have a nodal basis consisting of values at the nodes x_i together with the midpoints $\frac{1}{2}(x_i + x_{i+1})$. Calculate the stiffness matrix for these elements.

0.x.5 Verify (0.5.2).

0.x.6 Under the same assumptions as in Theorem 0.4.5, prove that

$$\|u - u_I\| \le Ch^2 \|u''\|.$$

(Hint: use a homogeneity argument as in the proof of Theorem 0.4.5. Using the notation of that proof, show further that

$$\int_0^1 w(x)^2 \, dx \le \tilde{c} \int_0^1 w'(x)^2 \, dx,$$

by utilizing the fact that $w(0) = 0$. How small can you make $\tilde{c}$ if you use both $w(0) = 0$ and $w(1) = 0$?)

0.x.7 Using only Theorems 0.3.5 and 0.4.5, prove that

$$\inf_{v \in S} \|u - v\| \le Ch^2 \|u''\|.$$

Exercise 0.x.6 also would imply this result independently. Compare the different constants, C, derived with the different approaches.

0.x.8 Prove that (0.1.1) has a solution $u \in C^2([0,1])$ provided $f \in C^0([0,1])$. (Hint: write

$$u(x) = \int_0^x \left(\int_s^1 f(t) \, dt \right) ds$$

and verify the equations.)

0.x.9 Let V denote the space, and $a(\cdot, \cdot)$ the bilinear form, defined in Sect. 0.1. Prove the following *coercivity* result

$$\|v\|^2 + \|v'\|^2 \le Ca(v, v) \quad \forall v \in V.$$

Give a value for C. (Hint: see the hint in exercise 0.x.6. For simplicity, restrict the result to $v \in V \cap C^1(0,1)$.)

0.x.10 Let V denote the space, and $a(\cdot,\cdot)$ the bilinear form, defined in Sect. 0.1. Prove the following version of Sobolev's inequality:

$$\|v\|_{\max}^2 \leq Ca(v,v) \quad \forall v \in V.$$

Give a value for C. (Hint: see the hint in exercise 0.x.6. For simplicity, restrict the result to $v \in V \cap C^1(0,1)$.)

0.x.11 Consider the difference method represented by (0.5.3), namely

$$\frac{-2}{h_i + h_{i+1}} \left(\frac{U_{i+1} - U_i}{h_{i+1}} - \frac{U_i - U_{i-1}}{h_i} \right) = f(x_i).$$

Prove $\tilde{u}_S := \sum U_i \phi_i$ satisfies the following modification to (0.2.1):

$$a(\tilde{u}_S, v) = Q(fv) \quad \forall v \in S$$

where $a(\cdot,\cdot)$ is the bilinear form defined in Sect. 0.1, S consists of piecewise linears as defined in Sect. 0.4 and Q denotes the quadrature approximation based on the trapezoidal rule

$$Q(w) := \sum_{i=0}^{n} \frac{h_i + h_{i+1}}{2} w(x_i).$$

Here ϕ_i, x_i and h_i are as defined in Sect. 0.4; we further define $h_0 = h_{n+1} = 0$ for simplicity of notation.

0.x.12 Let Q be defined as in exercise 0.x.11. Prove that

$$\left| Q(w) - \int_0^1 w(x)\,dx \right| \leq Ch^2 \sum_{i=1}^{n} \int_{x_{i-1}}^{x_i} |w''(x)|\,dx.$$

(Hint: observe that the trapezoidal rule is exact for piecewise linears and refer to the hint in exercise 0.x.6.)

0.x.13 Let u_S solve (0.2.1) where S consists of piecewise linears as defined in Sect. 0.4 and let $\tilde{u}_S$ be as in exercise 0.x.11. Prove that

$$|a(u_S - \tilde{u}_S, v)| \leq Ch^2 \left(\|f'\| + \|f''\| \right) \left(\|v\| + \|v'\| \right) \quad \forall v \in S.$$

(Hint: apply exercise 0.x.12 and Schwarz' inequality.)

0.x.14 Let u_S and $\tilde{u}_S$ be as in exercise 0.x.13. Prove that

$$\|u_S - \tilde{u}_S\|_E \leq Ch^2 \left(\|f'\| + \|f''\| \right).$$

(Hint: apply exercise 0.x.13, pick $v = u_S - \tilde{u}_S$ and apply exercise 0.x.9.)

0.x.15 Let $\tilde{u}_S$ be as in exercise 0.x.11 and let u solve (0.1.1). Prove that

$$\|u - \tilde{u}_S\|_{\max} \leq Ch^2 \left(\|f\|_{\max} + \|f'\| + \|f''\| \right).$$

(Hint: apply exercise 0.x.14 and Theorem 0.7.2.)

Chapter 1

Sobolev Spaces

This chapter is devoted to developing function spaces that are used in the variational formulation of differential equations. We begin with a review of Lebesgue integration theory, upon which our notion of "variational" or "weak" derivative rests. Functions with such "generalized" derivatives make up the spaces commonly referred to as Sobolev spaces. We develop only a small fraction of the known theory for these spaces—just enough to establish a foundation for the finite element method.

1.1 Review of Lebesgue Integration Theory

We will now review the basic concepts of Lebesgue integration theory, cf. (Halmos 1991), (Royden 1988) or (Rudin 1987). By "domain" we mean a Lebesgue-measurable (usually either open or closed) subset of $\mathbb{R}^n$ with non-empty interior. We restrict our attention for simplicity to real-valued functions, f, on a given domain, Ω, that are Lebesgue measurable; by

$$\int_\Omega f(x)\,dx$$

we denote the Lebesgue integral of f (dx denotes Lebesgue measure). For $1 \le p < \infty$, let

$$\|f\|_{L^p(\Omega)} := \left(\int_\Omega |f(x)|^p\,dx \right)^{1/p},$$

and for the case $p = \infty$ set

$$\|f\|_{L^\infty(\Omega)} := \text{ess sup}\,\{|f(x)| \, : \, x \in \Omega\}.$$

In either case, we define the *Lebesgue spaces*

(1.1.1) $$L^p(\Omega) := \{f \, : \, \|f\|_{L^p(\Omega)} < \infty\}.$$

To avoid trivial differences between functions, we identify two functions, f and g, that satisfy $\|f - g\|_{L^p(\Omega)} = 0$. For example, take $n = 1$, $\Omega = [-1, 1]$ and

$$(1.1.2) \qquad f(x) := \begin{cases} 1 & x \geq 0 \\ 0 & x < 0 \end{cases} \quad \text{and} \quad g(x) := \begin{cases} 1 & x > 0 \\ 0 & x \leq 0. \end{cases}$$

Since f and g differ only on a set of measure zero (one point, in this case), we view them as representing the same function. With a small ambiguity of notation, we then think of $L^p(\Omega)$ as a set of *equivalence classes* of functions with respect to this identification. There are some famous (and useful) inequalities that hold for the functionals defined above:

(1.1.3) **Minkowski's Inequality**   For $1 \leq p \leq \infty$ and $f, g \in L^p(\Omega)$, we have

$$\|f + g\|_{L^p(\Omega)} \leq \|f\|_{L^p(\Omega)} + \|g\|_{L^p(\Omega)}.$$

(1.1.4) **Hölder's Inequality**   For $1 \leq p, q \leq \infty$ such that $1 = 1/p + 1/q$, if $f \in L^p(\Omega)$ and $g \in L^q(\Omega)$, then $f\,g \in L^1(\Omega)$ and

$$\|f\,g\|_{L^1(\Omega)} \leq \|f\|_{L^p(\Omega)} \|g\|_{L^q(\Omega)}.$$

(1.1.5) **Schwarz' Inequality**   This is simply *Hölder's inequality in the special case* $p = q = 2$, viz. if $f, g \in L^2(\Omega)$ then $f\,g \in L^1(\Omega)$ and

$$\int_\Omega |f(x)g(x)|\, dx \leq \|f\|_{L^2(\Omega)} \|g\|_{L^2(\Omega)}.$$

In view of Minkowski's inequality and the definitions of $\|\cdot\|_{L^p(\Omega)}$, the space $L^p(\Omega)$ is closed under linear combinations, i.e., it is a *linear* (or *vector*) space. Moreover, the functionals $\|\cdot\|_{L^p(\Omega)}$ have properties that classify them as *norms*.

(1.1.6) Definition. *Given a linear (vector) space V, a **norm**, $\|\cdot\|$, is a function on V with values in the non-negative reals having the following properties:*

$$
\begin{aligned}
i) \quad & \|v\| \geq 0 \quad \forall v \in V \\
& \|v\| = 0 \iff v = 0 \\
ii) \quad & \|c \cdot v\| = |c| \cdot \|v\| \quad \forall c \in \mathbb{R}, v \in V, \quad \text{and} \\
iii) \quad & \|v + w\| \leq \|v\| + \|w\| \quad \forall\, v, w \in V \quad \textbf{(the triangle inequality)}.
\end{aligned}
$$

A norm, $\|\cdot\|$, can be used to define a notion of distance, or *metric*, $d(v, w) = \|v - w\|$ for points $v, w \in V$. A vector space endowed with the topology induced by this metric is called a *normed linear space*. Recall that a metric space, V, is called *complete* if every *Cauchy* sequence $\{v_j\}$ of elements of V has a limit $v \in V$. For a normed linear space, a Cauchy sequence is one such that $\|v_j - v_k\| \to 0$ as $j, k \to \infty$, and completeness means that $\|v - v_j\| \to 0$ as $j \to \infty$. The following definition encapsulates some key features of linear spaces of infinite dimensions needed for theoretical development.

(1.1.7) Definition. *A normed linear space* $(V, \|\cdot\|)$ *is called a* **Banach space** *if it is complete with respect to the metric induced by the norm,* $\|\cdot\|$.

(1.1.8) Theorem. *For* $1 \leq p \leq \infty$, $L^p(\Omega)$ *is a Banach space.*

This theorem (whose proof may be found in the references at the beginning of this section) is a cornerstone of Lebesgue integration theory. Note that it incorporates both Minkowski's inequality and a limit theorem for the Lebesgue integral; that is, if $f_j \to f$ in $L^p(\Omega)$ then (cf. exercise 1.x.3)

$$(1.1.9) \qquad \int_\Omega |f_j(x)|^p \, dx \to \int_\Omega |f(x)|^p \, dx \quad \text{as} \quad j \to \infty.$$

However, Hölder's inequality and more subtle limit theorems are not reflected in this characterization of the Lebesgue spaces.

A key reason that the Lebesgue integral is preferred over the Riemann integral is the aspect of "completeness" that it enjoys, i.e., that appropriate limits of integrable functions are integrable, a property that the Riemann integral does not have. For example, we can easily evaluate an improper integral to determine that the function $\log x$ has a finite integral on any finite interval of the form $[0, a]$. Correspondingly, if $\{r_n : n = 1, 2, \ldots\}$ is dense in the interval $[0, 1]$, then the functions

$$(1.1.10) \qquad f_j(x) := \sum_{n=1}^{j} 2^{-n} \log |x - r_n|$$

all have improper integrals, and one easily sees that

$$(1.1.11) \qquad \left| \int_0^1 f_j(x) \, dx \right| \leq 2 \int_0^1 |\log x| \, dx.$$

Therefore, the "limit" function

$$(1.1.12) \qquad f(x) := \sum_{n=1}^{\infty} 2^{-n} \log |x - r_n|$$

should have a finite "integral" on $[0, 1]$, again satisfying

$$(1.1.13) \qquad \left| \int_0^1 f(x) \, dx \right| \leq 2 \int_0^1 |\log x| \, dx.$$

However, f is infinite at some point in any open sub-interval of $[0, 1]$ and so it is not Riemann integrable on any sub-interval of $[0, 1]$. Thus, it is not possible, even via "improper" Riemann integrals, to determine if f has a finite integral. On the other hand, one can show (see exercise 1.x.6) that it is Lebesgue integrable and that (1.1.13) holds.

1.2 Generalized (Weak) Derivatives

There are several definitions of derivative that are useful in different situations. The "calculus" definition, viz.

$$u'(x) = \lim_{h \to 0} \frac{u(x+h) - u(x)}{h},$$

is a "local" definition, involving information about the function u only near the point x. The variational formulation developed in Chapter 0 takes a more global view, because pointwise values of derivatives are not needed; only derivatives that can be interpreted as functions in the Lebesgue space $L^2(\Omega)$ occur. In the previous section, we have seen that pointwise values of functions in Lebesgue spaces are irrelevant (cf. (1.1.2)); a function in one of these spaces is determined only by its global behavior. Thus, it is natural to develop a global notion of *derivative* more suited to the Lebesgue spaces. We do so using a "duality" technique, defining derivatives for a class of not-so-smooth functions (see Definition 1.2.3) by comparing them with very-very-smooth functions (introduced in Definition 1.2.1).

First, let us introduce some short-hand notation for (calculus) partial derivatives, the *multi-index* notation. A multi-index, α, is an n-tuple of non-negative integers, α_i. The length of α is given by

$$|\alpha| := \sum_{i=1}^{n} \alpha_i.$$

For $\phi \in C^\infty$, denote by

$$D^\alpha \phi, \quad D^\alpha_x \phi, \quad \left(\frac{\partial}{\partial x}\right)^\alpha \phi, \quad \phi^{(\alpha)}, \quad \text{and} \quad \partial^\alpha_x \phi$$

the usual (pointwise) partial derivative

$$\left(\frac{\partial}{\partial x_1}\right)^{\alpha_1} \cdots \left(\frac{\partial}{\partial x_n}\right)^{\alpha_n} \phi.$$

Given a vector $(x_1, \ldots, x_n)$, we define $x^\alpha := x_1^{\alpha_1} \cdot x_2^{\alpha_2} \cdots x_n^{\alpha_n}$. Note that if x is replaced formally by the symbol $\frac{\partial}{\partial x} := \left(\frac{\partial}{\partial x_1}, \ldots, \frac{\partial}{\partial x_n}\right)$, then this definition of x^α is consistent with the previous definition of $\left(\frac{\partial}{\partial x}\right)^\alpha$. Note that the *order* of this derivative is given by $|\alpha|$.

Next, let us introduce the concept of the *support* of a function defined on some domain in $\mathbb{R}^n$. For a continuous function, u, this is the closure of the (open) set $\{x : u(x) \neq 0\}$. If this is a compact set (i.e., if it is bounded) and it is a subset of the *interior* of a set, Ω, then u is said to have "compact support" with respect to Ω. (Outside the support of a function, it is natural to define it to be zero, thus extending it to be defined on all of $\mathbb{R}^n$.) When

Ω is a bounded set, it is equivalent to say that u vanishes in a neighborhood of $\partial\Omega$.

(1.2.1) Definition. *Let Ω be a domain in $\mathbb{R}^n$. Denote by $\mathcal{D}(\Omega)$ or $C_0^\infty(\Omega)$ the set of $C^\infty(\Omega)$ functions with compact support in Ω.*

Before proceeding any further, it would be wise to verify that we have not just introduced a vacuous definition, which we do in the following:

(1.2.2) Example. Define

$$\phi(x) := \begin{cases} e^{1/(|x|^2-1)} & |x| < 1 \\ 0 & |x| \geq 1 . \end{cases}$$

We claim that, for any multi-index α, $\phi^{(\alpha)}(x) = P_\alpha(x)\phi(x)/(1 - |x|^2)^{|\alpha|}$ for some polynomial P_α, as we now show. For $|x| < 1$, we can differentiate and determine, inductively in α, that $\phi^{(\alpha)}(x) = P_\alpha(x)e^{-t}t^{|\alpha|}$ for some polynomial P_α, where $t = 1/(1 - |x|^2)$. Further, $\phi^{(\alpha)}(x) = 0$ for $|x| > 1$. Thus, the formula above for $\phi^{(\alpha)}$ is verified in the case $|x| \neq 1$. Since the exponential increases faster than any finite power, $\phi^{(\alpha)}(x)/\left(1 - |x|^2\right)^k = P_\alpha(x)t^{|\alpha|+k}/e^t \to 0$ as $|x| \to 1$ (i.e., as $t \to \infty$) for any integer k. Applying (inductively) these facts with $k = 0$ shows that $\phi^{(\alpha)}$ is continuous at $|x| = 1$, and using $k = 1$ shows it is also differentiable there, and has derivative zero. Thus, the claimed formula holds for all x. Moreover, we also see from the argument that $\phi^{(\alpha)}$ is bounded and continuous for all α. Thus, $\phi \in \mathcal{D}(\Omega)$ for any open set Ω containing the closed unit ball. By scaling variables appropriately, we see that $\mathcal{D}(\Omega) \neq \emptyset$ for any Ω with non-empty interior.

We now use the space $\mathcal{D}$ to extend the notion of pointwise derivative to a class of functions larger than C^∞. For simplicity, we restrict our notion of derivatives to the following space of functions (see (Schwartz 1957) for a more general definition).

(1.2.3) Definition. *Given a domain Ω, the set of **locally integrable** functions is denoted by*

$$L^1_{loc}(\Omega) := \left\{ f \; : \; f \in L^1(K) \quad \forall \text{ compact } K \subset \text{interior } \Omega \right\}.$$

Functions in $L^1_{loc}(\Omega)$ can behave arbitrarily badly near the boundary, e.g., the function $e^{e^{1/\text{dist}(x,\partial\Omega)}} \in L^1_{loc}(\Omega)$, although this aspect is somewhat tangential to our use of the space. One notational convenience is that $L^1_{loc}(\Omega)$ contains all of $C^0(\Omega)$, without growth restrictions. Finally, we come to our new definition of derivative.

(1.2.4) Definition. *We say that a given function* $f \in L^1_{loc}(\Omega)$ *has a* **weak derivative,** $D^\alpha_w f$, *provided there exists a function* $g \in L^1_{loc}(\Omega)$ *such that*

$$\int_\Omega g(x)\phi(x)\, dx = (-1)^{|\alpha|} \int_\Omega f(x)\phi^{(\alpha)}(x)\, dx \quad \forall \phi \in \mathcal{D}(\Omega).$$

If such a g *exists, we define* $D^\alpha_w f = g$.

(1.2.5) Example. Take $n = 1$, $\Omega = [-1, 1]$, and $f(x) = 1 - |x|$. We claim that $D^1_w f$ exists and is given by

$$g(x) := \begin{cases} 1 & x < 0 \\ -1 & x > 0. \end{cases}$$

To see this, we break the interval $[-1, 1]$ into the two parts in which f is smooth, and we integrate by parts. Let $\phi \in \mathcal{D}(\Omega)$. Then

$$\int_{-1}^1 f(x)\phi'(x)\, dx = \int_{-1}^0 f(x)\phi'(x)\, dx + \int_0^1 f(x)\phi'(x)\, dx$$

$$= -\int_{-1}^0 (+1)\phi(x)\, dx + f\phi|^0_{-1} - \int_0^1 (-1)\phi(x)\, dx + f\phi|^1_0$$

$$= -\int_{-1}^1 g(x)\phi(x)\, dx + (f\phi)(0-) - (f\phi)(0+)$$

$$= -\int_{-1}^1 g(x)\phi(x)\, dx$$

because f is continuous at 0. One may check (cf. exercise 1.x.10) that $D^j_w f$ does not exist for $j > 1$.

One can see that, roughly speaking, the new definition of derivative is the same as the old one wherever the function being differentiated is regular enough. In particular, continuity of f in the example was enough to insure existence of a first-order weak derivative, but not second-order. This phenomenon depends on the dimension n as well, precluding a simple characterization of the relation between the calculus and weak derivatives, as the following example shows.

(1.2.6) Example. Let ρ be a smooth function defined for $0 < r \leq 1$ satisfying

$$\int_0^1 |\rho'(r)| r^{n-1}\, dr < \infty.$$

Define f on $\Omega = \{x \in \mathbb{R}^n : |x| < 1\}$ via $f(x) = \rho(|x|)$. Then $D^\alpha_w f$ exists for all $|\alpha| = 1$ and is given by

$$g(x) = \rho'(|x|)x^\alpha/|x|.$$

The verification of this is left to the reader in exercise 1.x.12.

This example shows that the relationship between the calculus and weak derivatives depends on dimension. That is, whether a function such as $|x|^r$ has a weak derivative depends not only on r but also on n (cf. exercise 1.x.13). However, the following fact (whose proof is left as an exercise) shows that the latter is a generalization of the former.

(1.2.7) Proposition. *Let α be arbitrary and let $\psi \in C^{|\alpha|}(\Omega)$. Then the weak derivative $D_w^\alpha \psi$ exists and is given by $D^\alpha \psi$.*

As a consequence of this proposition, we ignore the differences in definition of D and D_w from now on. That is, differentiation symbols will refer to weak derivatives in general, but we will also use classical properties of derivatives of smooth functions as appropriate.

1.3 Sobolev Norms and Associated Spaces

Using the notion of weak derivative, we can generalize the Lebesgue norms and spaces to include derivatives.

(1.3.1) Definition. *Let k be a non-negative integer, and let $f \in L^1_{loc}(\Omega)$. Suppose that the weak derivatives $D_w^\alpha f$ exist for all $|\alpha| \leq k$. Define the* **Sobolev norm**

$$\|f\|_{W_p^k(\Omega)} := \left(\sum_{|\alpha| \leq k} \|D_w^\alpha f\|^p_{L^p(\Omega)} \right)^{1/p}$$

in the case $1 \leq p < \infty$, and in the case $p = \infty$

$$\|f\|_{W_\infty^k(\Omega)} := \max_{|\alpha| \leq k} \|D_w^\alpha f\|_{L^\infty(\Omega)} \, .$$

In either case, we define the **Sobolev spaces** *via*

$$W_p^k(\Omega) := \left\{ f \in L^1_{loc}(\Omega) \, : \, \|f\|_{W_p^k(\Omega)} < \infty \right\}.$$

The Sobolev spaces can be related in special cases to other spaces. For example, recall the *Lipschitz norm*

$$\|f\|_{Lip(\Omega)} = \|f\|_{L^\infty(\Omega)} + \sup \left\{ \frac{|f(x) - f(y)|}{|x - y|} \, : \, x, y \in \Omega; x \neq y \right\},$$

and the corresponding space of *Lipschitz functions*

$$Lip(\Omega) = \left\{ f \in L^\infty(\Omega) : \|f\|_{Lip(\Omega)} < \infty \right\}.$$

Then for all dimensions n, we have $Lip(\Omega) = W^1_\infty(\Omega)$ with equivalent norms, at least under certain conditions on the domain Ω (cf. exercises 1.x.15 and 1.x.14). (Two norms, $\|\cdot\|_1$ and $\|\cdot\|_2$, on a linear space V are said to be *equivalent* provided there is a positive constant $C < \infty$ such that $\|v\|_1/C \le \|v\|_2 \le C\|v\|_1 \quad \forall v \in V$.) Moreover, for $k > 1$

$$W^k_\infty(\Omega) = \left\{ f \in C^{k-1}(\Omega) : f^{(\alpha)} \in Lip(\Omega) \; \forall \, |\alpha| \le k - 1 \right\}.$$

In one dimension ($n = 1$), the space $W^1_1(\Omega)$ can be characterized as the set of *absolutely continuous* functions on an interval Ω (cf. (Hartman & Mikusinski 1961) and exercises 1.x.17 and 1.x.22).

It is easy to see that $\|\cdot\|_{W^k_p(\Omega)}$ is a norm. Thus, $W^k_p(\Omega)$ is by definition a normed linear space. The following theorem shows that it is complete.

(1.3.2) Theorem. *The Sobolev space $W^k_p(\Omega)$ is a Banach space.*

Proof. Let $\{v_j\}$ be a Cauchy sequence with respect to the norm $\|\cdot\|_{W^k_p(\Omega)}$. Since the $\|\cdot\|_{W^k_p(\Omega)}$ norm is just a combination of $\|\cdot\|_{L^p(\Omega)}$ norms of weak derivatives, it follows that, for all $|\alpha| \le k$, $\{D^\alpha_w v_j\}$ is a Cauchy sequence with respect to the norm $\|\cdot\|_{L^p(\Omega)}$. Thus, Theorem 1.1.8 implies the existence of $v^\alpha \in L^p(\Omega)$ such that $\|D^\alpha_w v_j - v^\alpha\|_{L^p(\Omega)} \to 0$ as $j \to \infty$. In particular, $v_j \to v^{(0,\dots,0)} =: v$ in $L^p(\Omega)$. What remains to check is that $D^\alpha_w v$ exists and is equal to v^α.

First, note that if $w_j \to w$ in $L^p(\Omega)$, then for all $\phi \in \mathcal{D}(\Omega)$

$$(1.3.3) \qquad \int_\Omega w_j(x)\phi(x)\,dx \to \int_\Omega w(x)\phi(x)\,dx.$$

This follows from (1.1.9) and Hölder's inequality:

$$\|w_j\phi - w\phi\|_{L^p(\Omega)} \le \|w_j - w\|_{L^p(\Omega)}\|\phi\|_{L^\infty(\Omega)} \to 0 \quad \text{as} \quad j \to \infty.$$

To show that $D^\alpha_w v = v^\alpha$, we must show that

$$\int_\Omega v^\alpha \phi\,dx = (-1)^{|\alpha|} \int_\Omega v\phi^{(\alpha)}\,dx \quad \forall \phi \in \mathcal{D}(\Omega).$$

This follows from the definition of the weak derivative, $D^\alpha_w v_j$, and two applications of (1.3.3):

$$\int_\Omega v^\alpha \phi\,dx = \lim_{j\to\infty} \int_\Omega (D^\alpha_w v_j)\,\phi\,dx$$

$$= \lim_{j\to\infty} (-1)^{|\alpha|} \int_\Omega v_j \phi^{(\alpha)}\,dx = (-1)^{|\alpha|} \int_\Omega v\phi^{(\alpha)}\,dx. \qquad \square$$

There is another potential definition of Sobolev space that could be made. Let $H_p^k(\Omega)$ denote the closure of $C^k(\Omega)$ with respect to the Sobolev norm $\|\cdot\|_{W_p^k(\Omega)}$. In the case $p = \infty$, we have $H_\infty^k = C^k$, and this is not the same as $W_\infty^k(\Omega)$. Indeed, we have already identified the latter as being related to certain Lipschitz spaces. However, for $1 \le p < \infty$, it turns out that $H_p^k(\Omega) = W_p^k(\Omega)$. The following result was proved in a paper (Meyers & Serrin 1964) that is celebrated both for the importance of the result and the brevity of its title.

(1.3.4) Theorem. *Let Ω be any open set. Then $C^\infty(\Omega) \cap W_p^k(\Omega)$ is dense in $W_p^k(\Omega)$ for $p < \infty$.*

(1.3.5) *Remark.* This result should be contrasted with another kind of density result, namely, the density of $C^\infty(\overline{\Omega})$ in Sobolev spaces. The latter cannot happen whenever part of the domain lies on both sides of part of its boundary, as occurs with a slit domain that is frequently used to model crack propagation problems (cf. exercise 1.x.26). In order for this stronger density result to hold, some sort of regularity condition must also hold. For example, it is known (cf. Adams 1975) to be valid if Ω satisfies the *segment* condition, i.e., if for all $x \in \partial\Omega$ there is an open ball B_x containing x and a non-trivial vector n_x such that for all $z \in \overline{\Omega} \cap B_x$ the segment $\{z + tn_x : 0 < t < 1\} \subset \Omega$. The vector n_x plays the role of an inward-directed normal to $\partial\Omega$ at x. Many of the domains considered here satisfy this condition (cf. exercise 1.x.25).

(1.3.6) *Remark.* The validity of the theorem can be seen in the case $\Omega = \mathbb{R}^n$ as follows. (For a more general case, see exercise 1.x.19.) From Example 1.2.2, we see that there exist non-negative $\phi \in \mathcal{D}(\Omega)$, and so by normalizing appropriately, we can assume that there is such a $\phi \in \mathcal{D}(\Omega)$ further satisfying $\int_\Omega \phi(x)\, dx = 1$. For any $f \in L_{loc}^1(\Omega)$ and $\epsilon > 0$, define

$$f_\epsilon(x) = \epsilon^{-n} \int_\Omega f(y)\phi\left((x-y)/\epsilon\right) dy,$$

that is, we have $f_\epsilon = f * \phi^\epsilon$ where $\phi^\epsilon(y) := \epsilon^{-n}\phi(y/\epsilon) \quad \forall y \in \Omega$ and "$*$" denotes convolution. From the dominated convergence theorem, it is easy to see that f_ϵ is C^∞ and that $f \in L^p(\Omega)$ implies $f_\epsilon \to f$ in $L^p(\Omega)$. Moreover, if $f \in W_p^k(\Omega)$ then the same can be said of the derivatives of f_ϵ and f, by differentiating under the integral sign and using the definition of weak derivative, thus proving the result. The function f_ϵ is called a "mollification" of f.

For technical reasons it is useful to introduce the following notation for the Sobolev *semi-norms*.

(1.3.7) **Definition.** *For k a non-negative integer and $f \in W_p^k(\Omega)$, let*

$$|f|_{W_p^k(\Omega)} = \left(\sum_{|\alpha|=k} \|D_w^\alpha f\|_{L^p(\Omega)}^p \right)^{1/p}$$

in the case $1 \le p < \infty$, and in the case $p = \infty$

$$|f|_{W_\infty^k(\Omega)} = \max_{|\alpha|=k} \|D_w^\alpha f\|_{L^\infty(\Omega)} .$$

1.4 Inclusion Relations and Sobolev's Inequality

Given the number of indices defining Sobolev spaces, it is natural to hope that there are inclusion relations to provide some sort of ordering among them. Using the Definition 1.3.1 and exercise 1.x.1, it is easy to derive the following propositions.

(1.4.1) **Proposition.** *Suppose that Ω is any domain, k and m are non-negative integers satisfying $k \le m$, and p is any real number satisfying $1 \le p \le \infty$. Then $W_p^m(\Omega) \subset W_p^k(\Omega)$.*

(1.4.2) **Proposition.** *Suppose that Ω is a bounded domain, k is a non-negative integer, and p and q are real numbers satisfying $1 \le p \le q \le \infty$. Then $W_q^k(\Omega) \subset W_p^k(\Omega)$.*

However, there are more subtle relations among the Sobolev spaces. For example, there are cases when $k < m$ and $p > q$ and $W_q^m(\Omega) \subset W_p^k(\Omega)$. The existence of Sobolev derivatives imply a stronger integrability condition of a function. To set the stage, let us consider an example to give us guidance as to possible relations among k, m, p, and q for such a result to hold.

(1.4.3) **Example.** Let $n \ge 2$, let $\Omega = \{x \in \mathbb{R}^n : |x| < 1/2\}$ and consider the function $f(x) = \log |\log |x||$. From Example 1.2.6 (and exercise 1.x.12), we see that f has first-order weak derivatives

$$D^\alpha f(x) = x^\alpha / \left(|x|^2 \log |x| \right)$$

$(|\alpha| = 1)$. From exercise 1.x.5, we see that $D^\alpha f \in L^p(\Omega)$ provided $p \le n$. For example,

$$|D^\alpha f(x)|^n \le \rho(|x|) := 1 / \left(|x|^n |\log |x||^n \right)$$

satisfies the condition of exercise 1.x.5 because $\rho(r)r^{n-1} = 1/\left(r|\log r|^n\right)$ is integrable for all $n \geq 2$ on $[0, 1/2]$. In fact, the change of variables $r = e^{-t}$ gives

$$\int_0^{1/2} \frac{dr}{r|\log r|^n} = \int_{\log 2}^\infty \frac{dt}{t^n} < \infty. \qquad (n \geq 2)$$

Similarly, it is easy to see that $f \in L^p(\Omega)$ for $p < \infty$. Thus, $f \in W_p^1(\Omega)$ for $p \leq n$. Note, however, that in no case is $f \in L^\infty(\Omega)$.

This example shows that there are functions that are essentially infinite at points (such points could be chosen as in (1.1.12) to be everywhere dense), yet which have p-th power integrable weak derivatives. Moreover, as the dimension n increases, the integrability power p increases as well. On the other hand, the following result, which will be proved in Chapter 4, shows that if a function has p-th power integrable weak derivatives for sufficiently large p (with n fixed), it must be bounded (and, in fact, can be viewed as being continuous). But before we state the result, we must introduce a regularity condition on the domain boundary for the result to be true.

(1.4.4) Definition. *We say Ω has a **Lipschitz boundary** $\partial\Omega$ provided there exists a collection of open sets O_i, a positive parameter ϵ, an integer N and a finite number M, such that for all $x \in \partial\Omega$ the ball of radius ϵ centered at x is contained in some O_i, no more than N of the sets O_i intersect nontrivially, and each domain $O_i \cap \Omega = O_i \cap \Omega_i$ where Ω_i is a domain whose boundary is a graph of a Lipschitz function ϕ_i (i.e., $\Omega_i = \left\{(x, y) \in \mathbb{R}^n : x \in \mathbb{R}^{n-1},\ y < \phi_i(x)\right\}$) satisfying $\|\phi_i\|_{Lip(\mathbb{R}^{n-1})} \leq M$.*

One consequence of this definition is that we can now relate Sobolev spaces on a given domain to those on all of $\mathbb{R}^n$.

(1.4.5) Theorem. *Suppose that Ω has a Lipschitz boundary. Then there is an **extension** mapping $E : W_p^k(\Omega) \to W_p^k(\mathbb{R}^n)$ defined for all non-negative integers k and real numbers p in the range $1 \leq p \leq \infty$ satisfying $Ev|_\Omega = v$ for all $v \in W_p^k(\Omega)$ and*

$$\|Ev\|_{W_p^k(\mathbb{R}^n)} \leq C\,\|v\|_{W_p^k(\Omega)}$$

where C is independent of v.

For a proof of this result, as well as more details concerning other material in this section, see (Stein 1970). Of course, the complementary result is true for *any* domain, namely, that the natural *restriction* allows us to view functions in $W_p^k(\mathbb{R}^n)$ as well defined in $W_p^k(\Omega)$. We now return to the question regarding the relationship between Sobolev spaces with different indices.

(1.4.6) Theorem. (Sobolev's Inequality) *Let Ω be an n-dimensional domain with Lipschitz boundary, let k be a positive integer and let p be a real number in the range $1 \leq p < \infty$ such that*

$$k \geq n \quad \text{when} \quad p = 1$$
$$k > n/p \quad \text{when} \quad p > 1.$$

Then there is a constant C such that for all $u \in W_p^k(\Omega)$

$$\|u\|_{L^\infty(\Omega)} \leq C \|u\|_{W_p^k(\Omega)}.$$

Moreover, there is a continuous function in the $L^\infty(\Omega)$ equivalence class of u.

This result says that any function with suitably regular weak derivatives may be viewed as a continuous, bounded function. Note that Example 1.4.3 shows that the result is sharp, namely, that the condition $k > n/p$ cannot be relaxed (unless $p = 1$), at least when $n \geq 2$. When $n = 1$, Sobolev's inequality says that integrability of first-order derivatives to *any* power $p \geq 1$ is sufficient to guarantee continuity. The result will be proved as a corollary to our polynomial approximation theory to be developed in Chapter 4. Note that we can apply it to derivatives of functions in Sobolev spaces to derive the following:

(1.4.7) Corollary. *Let Ω be an n-dimensional domain with Lipschitz boundary, and let k and m be positive integers satisfying $m < k$ and let p be a real number in the range $1 \leq p < \infty$ such that*

$$k - m \geq n \quad \text{when} \quad p = 1$$
$$k - m > n/p \quad \text{when} \quad p > 1.$$

Then there is a constant C such that for all $u \in W_p^k(\Omega)$

$$\|u\|_{W_\infty^m(\Omega)} \leq C \|u\|_{W_p^k(\Omega)}.$$

Moreover, there is a C^m function in the $L^p(\Omega)$ equivalence class of u.

(1.4.8) Remark. If $\partial\Omega$ is not Lipschitz continuous, then neither Theorem 1.4.5 nor Theorem 1.4.6 need hold. For example, let

$$\Omega = \left\{ (x,y) \in \mathbb{R}^2 \ : \ 0 < x < 1, \ |y| < x^r \right\}$$

where $r > 1$, and let $u(x,y) = x^{-\epsilon/p}$, where $0 < \epsilon < r$. Then

$$\sum_{|\alpha|=1} \int_\Omega |D^\alpha u|^p \, dx dy = c_{\epsilon,p} \int_0^1 x^{-\epsilon-p+r} \, dx < \infty,$$

provided $p < 1 + r - \epsilon$. In this case, $u \in W_p^1(\Omega)$ but u is not essentially bounded on Ω if $\epsilon > 0$. Choosing ϵ so that it is possible to have $p > 2$, we find that a Lipschitz boundary is necessary for Sobolev's inequality to hold. Since Sobolev's inequality does hold on $\mathbb{R}^n$, it is not possible to extend u to an element of $W_2^1(\mathbb{R}^2)$. Thus, the extension theorem for Sobolev functions can not hold if $\partial\Omega$ is not Lipschitz continuous.

1.5 Review of Chapter 0

At this point we can tie up many of the loose ends from the previous chapter. We see that the space V introduced there can now be rigorously defined as

$$V = \left\{v \in W_2^1(\Omega) \, : \, v(0) = 0\right\}$$

where $\Omega = [0,1]$, and that this makes sense because Sobolev's inequality guarantees that pointwise values are well defined for functions in $W_2^1(\Omega)$. (In fact, we are allowed to view $W_2^1(\Omega)$ as a subspace of $C_b(\Omega)$, the Banach space of bounded continuous functions.)

The derivation of the variational formulation (0.1.3) for solution of (0.1.1) can now be made rigorous (cf. exercise 1.x.24). We have stated in Sect. 1.3 (see the related exercises) that functions in $W_1^1(\Omega)$, and *a fortiori* those in $W_2^1(\Omega)$, are absolutely continuous, implying that the Cantor function is not among them. However, we saw in Example 1.2.5 that piecewise linear functions have weak derivatives that are piecewise constant. Thus, we can assert that the spaces S constructed in Sect. 0.5 satisfy $S \subset W_\infty^1(\Omega)$, and therefore that $S \subset V$.

Since Sobolev's inequality implies that $V \subset C_b(\Omega)$, exercise 0.x.8 shows that w in the duality argument leading to Theorem 0.3.5 is well defined (also see exercise 1.x.23).

In the error estimates for $u - u_S$ we make reference to the $L^2(\Omega)$ norm, $\|\cdot\|$, of second derivatives of functions. We can now make those expressions rigorous by interpreting them in the context of functions in $W_2^2(\Omega)$. In particular, we can re-state the approximation assumption (0.3.4) as

$$(0.3.4\text{bis}) \quad \exists \epsilon < \infty \quad \text{such that} \quad \inf_{v \in S} \|w - v\|_E \le \epsilon \|w''\| \quad \forall w \in W_2^2(\Omega),$$

and the only condition needed for Theorem 0.4.5 to hold is $f \in L^2(\Omega)$ (cf. exercise 1.x.23).

Finally, in the proofs in Chapter 0 we argued that $a(v,v) = 0$ implied $v \equiv 0$. While it certainly implies that the weak derivative of v is zero as a function in $L^2(\Omega)$, to conclude that v must be constant requires a notion of *coercivity* that will subsequently be developed in detail. For now, consider the following simple case. From exercise 1.x.16, we know that, for all $v \in V$, we can write

$$v(x) = \int_0^x D_w^1 v(s)\, ds$$

and use Schwarz' inequality (cf. the proof of Theorem 0.4.5) to estimate

$$|v(x)|^2 \le x \cdot \int_0^x D_w^1 v(s)^2\, ds.$$

Integrating with respect to x yields

$$\|v\|_{L^2(\Omega)}^2 \le \frac{1}{2} a(v,v).$$

In particular, this shows that if $v \in V$ satisfies $a(v,v) = 0$ then $v = 0$ as an element of $L^2(\Omega)$. Moreover, recalling the definition of the $W_2^1(\Omega)$ norm, we see that

(1.5.1) $$\|v\|_{W_2^1(\Omega)}^2 \le \frac{3}{2} a(v,v) \quad \forall v \in V.$$

Thus, we can conclude that vanishing of $a(v,v)$ for $v \in V$ implies v is the zero element in V (or $W_2^1(\Omega)$). Inequality (1.5.1) is a *coercivity* inequality for the bilinear form $a(\cdot,\cdot)$ on the space V. Note that this inequality is only valid on the subspace V of $W_2^1(\Omega)$, not all of $W_2^1(\Omega)$, since it fails if we take v to be a non-zero constant function (regardless of what constant would be substituted for $\frac{3}{2}$).

1.6 Trace Theorems

In the previous section we saw that it was possible to interpret the "boundary condition" $v(0) = 0$ in the definition of the space V using Sobolev's inequality. As a guide to the higher-dimensional cases of interest later, this is somewhat misleading, in that Sobolev's inequality, as presented in Sect. 1.4, will not suffice to interpret boundary conditions for higher-dimensional problems. For example, we have already seen in Example 1.4.3 that, when $n \ge 2$, the analogue of the space V, namely $W_2^1(\Omega)$, contains unbounded functions. Thus, we cannot interpret the boundary conditions in a pointwise sense, and Sobolev's inequality will have to be augmented in a substantial way to apply when $n \ge 2$. On the other hand, the function in Example 1.4.3 can be interpreted as an L^p function on any line in $\mathbb{R}^2$ since the function $\log |\log |\cdot||$ is p-th power integrable in one dimension.

     The correct interpretation of the situation is as follows. The boundary $\partial\Omega$ of an n-dimensional domain Ω can be interpreted as an $n-1$-dimensional object, a *manifold*. When $n = 1$ it consists of distinct points—the zero-dimensional case of a manifold. Sobolev's inequality gives conditions under which point values are well defined for functions in a Sobolev space, and thus for boundary values in the one-dimensional case. For higher dimensional problems, we must seek an interpretation of restrictions of Sobolev-class functions to manifolds of dimension $n - 1$, and in particular it should make good sense (say, in a Lebesgue class) for functions in $W_2^1(\Omega)$.

We begin with a simple example to explain the ideas. Let Ω denote the unit disk in $\mathbb{R}^2$:

$$\Omega = \{(x, y) : x^2 + y^2 < 1\} = \{(r, \theta) : 0 \le r < 1, 0 \le \theta < 2\pi\}.$$

Let $u \in C^1(\overline{\Omega})$, and consider its restriction to $\partial\Omega$ as follows:

$$
\begin{aligned}
u(1, \theta)^2 &= \int_0^1 \frac{\partial}{\partial r} \left(r^2 u(r, \theta)^2 \right) \, dr \\
&= \int_0^1 2 \left(r^2 u u_r + r u^2 \right) (r, \theta) \, dr \\
&= \int_0^1 2 \left(r^2 u \nabla u \cdot \frac{(x, y)}{r} + r u^2 \right) (r, \theta) \, dr \\
&\le \int_0^1 2 \left(r^2 |u| |\nabla u| + r u^2 \right) (r, \theta) \, dr \\
&\le \int_0^1 2 \left(|u| |\nabla u| + u^2 \right) (r, \theta) \, r \, dr. \qquad (r \le 1)
\end{aligned}
$$

Integrating with respect to θ and using polar coordinates (cf. exercise 1.x.4), we find

$$\int_{\partial\Omega} u^2 \, d\theta \le 2 \int_\Omega \left(|u| |\nabla u| + u^2 \right) \, dx \, dy,$$

where we define the boundary integral (and corresponding norm) in the obvious way:

(1.6.1) $$\int_{\partial\Omega} u^2 \, d\theta := \int_0^{2\pi} u(1, \theta)^2 \, d\theta =: \|u\|_{L^2(\partial\Omega)}^2.$$

Using Schwarz' inequality, we have

$$\|u\|_{L^2(\partial\Omega)}^2 \le 2 \|u\|_{L^2(\Omega)} \left(\int_\Omega |\nabla u|^2 \, dx \, dy \right)^{1/2} + 2 \int_\Omega u^2 \, dx \, dy.$$

The *arithmetic-geometric mean inequality* (cf. exercise 1.x.32), implies that

$$\left(\int_\Omega |\nabla u|^2 \, dx \, dy \right)^{1/2} + \left(\int_\Omega u^2 \, dx \, dy \right)^{1/2} \le \left(2 \int_\Omega \left(|\nabla u|^2 + u^2 \right) \, dx \, dy \right)^{1/2}.$$

Therefore,

(1.6.2) $$\|u\|_{L^2(\partial\Omega)} \le \sqrt[4]{8} \, \|u\|_{L^2(\Omega)}^{1/2} \|u\|_{W_2^1(\Omega)}^{1/2}.$$

This is an inequality analogous to Sobolev's inequality, Theorem 1.4.6, except that the $L^\infty(\Omega)$ norm on the left-hand side of the inequality has been replaced by $\|u\|_{L^2(\partial\Omega)}$. Although we have only proved the inequality for smooth u, we will see that it makes sense for all $u \in W_2^1(\Omega)$, and correspondingly that for such u the restriction $u|_{\partial\Omega}$ makes sense as a function in $L^2(\partial\Omega)$. But first, we should say what we mean by the latter space. Using Definition 1.6.1, we can identify it simply with $L^2([0, 2\pi])$ using the coordinate mapping $\theta \to (\cos\theta, \sin\theta)$. (More general boundaries will be discussed shortly.) We can now use inequality (1.6.2) to prove the following result.

(1.6.3) Proposition. *Let Ω denote the unit disk in $\mathbb{R}^2$. For all $u \in W_2^1(\Omega)$, the restriction $u|_{\partial\Omega}$ may be interpreted as a function in $L^2(\partial\Omega)$ satisfying* (1.6.2).

Proof. We will use (1.6.2) three times in the proof, which so far has only been derived for smooth functions. However, in view of Remark 1.3.5, such functions are dense in $W_2^1(\Omega)$, so we may pick a sequence $u_j \in C^1(\overline{\Omega})$ such that $\|u - u_j\|_{W_2^1(\Omega)} \leq 1/j$ for all j. By (1.6.2) and the triangle inequality,

$$\begin{aligned}
\|u_k - u_j\|_{L^2(\partial\Omega)} &\leq \sqrt[4]{8}\, \|u_k - u_j\|_{L^2(\Omega)}^{1/2} \|u_k - u_j\|_{W_2^1(\Omega)}^{1/2} \\
&\leq \sqrt[4]{8}\, \|u_k - u_j\|_{W_2^1(\Omega)} \leq \sqrt[4]{8}\left(\frac{1}{j} + \frac{1}{k}\right)
\end{aligned}$$

for all j and k, so that $\{u_j\}$ is a Cauchy sequence in $L^2(\partial\Omega)$. Since this space is complete, there must be a limit $v \in L^2(\partial\Omega)$ such that $\|v - u_j\|_{L^2(\partial\Omega)} \to 0$ as $j \to \infty$. We *define*

$$u|_{\partial\Omega} := v.$$

The first thing we need to check is that this definition does not depend on the particular sequence that we chose. So suppose that v_j is another sequence of $C^1(\overline{\Omega})$ functions that satisfy $\|u - v_j\|_{W_2^1(\Omega)} \to 0$ as $j \to \infty$. Using the triangle inequality a few times and (1.6.2) again, we see that

$$\begin{aligned}
&\|v - v_j\|_{L^2(\partial\Omega)} \\
&\leq \|v - u_j\|_{L^2(\partial\Omega)} + \|u_j - v_j\|_{L^2(\partial\Omega)} \\
&\leq \|v - u_j\|_{L^2(\partial\Omega)} + \sqrt[4]{8}\, \|u_j - v_j\|_{W_2^1(\Omega)} \\
&\leq \|v - u_j\|_{L^2(\partial\Omega)} + \sqrt[4]{8}\left(\|u_j - u\|_{W_2^1(\Omega)} + \|u - v_j\|_{W_2^1(\Omega)}\right) \\
&\to 0
\end{aligned}$$

as $j \to \infty$. Thus, $u|_{\partial\Omega}$ is well defined in $L^2(\partial\Omega)$. All that remains is to check that (1.6.2) holds for u. Again, we use the validity of it for smooth functions:

$$\begin{aligned}
\|u\|_{L^2(\partial\Omega)} &= \|v\|_{L^2(\partial\Omega)} = \lim_{j\to\infty} \|u_j\|_{L^2(\partial\Omega)} \\
&\leq \lim_{j\to\infty} \sqrt[4]{8}\, \|u_j\|_{L^2(\Omega)}^{1/2} \|u_j\|_{W_2^1(\Omega)}^{1/2} = \sqrt[4]{8}\, \|u\|_{L^2(\Omega)}^{1/2} \|u\|_{W_2^1(\Omega)}^{1/2}.
\end{aligned}$$

This completes the proof of the proposition. Note that (1.6.2) was used repeatedly to extend its validity on a dense subspace to all of $W_2^1(\Omega)$; this is a prototypical example of a *density argument*.   □

(1.6.4) Remark. Note that this proposition does *not* assert that *pointwise* values of u on $\partial\Omega$ make sense, only that $u|_{\partial\Omega}$ is square integrable on $\partial\Omega$. This leaves open the possibility (cf. exercise 1.x.28) that u could be infinite at a dense set of points on $\partial\Omega$. For smooth functions, the trace defined here is the same as the ordinary pointwise restriction to the boundary.

(1.6.5) Remark. The proposition, at first glance, says that functions in $W_2^1(\Omega)$ have boundary values in $L^2(\partial\Omega)$, and this is true. However, this by itself would not be a sharp result (not every element of $L^2(\partial\Omega)$ is the trace of some element of $W_2^1(\Omega)$). But on closer inspection it says something which is sharp, namely, that the $L^2(\partial\Omega)$ norm of a function can be bounded by just part of the $W_2^1(\Omega)$ norm (the square root of it), together with the $L^2(\Omega)$ norm. This result might seem strange until we see that it is dimensionally correct. That is, suppose that functions are measured in some unit U, and that L denotes the length unit. Then the units of the $W_2^1(\Omega)$ norm (ignoring lower order terms) equal U, and those of the $L^2(\Omega)$ norm equal $U \cdot L$. Neither of these matches the units of the square root of the left-hand side of (1.6.2), $U\sqrt{L}$, but the square root of their product does. Such a *dimensionality argument* can not prove an inequality such as (1.6.2), but it can be used to disprove one, or simplify its proof (cf. exercise 1.x.31).

Now let us describe a generalization of Proposition 1.6.3 to more complex domains. One natural approach is to work in the class of Lipschitz domains. If $\partial\Omega$ is given as the graph of a function ϕ (cf. (1.4.4)), we can define the integral on $\partial\Omega$ as

$$\int_{\partial\Omega} f\, dS := \int_{\mathbb{R}^{n-1}} f(x, \phi(x))\sqrt{1 + |\nabla\phi(x)|^2}\, dx.$$

If ϕ is Lipschitz, then the weight $\sqrt{1 + |\nabla\phi(x)|^2}$ is an L^∞ function (cf. exercise 1.x.14). In this way (cf. Grisvard 1985), we can define the integral on any Lipschitz boundary, and correspondingly associated Lebesgue spaces. Moreover, the following result holds.

(1.6.6) Theorem. *Suppose that Ω has a Lipschitz boundary, and that p is a real number in the range $1 \le p \le \infty$. Then there is a constant, C, such that*

$$\|v\|_{L^p(\partial\Omega)} \le C\, \|v\|_{L^p(\Omega)}^{1-1/p} \|v\|_{W_p^1(\Omega)}^{1/p} \quad \forall v \in W_p^1(\Omega).$$

We will use the notation $\mathring{W}_p^1(\Omega)$ to denote the subset of $W_p^1(\Omega)$, consisting of functions whose trace on $\partial\Omega$ is zero, that is

(1.6.7) $\mathring{W}_p^1(\Omega) = \left\{ v \in W_p^1(\Omega) \; : \; v|_{\partial\Omega} = 0 \text{ in } L^2(\partial\Omega) \right\}.$

Similarly, we let $\mathring{W}_p^k(\Omega)$ denote the subset of $W_p^k(\Omega)$ consisting of functions whose derivatives of order $k-1$ are in $\mathring{W}_p^1(\Omega)$, i.e.

(1.6.8) $\mathring{W}_p^k(\Omega) = \left\{ v \in W_p^k(\Omega) \; : \; v^{(\alpha)}|_{\partial\Omega} = 0 \text{ in } L^2(\partial\Omega) \quad \forall |\alpha| < k \right\}.$

1.7 Negative Norms and Duality

In this section we introduce ideas that lead to the definition of Sobolev spaces W_p^k for negative integers k. This definition is based on the concept of *duality* in Banach spaces. The *dual space*, B', to a Banach space, B, is a set of linear functionals on B. (A *linear functional* on a linear space B is simply a linear function from B into the reals, $\mathbb{R}$, i.e., a function $L : B \to \mathbb{R}$ such that

$$L(u + av) \; = \; L(u) + aL(v) \quad \forall u, v \in B, \, a \in \mathbb{R}.)$$

More precisely, we distinguish between the linear space, B^*, of *all* linear functionals on B (cf. exercise 1.x.33), and the subspace $B' \subset B^*$ of *continuous* linear functionals on B. The following observation simplifies the characterization of such functionals.

(1.7.1) Proposition. *A linear functional, L, on a Banach space, B, is continuous if and only if it is bounded, i.e., if there is a finite constant C such that $|L(v)| \le C\|v\|_B \quad \forall v \in B$.*

Proof. A bounded linear function is actually Lipschitz continuous, i.e.,

$$|L(u) - L(v)| \; = \; |L(u - v)| \; \le \; C\|u - v\|_B \quad \forall u, v \in B.$$

Conversely, suppose L is continuous. If it is not bounded, then there must be a sequence $\{v_n\}$ in B such that $|L(v_n)|/\|v_n\|_B \ge n$. Renormalizing by setting $w_n = v_n/n\|v_n\|_B$ gives $|L(w_n)| \ge 1$ but $\|w_n\|_B \le 1/n$, and thus $w_n \to 0$. But, by continuity of L, we should have $L(w_n) \to 0$, the desired contradiction.   □

For a continuous linear functional, L, on a Banach space, B, the proposition states that the following quantity is always finite:

(1.7.2) $\|L\|_{B'} \; := \; \sup_{0 \ne v \in B} \frac{L(v)}{\|v\|_B}.$

Exercise 1.x.34 shows that this forms a norm on B', called the *dual norm*, and one can show (cf. Trèves 1967) that B' is complete with respect to it, i.e., that B' is also a Banach space.

(1.7.3) Example. The dual space of a Banach space need not be a mysterious object. One of the key results of Lebesgue integration theory is that the dual spaces of L^p can be easily identified, for $1 \le p < \infty$. From Hölder's inequality, any function $f \in L^q(\Omega)$ (where $\frac{1}{q} + \frac{1}{p} = 1$ defines the dual index, q, to p) can be viewed as a continuous linear functional via

$$L^p(\Omega) \ni v \to \int_\Omega v(x) f(x) \, dx.$$

One version of the Riesz Representation Theorem states that *all* continuous linear functionals on $L^p(\Omega)$ arise in this way, i.e., that $(L^p(\Omega))'$ is isomorphic to $L^q(\Omega)$.

(1.7.4) Example. The dual space of a Banach space can also contain totally new objects. For example, Sobolev's inequality shows that the Dirac δ-function is a continuous linear functional on W_p^k, provided k and p satisfy the appropriate relation given in (1.4.6). Specifically, the Dirac δ-function is the linear functional

$$W_p^k(\Omega) \ni v \to v(y) =: \delta_y(v),$$

where y denotes a given point in the domain Ω. It can be seen that this can not arise via an integration process using any locally integrable function (cf. exercise 1.x.36), i.e., it can not be viewed as a member of any of the spaces introduced so far.

(1.7.5) Definition. *Let p be in the range $1 \le p \le \infty$, and let k be a negative integer. Let q be the dual index to p, i.e., $\frac{1}{q} + \frac{1}{p} = 1$. Then the Sobolev space $W_p^k(\Omega)$ is defined to be the dual space $\left(W_q^{-k}(\Omega)\right)'$ with norm given by the dual norm (cf. (1.7.2)).*

(1.7.6) Remark. Note that we have defined the negative-index Sobolev spaces so that, if the same definition were used as well for $k = 0$, then the two definitions (cf. (1.3.1)) would agree for $1 < p < \infty$, in view of Remark 1.7.3. Note also that different dual spaces are used to define negative-index Sobolev spaces, in particular, it is frequently useful to use the dual of a subspace of $W_q^{-k}(\Omega)$, and in view of exercise 1.x.35, this leads to a slightly larger space. In either case, the negative Sobolev spaces are big enough to include interesting new objects, such as the Dirac δ-function. Example 1.7.4 shows that $\delta \in W_p^k(\Omega)$ provided $k < -n + n/p$ (or $k \le -n$ if $p = \infty$).

1.x Exercises

1.x.1 Suppose that Ω is bounded and that $1 \leq p \leq q \leq \infty$. Prove that $L^q(\Omega) \subset L^p(\Omega)$. (Hint: use Hölder's inequality.) Give examples to show that the inclusion is strict if $p < q$ and false if Ω is not bounded.

1.x.2 Show that the set of bounded, continuous functions on a domain Ω forms a Banach space with norm $\|\cdot\|_{L^\infty(\Omega)}$.

1.x.3 Suppose that Ω is bounded and that $f_j \to f$ in $L^p(\Omega)$. Using Hölder's inequality prove that

$$\int_\Omega f_j(x)\, dx \to \int_\Omega f(x)\, dx \quad \text{as} \quad j \to \infty.$$

1.x.4 *Just in case you have not seen polar coordinates in n dimensions.* Define, inductively, mappings x^k from subsets $\Omega_k \subset \mathbb{R}^k$ into (the unit sphere in) $\mathbb{R}^{k+1}$ via $(x_1^k, \ldots, x_{k+1}^k)(\omega, \phi) := \big((\sin \phi)x^{k-1}(\omega),\ \cos \phi\big)$ on

$$\Omega_k := \{(\omega, \phi)\ :\ \omega \in \Omega_{k-1},\ 0 \leq \phi < \pi\},$$

with $x^1(\omega) = (\cos \omega, \sin \omega)$ and $\Omega_1 = \{\omega\ :\ 0 \leq \omega < 2\pi\}$. Define the *polar coordinate mapping* $X(\omega, r) := rx^{n-1}(\omega)$ for $r \geq 0$ and $\omega \in \Omega_{n-1}$. Prove that

$$\left| \det \frac{\partial X(\omega, r)}{\partial(\omega, r)} \right| = r^{n-1} \prod_{j=2}^{n-1} (\sin \omega_j)^{j-1}$$

for $\omega = (\omega_1, \ldots, \omega_{n-1}) \in \Omega_{n-1}$ and $r \geq 0$. (Hint: do the calculation for $n = 2$ and then show that

$$\det \begin{pmatrix} \dfrac{\partial x^{k+1}}{\partial(\omega, \phi)} \\ x^{k+1}(\omega, \phi) \end{pmatrix} = (\sin \phi)^k \det \begin{pmatrix} \dfrac{\partial x^k}{\partial \omega} \\ x^k(\omega) \end{pmatrix}$$

by induction on k.)

1.x.5 Let n be a positive integer, and suppose that ρ is a non-negative, smooth function defined for $0 < r \leq 1$ satisfying

$$\lim_{\epsilon \to 0+} \int_\epsilon^1 \rho(r) r^{n-1}\, dr < \infty.$$

Define f on $\Omega = \{x \in \mathbb{R}^n\ :\ |x| < 1\}$ via $f(x) = \rho(|x|)$. Show that $f \in L^1(\Omega)$. (Hint: use the Monotone Convergence Theorem, polar coordinates, cf. exercise 1.x.4, and Fubini's Theorem.)

1.x.6 Let $\Omega = [0, 1]$ and $1 \leq p < \infty$. Show that the function f defined in (1.1.12) is in $L^p(\Omega)$, and moreover, that $\|f - f_j\|_{L^p(\Omega)} \to 0$ as

$j \to \infty$. (Hint: first show that log $x \in L^p(\Omega)$, and then use the fact that $L^p(\Omega)$ is a Banach space.)

1.x.7 Pick your favorite dense sequence $\{r_n\}$. Graph the function f defined in (1.1.10) for various values of j.

1.x.8 Prove that $\left(\frac{\partial}{\partial x}\right)^\alpha |x| = x^\alpha / |x|$ for all $x \neq 0$ and $|\alpha| = 1$.

1.x.9 Prove Proposition 1.2.7. (Hint: use integration by parts and the fact that $C^0(\Omega) \subset L^1_{loc}(\Omega)$.)

1.x.10 Prove that weak derivatives of order greater than one of the function, f, in Example 1.2.5 do not exist.

1.x.11 Show that the condition in Example 1.2.6 implies that

$$\lim_{r \to 0} r^{n-1}|\rho(r)| = 0.$$

(Hint: First show that $r^{n-1}|\rho(r)| \leq \int_r^1 t^{n-1}|\rho'(t)|\,dt + \tilde{C} \leq C$. If the limit is not zero, then there are points r_j tending to zero such that

$$\left| r_j^{n-1} \int_{r_j}^{r_j+\gamma r_j} \rho'(t)\,dt \right| = \left| r_j^{n-1}\left(\rho(r_j+\gamma r_j) - \rho(r_j)\right) \right|$$

$$\geq \left| r_j^{n-1}\left(C(r_j+\gamma r_j)^{1-n} - c r_j^{1-n}\right) \right|$$

$$= \delta > 0$$

for appropriate $\gamma > 0$. Show this contradicts the condition. Or apply Hardy's inequality (Stein 1970).)

1.x.12 Verify the existence of weak derivatives claimed in Example1.2.6. (Hint: use exercises 1.x.11, 1.x.8, 1.x.5, the Divergence Theorem and a limit theorem for the Lebesgue integral.)

1.x.13 Let $f(x) = |x|^r$ for a given real number r. Prove that f has first-order weak derivatives on the unit ball provided that $r > 1 - n$.

1.x.14 Prove that $Lip(\Omega) \subset W^1_\infty(\Omega)$. (Hint: $f \in Lip(\Omega)$ is a fortiori Lipschitz continuous in each variable separately, so has partial derivatives a.e. Using Fubini's Theorem, show by contradiction that these must be essentially bounded on Ω. Show that these derivatives are actually *weak* derivatives by using the fact that Lipschitz functions are absolutely continuous and that integration by parts is justified for absolutely continuous functions.) What is the constant in the equivalence relation between the Lipschitz norm and the Sobolev $W^1_\infty(\Omega)$ norm corresponding to this inclusion?

1.x.15 Suppose that Ω is convex. Prove that $W^1_\infty(\Omega) \subset Lip(\Omega)$. (Hint: given $f \in W^1_\infty(\Omega), \phi \in \mathcal{D}(\Omega)$ and $y, z \in \Omega$, write

$$f(y) - f(z) = \lim_{\epsilon \to 0} \int_\Omega f(x) D\phi^\epsilon(x)\, dx$$

where D denotes the directional derivative in the direction $n_{y,z} = (y - z)/|y - z|$ and

$$\phi^\epsilon(x) = \epsilon^{-n} \int_0^\infty \phi\left(\frac{x - y - tn_{y,z}}{\epsilon}\right) \phi\left(\frac{x - z - tn_{y,z}}{\epsilon}\right) dt.$$

Note: the identity above must be proved, cf. the mollification argument in Remark 1.3.6 and exercise 1.x.18. Show that $\phi^\epsilon \in \mathcal{D}(\Omega)$ and that $\|\phi^\epsilon\|_{L^1(\mathbb{R}^n)} = |y - z|$ by verifying and using the identity

$$\phi^\epsilon(x) = -\epsilon^{-n}|y - z| \int_{-1}^0 \phi\left(\frac{x - y - \tau(y - z)}{\epsilon}\right) d\tau.$$

Apply the definition of weak derivative and Hölder's inequality.)

1.x.16 Let $n = 1$, $\Omega = [a, b]$ and $f \in W_1^1(\Omega)$. Prove that

$$\int_a^b D_w^1 f(x)\, dx = f(b) - f(a)$$

under the assumption that f is continuous at a and b. (Hint: use the "integration by parts" formula that defines the weak derivative and choose an appropriate sequence $\phi_j \in \mathcal{D}(\Omega)$ such that $\phi_j \to 1$ on Ω, cf. exercise 1.x.15.)

1.x.17 Prove that absolutely continuous functions on an interval $[a, b]$ are in $W_1^1([a, b])$. (Hint: use the fact that integration by parts makes sense for absolutely continuous functions.)

1.x.18 Verify the statements of Remark 1.3.6. (Hints: show that $D^\alpha f_\epsilon = \epsilon^{-|\alpha|} f * (D^\alpha \phi)^\epsilon$ for all α; prove that $f_\epsilon \to f$ in $L^p(\Omega)$; show also that $D^\alpha f_\epsilon = (D_w^\alpha f) * \phi^\epsilon = (D_w^\alpha f)_\epsilon$ under appropriate conditions.)

1.x.19 A domain Ω is said to be *star-shaped* with respect to a point x if $\forall y \in \Omega$ the line-segment connecting x and y lies within Ω. (Here, we mean the closed line-segment, so necessarily $x \in \Omega$.) Prove that $C^\infty(\overline{\Omega})$ is dense in $W_p^k(\Omega)$ for $1 \le p < \infty$ under the assumption that Ω is star-shaped. (Hint: see (Dupont and Scott 1979); assume $x = 0$ and show that the "dilation" $u_\rho(y) = u(\rho y)$ defined for $\rho < 1$ and $y \in \Omega$ satisfies $u_\rho \to u$ in $W_p^k(\Omega)$ as $\rho \to 1$. Mollify u_ρ as in Remark 1.3.6 to get a smooth function $u_{\rho,\epsilon} \to u$ as $\rho \to 1$ and $\epsilon \to 0$.)

1.x.20 Prove Sobolev's inequality in the case of $n = 1$, i.e., let $\Omega = [a, b]$ and show that

$$\|u\|_{L^\infty(\Omega)} \le C \|u\|_{W_1^1(\Omega)}.$$

(Hint: use the fundamental theorem of calculus and Theorem 1.3.4. See the proof of the trace inequality (1.6.2) for technical help.) How does the constant, C, depend on a and b?

1.x.21 Let $\Omega = [a, b]$ (here $n = 1$). Prove that all functions in $W_1^1(\Omega)$ are continuous (have a continuous representative). (Hint: use exercise 1.x.20 and Theorem 1.3.4.)

1.x.22 Prove that functions in $W_1^1([a, b])$ are absolutely continuous on $[a, b]$. (Hint: use exercises 1.x.16 and 1.x.21 and the "continuity" of the Lebesgue integral.)

1.x.23 Using the hint in exercise 0.x.8, show that, given $f \in L^2(\Omega)$, there is a solution $u \in W_2^2(\Omega)$ satisfying (0.1.1). Verify that all boundary conditions make sense and are satisfied and explain in what sense the differential equation is to be interpreted.

1.x.24 Let u be the solution to (0.1.1) given either by exercise (0.x.8) or exercise 1.x.23. Justify rigorously that (0.1.3) holds.

1.x.25 Show that any Lipschitz domain satisfies the segment condition (1.3.5).

1.x.26 Let $\Omega = \{(x, y) \in \mathbb{R}^2 : |x| < 1, |y| < 1, y \neq 0 \text{ for } x \geq 0\}$. Show that there are functions in $W_p^1(\Omega)$ that cannot be limits of functions in $C^0(\overline{\Omega})$.

1.x.27 Does a "cusp" domain, $\Omega = \{(x, y) : 0 < x < 1, 0 < y < x^r\}$, with $r > 1$, satisfy the segment condition? Does its complement?

1.x.28 Let Ω be the unit disk in the plane. Show that there exist $u \in W_2^1(\Omega)$ that are infinite at a dense set of points on $\partial\Omega$. (Hint: see (1.1.12) and (1.4.3).)

1.x.29 Let Ω be a domain in the plane. Show that there exist $u \in W_2^1(\Omega)$ that are infinite at a dense set of points in Ω. (Hint: see (1.1.12) and (1.4.3).)

1.x.30 Suppose Ω is as in Proposition 1.6.3, and let p be a real number in the range $1 \leq p \leq \infty$. Prove that there is a constant C such that

$$\|v\|_{L^p(\partial\Omega)} \leq C \|v\|_{L^p(\Omega)}^{1-1/p} \|v\|_{W_p^1(\Omega)}^{1/p} \quad \forall v \in W_p^1(\Omega).$$

Explain what this means in the case $p = \infty$.

1.x.31 Let Ω denote the upper half-plane. Show that no inequality of the form

$$\|v\|_{L^2(\partial\Omega)} \leq C \|v\|_{L^2(\Omega)}^\lambda \|v\|_{W_2^1(\Omega)}^\mu$$

holds, unless $\lambda \leq 1/2$ and $\lambda + \mu = 1$. (Hint: suppose it holds and consider the functions $u(x) = Uv(Lx)$ for U and L arbitrary to reach a contradiction.) Show that, to prove it for $\lambda = \mu = 1/2$, it suffices to assume $\|v\|_{L^2(\Omega)} = \|v\|_{W_2^1(\Omega)} = 1$, i.e., deduce the general case from this special case. (Hint: use the same scalings.) Use the same idea to prove the inequality

$$\|v\|_{L^2(\partial\Omega)} \leq C \|v\|_{L^2(\Omega)}^{1/2} |v|_{W_2^1(\Omega)}^{1/2} .$$

1.x.32 Let $a, b \in \mathbb{R}$. Prove that $ab \leq \frac{\epsilon}{2}a^2 + \frac{1}{2\epsilon}b^2$ for any $\epsilon > 0$. (Hint: expand $(a+b)^2 \geq 0$.) Apply this to show that $a + b \leq \left((1+\epsilon)a^2 + (1+1/\epsilon)b^2\right)^{1/2}$.

1.x.33 Show that the set of linear functionals on a linear space is itself a linear space, where the operation of *addition* is given by $(L_1 + L_2)(v) := L_1(v) + L_2(v)$ and *scalar multiplication* is defined by $(aL)(v) := aL(v)$.

1.x.34 Show that the expression in (1.7.2) defines a norm on the dual space of a Banach space, i.e., verify the conditions of (1.1.6).

1.x.35 Let B and C be two Banach spaces with $B \subset C$ and with the inclusion being continuous. Show that $C' \subset B'$, with the inclusion being continuous.

1.x.36 Let Ω be an open set and $y \in \Omega$. Show that there is no function $f \in L_{loc}^1(\Omega)$ such that $\phi(y) = \int_\Omega f(x)\phi(x)\,dx \quad \forall \phi \in \mathcal{D}(\Omega)$.

1.x.37 Let $\Omega = [0, 1]$. Prove that $\{v \in C^1([0,1] \ : \ v(0) = 0\}$ is dense in the set $\{v \in W_2^1(\Omega) : v(0) = 0\}$. (Hint: use the density result in Remark 1.3.5 and Sobolev's inequality.)

1.x.38 Let $\Omega = [0, 1]$. Prove that $\{v \in C^1([0,1] \ : \ v'(1) = 0\}$ is dense in $W_2^1(\Omega)$. (Hint: use the density result in Remark 1.3.5 to get a sequence $v_n \in C^1([0,1]$ converging to v in $W_2^1(\Omega)$ and modify each v_n near 1 to make the derivative zero.)

1.x.39 Let Ω be an open set. Suppose that $f \in L_{loc}^1(\Omega)$ such that $\int_\Omega f\phi\,dx = 0 \quad \forall \phi \in \mathcal{D}(\Omega)$. Prove that $f = 0$ a.e. (Hint: By contradiction. Show otherwise that $\exists \epsilon > 0$ such that $A = \{x \in \Omega : \pm f(x) > \epsilon\}$ has positive measure. Approximate A in measure by a finite union of balls A_N and let $\tilde{A}_N$ denote concentric balls of slightly smaller size, also approximating A in measure. Choose $\phi_N \in \mathcal{D}(A_N)$ with the properties $0 \leq \phi_N \leq 1$ everywhere and $\phi_N \geq \frac{1}{2}$ on $\tilde{A}_N$ and integrate $f\phi_N$ over Ω to reach a contradiction. Construct ϕ_N as follows. Write $A_N = \cup_j B_j$ and $\tilde{A}_N = \cup_j \tilde{B}_j$ and choose $\psi_j \in \mathcal{D}(B_j)$, such that $0 \leq \psi_j \leq 1$ everywhere and $\psi_j = 1$ on $\tilde{B}_j$. Define

$$\phi_N(x) = \sum_j \psi_j(x) \Big/ \Big(1 + \sum_j \psi_j(x)\Big).$$

This can be viewed as an *approximate partition of unity* argument.)

1.x.40 Show that the weak derivative is unique. (Hint: use exercise 1.x.39.)

1.x.41 Suppose, as in exercise 1.x.39, that $\tilde{B}_j \subset B_j$ are concentric balls, and let $\psi_j \in \mathcal{D}(B_j)$ such that $0 \le \psi_j \le 1$ everywhere and $\psi_j = 1$ on $\tilde{B}_j$. Let $\epsilon > 0$. Define

$$\phi^\epsilon(x) = \sum_j \psi_j(x) \Big/ \Big(\frac{\epsilon}{1-\epsilon} + \sum_j \psi_j(x) \Big).$$

Show that $0 \le \phi^\epsilon \le 1$ everywhere and $\phi^\epsilon \ge 1 - \epsilon$ on $\cup_j \tilde{B}_j$. Can you construct $\phi \in \mathcal{D}(\cup_j B_j)$ such that $0 \le \phi \le 1$ everywhere and $\phi = 1$ on $\cup_j \tilde{B}_j$? What can you say if the finite unions are replaced by infinite ones?

1.x.42 Let $v_\beta(r,\theta) = r^\beta \sin \beta\theta$ on $\Omega_\beta = \{(r,\theta) : r < 1, 0 < \theta < \pi/\beta\}$ where $1/2 \le \beta < \infty$. Determine the optimal values of k, p such that $v_\beta \in W_p^k(\Omega_\beta)$ as a function of β. The case $\beta = 1/2$ is called a *slit domain* and $\beta = 2/3$ a *re-entrant corner*. What does the case $\beta = 1$ correspond to?

1.x.43 The type of *finite cone* C used for ice cream can be represented as the union of the sets $t\xi + B_{\alpha t}$ for $0 < t < h$, where ξ is the *axis* of the cone, α measures the angle, and h is the height. Here B_t is the ball of radius t. A Lipschitz domain Ω satisfies a *cone property*: there is a finite set of finite cones C_i such that for all $x \in \Omega$, there is some i such that $x + C_i \subset \Omega$. Prove that smooth functions are dense in $W_p^k(\Omega)$ for such a domain. (Hint: Use a mollifier to average over the set $x + t\xi + B_{\alpha t}$ in defining a smoothed value at x, that is

$$v_t(x) := \int_{B_{\alpha t}} v(x + y + t\xi)\rho_t(y)\, dy.$$

Prove that v_t tends to v in $W_p^k(\Omega)$.)

1.x.44 A Cantor function can be defined by a limiting sequence of continuous, piecewise linear functions. Let $0 < \epsilon < 1$ and start with $f_0(x) = x$. Next, let $f_1(x)$ be equal to $\frac{1}{2}$ in the interval of width ϵ around $x = \frac{1}{2}$, and equal to f_0 at the ends of the interval $[0,1]$. Given f_i, define f_{i+1} inductively. In each interval where f_i is not constant, divide this interval in the way we did for the original one: let f_{i+1} be constant in a sub-interval of length $2^{-i}\epsilon$ in the center, and connect the remaining parts linearly. The constant is chosen to be the middle value. Show that the sequence f_i stays bounded in the space Lip^α of functions which satisfy a uniform bound of the form

$$|f(x) - f(y)| \le C|x-y|^\alpha$$

for some $1 > \alpha > 0$. Show that the best α can tend to one as ϵ tends to zero.

Chapter 2

Variational Formulation of Elliptic Boundary Value Problems

This chapter is devoted to the functional analysis tools required for developing the variational formulation of differential equations. It begins with an introduction to Hilbert spaces, including only material that is essential to later developments. The goal of the chapter is to provide a framework in which existence and uniqueness of solutions to variational problems may be established.

2.1 Inner-Product Spaces

(2.1.1) Definition. *A **bilinear form**, $b(\cdot, \cdot)$, on a linear space V is a mapping $b : V \times V \longrightarrow \mathbb{R}$ such that each of the maps $v \mapsto b(v, w)$ and $w \mapsto b(v, w)$ is a linear form on V. It is **symmetric** if $b(v, w) = b(w, v)$ for all $v, w \in V$. A (real) **inner product**, denoted by $(\cdot, \cdot)$, is a symmetric bilinear form on a linear space V that satisfies*

  (a) $(v, v) \geq 0 \; \forall\, v \in V$ *and*
  (b) $(v, v) = 0 \Longleftrightarrow v = 0$.

(2.1.2) Definition. *A linear space V together with an inner product defined on it is called an **inner-product space** and is denoted by $(V, (\cdot, \cdot))$.*

(2.1.3) Examples. The following are examples of inner-product spaces.

  (i) $V = \mathbb{R}^n, (x, y) := \sum_{i=1}^{n} x_i y_i$

  (ii) $V = L^2(\Omega), \Omega \subseteq \mathbb{R}^n, (u, v)_{L^2(\Omega)} := \int_\Omega u(x)v(x)dx$

  (iii) $V = W_2^k(\Omega), \Omega \subseteq \mathbb{R}^n, (u, v)_k := \sum_{|\alpha| \leq k} (D^\alpha u, D^\alpha v)_{L^2(\Omega)}$

   Notation. The inner-product space (iii) is often denoted by $H^k(\Omega)$. Thus, $H^k(\Omega) = W_2^k(\Omega)$.

(2.1.4) Theorem. (The Schwarz Inequality)   *If* $(V, (\cdot, \cdot))$ *is an inner-product space, then*

$$(2.1.5) \qquad |(u, v)| \leq (u, u)^{1/2}(v, v)^{1/2}.$$

The equality holds if and only if u and v are linearly dependent.

Proof. For $t \in \mathbb{R}$

$$(2.1.6) \qquad 0 \leq (u - tv, u - tv) = (u, u) - 2t(u, v) + t^2(v, v).$$

If $(v, v) = 0$, then $(u, u) - 2t(u, v) \geq 0$ $\forall t \in \mathbb{R}$, which forces $(u, v) = 0$, so the inequality holds trivially. Thus, suppose $(v, v) \neq 0$. Substituting $t = (u, v)/(v, v)$ into this inequality, we obtain

$$(2.1.7) \qquad 0 \leq (u, u) - |(u, v)|^2/(v, v)$$

which is equivalent to (2.1.5). Note that we did not use part (b) of Definition 2.1.1 to prove (2.1.5).

  If u and v are linearly dependent, one can easily see that equality holds in (2.1.5).

  Conversely, we assume that equality holds. If $v = 0$, then u and v are linearly dependent. If $v \neq 0$, take $\lambda = (u, v)/(v, v)$. It follows that $(u - \lambda v, u - \lambda v) = 0$, and property (b) of Definition 2.1.1 implies that $u - \lambda v = 0$, i.e. u and v are linearly dependent. □

(2.1.8) *Remark.* The Schwarz *inequality* (2.1.5) was proved without using property (b) of the inner product. An example where this might be useful is $a(u, v) = \int_\Omega \nabla u \cdot \nabla v \, dx$ on $H^1(\Omega)$. Thus, we have proved that

$$a(u, v)^2 \leq a(u, u) \, a(v, v) \qquad \text{for all } u, v \in H^1(\Omega),$$

even though $a(\cdot, \cdot)$ is not an inner product on $H^1(\Omega)$.

(2.1.9) Proposition. $\|v\| := \sqrt{(v, v)}$ *defines a norm in the inner-product space* $(V, (\cdot, \cdot))$.

Proof. One can easily show that $\|\alpha v\| = |\alpha| \, \|v\|$, $\|v\| \geq 0$, and $\|v\| = 0 \Longleftrightarrow v = 0$. It remains to prove the triangle inequality:

$$
\begin{aligned}
\|u + v\|^2 &= (u + v, u + v) \\
&= (u, u) + 2(u, v) + (v, v) \\
&= \|u\|^2 + 2(u, v) + \|v\|^2 \\
&\leq \|u\|^2 + 2\|u\| \, \|v\| + \|v\|^2 \qquad \text{(by Schwarz' inequality 2.1.5)} \\
&= (\|u\| + \|v\|)^2.
\end{aligned}
$$

Therefore, $\|u + v\| \leq \|u\| + \|v\|$. □

2.2 Hilbert Spaces

Proposition 2.1.9 says that, given an inner-product space $(V, (\cdot, \cdot))$, there is an associated norm defined on V, namely $\|v\| = \sqrt{(v, v)}$. Thus, an inner-product space can be made into a normed linear space.

(2.2.1) Definition. *Let $(V, (\cdot, \cdot))$ be an inner-product space. If the associated normed linear space $(V, \|\cdot\|)$ is complete, then $(V, (\cdot, \cdot))$ is called a* **Hilbert space**.

(2.2.2) Examples. The examples (i) - (iii) of (2.1.3) are all Hilbert spaces. In particular, the norm associated with the inner product $(\cdot, \cdot)_k$ on $W_2^k(\Omega)$ is the same as the norm $\|\cdot\|_{W_2^k(\Omega)}$ defined in Chapter 1 where $W_2^k(\Omega)$ was shown to be complete.

(2.2.3) Definition. *Let H be a Hilbert space and $S \subset H$ be a linear subset that is closed in H. (Recall that S* **linear** *means that $u, v \in S, \alpha \in \mathbb{R} \Longrightarrow u + \alpha v \in S$.) Then S is called a* **subspace of H**.

(2.2.4) Proposition.   *If S is a subspace of H, then $(S, (\cdot, \cdot))$ is also a Hilbert space.*

Proof. $(S, \|\cdot\|)$ is complete because S is closed in H under the norm $\|\cdot\|$.   □

(2.2.5) Examples of subspaces of Hilbert spaces.

  (i) H and $\{0\}$ are the obvious extreme cases. More interesting ones follow.
 (ii) Let $T : H \longrightarrow K$ be a continuous linear map of H into another linear space. Then $ker\, T$ is a subspace (see exercise 2.x.1).
(iii) Let $x \in H$ and define $x^\perp := \{v \in H : (v, x) = 0\}$. Then $x^\perp$ is a subspace of H. To see this, note that $x^\perp = ker\, L_x$, where L_x is the linear functional

$(2.2.6)$ $$L_x : v \mapsto (v, x).$$

By the Schwarz inequality (2.1.5),

$$|L_x(v)| \le \|x\|\, \|v\|$$

implying that L_x is bounded and therefore continuous. This proves that $x^\perp$ is a subspace of H in view of the previous example.

Note.   The overall objective of the next section is to prove that all $L \in H'$ are of the form L_x for $x \in H$, when H is a Hilbert space.

 (iv) Let $M \subset H$ be a subset and define

$$M^\perp := \{v \in H : (x, v) = 0\ \forall x \in M\}.$$

Note that

$$M^\perp = \bigcap_{x \in M} x^\perp$$

and each $x^\perp$ is a (closed) subspace of H. Thus, $M^\perp$ is a subspace of H.

(2.2.7) Proposition. *Let H be a Hilbert space.*

    (1) *For any subsets $M, N \subset H, M \subset N \implies N^\perp \subset M^\perp$.*
    (2) *For any subset M of H containing zero, $M \cap M^\perp = \{0\}$.*
    (3) *$\{0\}^\perp = H$.*
    (4) *$H^\perp = \{0\}$.*

Proof. For (2): Let $x \in M \cap M^\perp$. Then $x \in M \implies M^\perp \subset x^\perp$ and so

$$x \in M^\perp \implies x \in x^\perp \iff (x, x) = 0 \iff x = 0.$$

For (4): Since $H^\perp \subset H$, (2) implies that

$$H^\perp = H \cap H^\perp = \{0\}.$$

Parts (1) and (3) are left to the reader in exercise 2.x.3.   □

(2.2.8) Theorem. (Parallelogram Law)   *Let $\|\cdot\|$ be the norm associated with the inner product $(\cdot, \cdot)$ on H. We have*

$$(2.2.9) \qquad \|v + w\|^2 + \|v - w\|^2 = 2\left(\|v\|^2 + \|w\|^2\right).$$

Proof. A straight-forward calculation; see exercise 2.x.4.   □

2.3 Projections onto Subspaces

The following result establishes an essential geometric fact about Hilbert spaces.

(2.3.1) Proposition.   *Let M be a subspace of the Hilbert space H. Let $v \in H \setminus M$ and define $\delta := \inf\{\|v - w\| : w \in M\}$. (Note that $\delta > 0$ since M is closed in H.) Then there exists $w_0 \in M$ such that*
    (i) *$\|v - w_0\| = \delta$, i.e., there exists a closest point $w_0 \in M$ to v, and*
    (ii) *$v - w_0 \in M^\perp$.*

Proof. (i) Let $\{w_n\}$ be a minimizing sequence:

$$\lim_{n\to\infty} \|v - w_n\| = \delta.$$

We now show that $\{w_n\}$ is a Cauchy sequence. By the parallelogram law, $\|(w_n - v) + (w_m - v)\|^2 + \|(w_n - v) - (w_m - v)\|^2 = 2(\|w_n - v\|^2 + \|w_m - v\|^2)$, i.e.,

$$0 \le \|w_n - w_m\|^2 = 2(\|w_n - v\|^2 + \|w_m - v\|^2) - 4\left\|\tfrac{1}{2}(w_n + w_m) - v\right\|^2.$$

Since $\frac{1}{2}(w_n + w_m) \in M$, we have

$$\left\|\tfrac{1}{2}(w_n + w_m) - v\right\| \ge \delta$$

by the definition of δ. Thus,

$$0 \le \|w_n - w_m\|^2 \le 2(\|w_n - v\|^2 + \|w_m - v\|^2) - 4\delta^2.$$

Letting m, n tend towards infinity, we have

$$2(\|w_n - v\|^2 + \|w_m - v\|^2) - 4\delta^2 \to 2\delta^2 + 2\delta^2 - 4\delta^2 = 0.$$

Therefore, $\|w_n - w_m\|^2 \to 0$, proving that $\{w_n\}$ is Cauchy. Thus, there exists $w_0 \in \overline{M} = M$ such that $w_n \to w_0$. Continuity of the norm implies that $\|v - w_0\| = \delta$.

(ii) Let $z = v - w_0$, so that $\|z\| = \delta$. We will prove that $z \perp M$. Let $w \in M$ and $t \in \mathbb{R}$. Then $w_0 + tw \in M$ implies that $\|z - tw\|^2 = \|v - (w_0 + tw)\|^2$ has an absolute minimum at $t = 0$. Therefore,

$$0 = \frac{d}{dt}\|z - tw\|^2\Big|_{t=0} = -2(z, w).$$

This implies that, for all $w \in M$,

$$(v - w_0, w) = (z, w) = 0.$$

Since $w \in M$ was arbitrary, this implies $v - w_0 \in M^\perp$.   □

Proposition 2.3.1 says that, given a subspace M of H and $v \in H$, we can write $v = w_0 + w_1$, where $w_0 \in M$ and $w_1(= v - w_0) \in M^\perp$. Let us show that this decomposition of an element $v \in H$ is unique. In fact, from

$$w_0 + w_1 = v = z_0 + z_1, \qquad w_0, z_0 \in M, \qquad w_1, z_1 \in M^\perp,$$

we obtain

$$M \ni w_0 - z_0 = -(w_1 - z_1) \in M^\perp.$$

Since $M \cap M^\perp = \{0\}$, $w_0 = z_0$ and $w_1 = z_1$. This shows that the decomposition is unique. Therefore, we can define the following operators

(2.3.2) $$P_M : H \longrightarrow M, \qquad P_M^{\perp} : H \longrightarrow M^\perp$$

where the respective definitions of P_M and $P_M^{\perp}$ are given by

$$(2.3.3) \qquad P_M \, v \;=\; \begin{cases} v & \text{if } v \in M, \\ w_0 & \text{if } v \in H \setminus M; \end{cases}$$

$$(2.3.4) \qquad P_M{}^{\perp} v \;=\; \begin{cases} 0 & \text{if } v \in M, \\ v - w_0 & \text{if } v \in H \setminus M. \end{cases}$$

The uniqueness of the decomposition implies that $P_M{}^{\perp} = P_{M^{\perp}}$ (see exercise 2.x.5) so we need no longer be careful about where we put the " $\perp$ ". Summarizing the above observations, we state the following

(2.3.5) Proposition. *Given a subspace M of H and $v \in H$, there is a unique decomposition*

$$(2.3.6) \qquad v \;=\; P_M v + P_{M^{\perp}} v,$$

where $P_M : H \longrightarrow M$ and $P_{M^{\perp}} : H \longrightarrow M^{\perp}$. In other words,

$$(2.3.7) \qquad H \;=\; M \oplus M^{\perp}.$$

(2.3.8) Remark. The operators P_M and $P_{M^{\perp}}$ defined above are *linear* operators. To see this, note from the above proposition that

$$\alpha v_1 + \beta v_2 \;=\; P_M(\alpha v_1 + \beta v_2) + P_{M^{\perp}}(\alpha v_1 + \beta v_2),$$

where

$$v_1 \;=\; P_M v_1 + P_{M^{\perp}} v_1 \quad \text{and} \quad v_2 \;=\; P_M v_2 + P_{M^{\perp}} v_2.$$

That is

$$\begin{aligned} P_M(\alpha v_1 &+ \beta v_2) + P_{M^{\perp}}(\alpha v_1 + \beta v_2) \\ &= \alpha v_1 + \beta v_2 \\ &= (\alpha P_M v_1 + \beta P_M v_2) + (\alpha P_{M^{\perp}} v_1 + \beta P_{M^{\perp}} v_2). \end{aligned}$$

Uniqueness of decomposition of $\alpha v_1 + \beta v_2$ and the definitions of P_M and $P_{M^{\perp}}$ imply that

$$P_M(\alpha v_1 + \beta v_2) \;=\; \alpha P_M v_1 + \beta P_M v_2,$$

and

$$P_{M^{\perp}}(\alpha v_1 + \beta v_2) \;=\; \alpha P_{M^{\perp}} v_1 + \beta P_{M^{\perp}} v_2,$$

i.e., P_M and $P_{M^{\perp}}$ are linear.

(2.3.9) Definition. *An operator P on a linear space V is a projection if $P^2 = P$, i.e., $Pz = z$ for all z in the image of P.*

(2.3.10) Remark. The fact that P_M is a projection follows from its definition. That $P_M{}^{\perp}$ is also follows from the observation that $P_M{}^{\perp} = P_{M^{\perp}}$.

2.4 Riesz Representation Theorem

Given $u \in H$, recall that a continuous linear functional L_u can be defined on H by

$$(2.4.1) \qquad\qquad L_u(v) = (u, v).$$

The following theorem proves that the converse is also true.

(2.4.2) Theorem. (Riesz Representation Theorem) *Any continuous linear functional L on a Hilbert space H can be represented uniquely as*

$$(2.4.3) \qquad\qquad L(v) = (u, v)$$

for some $u \in H$. Furthermore, we have

$$(2.4.4) \qquad\qquad \|L\|_{H'} = \|u\|_H .$$

Proof. Uniqueness follows from the nondegeneracy of the inner product. For if u_1 and u_2 were two such solutions, we would have

$$
\begin{aligned}
0 &= L(u_1 - u_2) - L(u_1 - u_2) \\
  &= (u_1, u_1 - u_2) - (u_2, u_1 - u_2) \\
  &= (u_1 - u_2, u_1 - u_2)
\end{aligned}
$$

which implies $u_1 = u_2$. Now we prove existence.

Define $M := \{v \in H : L(v) = 0\}$. In view of Example 2.2.5.ii, M is a subspace of H. Therefore, $H = M \oplus M^\perp$ by Proposition 2.3.5.

Case (1): $\quad M^\perp = \{0\}$.
Thus, in this case $M = H$, implying that $L \equiv 0$. So take $u = 0$.

Case (2): $\quad M^\perp \neq \{0\}$.
Pick $z \in M^\perp$, $z \neq 0$. Then $L(z) \neq 0$. (Otherwise, $z \in M$, which implies that $z \in M^\perp \cap M = \{0\}$.) For $v \in H$ and $\beta = L(v)/L(z)$ we have

$$L(v - \beta z) = L(v) - \beta L(z) = 0,$$

i.e.

$$v - \beta z \in M.$$

Thus, $v - \beta z = P_M v$ and $\beta z = P_{M^\perp} v$. In particular, if $v \in M^\perp$, then $v = \beta z$ (that is, $v - \beta z = 0$), which proves that $M^\perp$ is one-dimensional. Now choose

$$(2.4.5) \qquad\qquad u := \frac{L(z)}{\|z\|_H^2} \, z .$$

Note that $u \in M^\perp$. We have

$$
\begin{aligned}
(u, v) &= \big(u, (v - \beta z) + \beta z\big) \\
&= (u, v - \beta z) + (u, \beta z) \\
&= (u, \beta z) && (u \in M^{\perp},\ v - \beta z \in M) \\
&= \beta \frac{L(z)}{\|z\|_H^2}(z, z) && \text{(definition of } u) \\
&= \beta L(z) \\
&= L(v). && \text{(definition of } \beta)
\end{aligned}
$$

Thus, $u := \big(L(z)/\|z\|^2\big)\, z$ is the desired element of H.

It remains to prove that $\|L\|_{H'} = \|u\|_H$. Let us first observe that

$$
\|u\|_H = \frac{|L(z)|}{\|z\|_H}
$$

from (2.4.5). Now, according to the definition (1.7.2) of the dual norm,

$$
\begin{aligned}
\|L\|_{H'} &= \sup_{0 \neq v \in H} \frac{|L(v)|}{\|v\|_H} && \text{(by 1.7.2)} \\
&= \sup_{0 \neq v \in H} \frac{|(u, v)|}{\|v\|_H} && \text{(by 2.4.3)} \\
&\leq \|u\|_H && \text{(Schwarz' inequality 2.1.5)} \\
&= \frac{|L(z)|}{\|z\|_H} && \text{(by 2.4.5)} \\
&\leq \|L\|_{H'} && \text{(by 1.7.2)}
\end{aligned}
$$

Therefore, $\|u\|_H = \|L\|_{H'}$.   □

(2.4.6) Remark. According to the Riesz Representation Theorem, there is a natural isometry between H and H' ($u \in H \longleftrightarrow L_u \in H'$). For this reason, H and H' are often identified. For example, we can write $W_2^m(\Omega) \cong W_2^{-m}(\Omega)$ (although they are completely different Hilbert spaces). We will use τ to represent the isometry from H' onto H.

2.5 Formulation of Symmetric Variational Problems

The purpose of the rest of this chapter is to apply the abstract Hilbert space theory developed in the previous sections to get existence and uniqueness results for variational formulations of boundary value problems.

(2.5.1) Example. Recall from Examples 2.1.3.iii and 2.2.2 that $H^1(0, 1) = W_2^1(0, 1)$ is a Hilbert space under the inner product

$$(u, v)_{H^1} = \int_0^1 uv \, dx + \int_0^1 u'v' \, dx.$$

In Chapter 0, we defined $V = \{v \in H^1(0,1) : v(0) = 0\}$. To see that V is a subspace of $H^1(0,1)$, let $\delta_0 : H^1(0,1) \longrightarrow \mathbb{R}$ by $\delta_0(v) = v(0)$. From Sobolev's inequality (1.4.6), δ_0 is a bounded linear functional on H^1, so it is continuous. Hence, $V = \delta_0^{-1}\{0\}$ is closed in H^1. We also defined $a(v, w) = \int_0^1 v'w' \, dx$. Note that $a(\cdot, \cdot)$ is a symmetric bilinear form on $H^1(0,1)$, but it is not an inner product on H^1 since $a(1, 1) = 0$. However, it does satisfy the *coercivity* property (1.5.1) on V. In view of the following, this shows that the variational problem in Chapter 0 is naturally expressed in a Hilbert-space setting.

(2.5.2) Definition. *A bilinear form $a(\cdot, \cdot)$ on a normed linear space H is said to be* **bounded** *(or* **continuous***) if $\exists\, C < \infty$ such that*

$$|a(v, w)| \le C \, \|v\|_H \, \|w\|_H \qquad \forall v, w \in H$$

and **coercive** *on $V \subset H$ if $\exists\, \alpha > 0$ such that*

$$a(v, v) \ge \alpha \, \|v\|_H^2 \qquad \forall v \in V.$$

(2.5.3) Proposition. *Let H be a Hilbert space, and suppose $a(\cdot, \cdot)$ is a symmetric bilinear form that is continuous on H and coercive on a subspace V of H. Then $\big(V, a(\cdot, \cdot)\big)$ is a Hilbert space.*

Proof. An immediate consequence of the coercivity of $a(\cdot, \cdot)$ is that if $v \in V$ and $a(v, v) = 0$, then $v \equiv 0$. Hence, $a(\cdot, \cdot)$ is an inner product on V.

Now let $\|v\|_E = \sqrt{a(v, v)}$, and suppose that $\{v_n\}$ is a Cauchy sequence in $(V, \|\cdot\|_E)$. By coercivity, $\{v_n\}$ is also Cauchy in $(H, \|\cdot\|_H)$. Since H is complete, $\exists\, v \in H$ such that $v_n \to v$ in the $\|\cdot\|_H$ norm. Since V is closed in H, $v \in V$. Now, $\|v - v_n\|_E \le \sqrt{c_1} \, \|v - v_n\|_H$ since $a(\cdot, \cdot)$ is bounded. Hence, $\{v_n\} \to v$ in the $\|\cdot\|_E$ norm, so $(V, \|\cdot\|_E)$ is complete. $\square$

In general, a symmetric variational problem is posed as follows. Suppose that the following three conditions are valid:

(2.5.4)
$$\begin{cases} (1) & \big(H, (\cdot, \cdot)\big) \text{ is a Hilbert space.} \\ (2) & V \text{ is a (closed) subspace of } H. \\ (3) & a(\cdot, \cdot) \text{ is a } bounded,\ symmetric \text{ bilinear form} \\ & \text{that is } coercive \text{ on } V. \end{cases}$$

Then the **symmetric variational problem** is the following.

(2.5.5)   Given $F \in V'$, find $u \in V$ such that $a(u, v) = F(v)$ $\forall v \in V$.

(2.5.6) Theorem. *Suppose that conditions* $(1) - (3)$ *of* $(2.5.4)$ *hold. Then there exists a unique* $u \in V$ *solving* $(2.5.5)$.

Proof. Proposition 2.5.3 implies that $a(\cdot, \cdot)$ is an inner product on V and that $(V, a(\cdot, \cdot))$ is a Hilbert space. Apply the Riesz Representation Theorem.
□

The **(Ritz-Galerkin) Approximation Problem** is the following.

> Given a finite-dimensional subspace $V_h \subset V$ and $F \in V'$, find $u_h \in V_h$ such that

(2.5.7) $a(u_h, v) = F(v)$ $\forall v \in V_h$.

(2.5.8) Theorem. *Under the conditions* $(2.5.4)$, *there exists a unique* u_h *that solves* $(2.5.7)$.

Proof. $(V_h, a(\cdot, \cdot))$ is a Hilbert space in its own right, and $F|_{V_h} \in V_h'$. Apply the Riesz Representation Theorem.
□

Error estimates for $u - u_h$ are a consequence of the following relationship.

(2.5.9) Proposition. (Fundamental Orthogonality) *Let u and u_h be solutions to* $(2.5.5)$ *and* $(2.5.7)$ *respectively. Then*

$$a(u - u_h, v) = 0 \quad \forall v \in V_h.$$

Proof. Subtract the two equations

$$
\begin{aligned}
a(u, v) &= F(v) & \forall v \in V \\
a(u_h, v) &= F(v) & \forall v \in V_h.
\end{aligned}
$$

□

(2.5.10) Corollary. $\|u - u_h\|_E = \min_{v \in V_h} \|u - v\|_E$.

Proof. Same as (0.3.3).
□

(2.5.11) *Remark.* **(The Ritz Method)** In the symmetric case, u_h minimizes the quadratic functional

$$Q(v) = a(v, v) - 2F(v)$$

over all $v \in V_h$ (see exercise 2.x.6).

Note that (2.5.10) and (2.5.11) are valid *only* in the symmetric case.

2.6 Formulation of Nonsymmetric Variational Problems

A nonsymmetric variational problem is posed as follows. Suppose that the following five conditions are valid:

(2.6.1) $\begin{cases} (1) & (H, (\cdot, \cdot)) \text{ is a Hilbert space.} \\ (2) & V \text{ is a (closed) subspace of } H. \\ (3) & a(\cdot, \cdot) \text{ is a bilinear form on } V, \; not \text{ necessarily symmetric.} \\ (4) & a(\cdot, \cdot) \text{ is continuous (bounded) on } V. \\ (5) & a(\cdot, \cdot) \text{ is coercive on } V. \end{cases}$

Then the **nonsymmetric variational problem** is the following.

(2.6.2)   Given $F \in V'$, find $u \in V$ such that $a(u, v) = F(v) \; \forall v \in V$.

The **(Galerkin) approximation problem** is the following.

Given a finite-dimensional subspace $V_h \subset V$ and $F \in V'$, find $u_h \in V_h$ such that

(2.6.3) $$a(u_h, v) = F(v) \qquad \forall v \in V_h.$$

The following questions arise.

1. Do there exist unique solutions u, u_h?
2. What are the error estimates for $u - u_h$?
3. Are there any interesting examples?

An Interesting Example. Consider the boundary value problem

(2.6.4) $\quad -u'' + u' + u = f \quad \text{on } [0, 1] \qquad u'(0) = u'(1) = 0.$

One variational formulation for this is: Take

$$V = H^1(0, 1)$$

(2.6.5) $$a(u, v) = \int_0^1 \left(u' v' + u' v + uv \right) dx$$

$$F(v) = (f, v)$$

and solve the variational equation (2.6.2). Note that $a(\cdot, \cdot)$ is *not* symmetric because of the $u' v$ term.

To prove $a(\cdot, \cdot)$ is continuous, observe that

$$|a(u, v)| \le |(u, v)_{H^1}| + \left| \int_0^1 u' v \, dx \right|$$

$$\le \|u\|_{H^1} \|v\|_{H^1} + \|u'\|_{L^2} \|v\|_{L^2} \qquad \text{(Schwarz' inequality 2.1.5)}$$

$$\le 2 \|u\|_{H^1} \|v\|_{H^1}.$$

Therefore, $a(\cdot,\cdot)$ is continuous (take $c_1 = 2$ in the definition).

To prove $a(\cdot,\cdot)$ is coercive, observe that

$$a(v,v) = \int_0^1 (v'^{\,2} + v'\,v + v^2)\,dx$$

$$= \frac{1}{2}\int_0^1 (v' + v)^2\,dx + \frac{1}{2}\int_0^1 (v'^2 + v^2)\,dx$$

$$\geq \frac{1}{2}\|v\|_{H^1}^2.$$

Therefore, $a(\cdot,\cdot)$ is coercive (take $c_2 = 1/2$ in the definition).   □

If the above differential equation is changed to

(2.6.6) $-u'' + ku' + u = f,$

then the corresponding $a(\cdot,\cdot)$ need *not* be coercive for large k.

(2.6.7) Remark. If $\big(H, (\cdot,\cdot)\big)$ is a Hilbert space, V is a subspace of H, and $a(\cdot,\cdot)$ is an inner product on V, then $\big(V, a(\cdot,\cdot)\big)$ need *not* be complete if $a(\cdot,\cdot)$ is not coercive. For example, let $H = H^1(0,1)$, $V = H, a(v,w) = \int_0^1 v\,w\,dx = (v,w)_{L^2(0,1)}$. Then $a(\cdot,\cdot)$ is an inner product in V, but convergence in the L^2 norm does not imply convergence in the H^1 norm since $H^1(0,1)$ is dense in $L^2(0,1)$.

2.7 The Lax-Milgram Theorem

We would like to prove the existence and uniqueness of the solution of the (nonsymmetric) variational problem:

Find $u \in V$ such that

(2.7.1) $a(u,v) = F(v) \qquad \forall v \in V,$

where V is a Hilbert space, $F \in V'$ and $a(\cdot,\cdot)$ is a continuous, coercive bilinear form that is not necessarily symmetric. The Lax-Milgram Theorem guarantees both existence and uniqueness of the solution to (2.7.1). First we need to prove the following lemma.

(2.7.2) Lemma. (Contraction Mapping Principle) *Given a Banach space V and a mapping $T : V \longrightarrow V$, satisfying*

(2.7.3) $\|Tv_1 - Tv_2\| \leq M\|v_1 - v_2\|$

for all $v_1, v_2 \in V$ and fixed M, $0 \leq M < 1$, there exists a unique $u \in V$ such that

(2.7.4) $u = Tu,$

i.e. the contraction mapping T has a unique fixed point u.

(2.7.5) *Remark.* We actually only need that V is a *complete metric space* in the lemma.

Proof. First, we show uniqueness. Suppose $Tv_1 = v_1$ and $Tv_2 = v_2$. Since T is a contraction mapping,

$$\|Tv_1 - Tv_2\| \le M\|v_1 - v_2\|$$

for some $0 \le M < 1$. But $\|Tv_1 - Tv_2\| = \|v_1 - v_2\|$. Therefore,

$$\|v_1 - v_2\| \le M\|v_1 - v_2\|.$$

This implies that $\|v_1 - v_2\| = 0$ (otherwise, we have $1 \le M$). Therefore, $v_1 = v_2$, i.e., the fixed point is unique.

Next, we show existence. Pick $v_0 \in V$ and define

$$v_1 = Tv_0, v_2 = Tv_1, \ldots, v_{k+1} = Tv_k, \ldots.$$

Note that $\|v_{k+1} - v_k\| = \|Tv_k - Tv_{k-1}\| \le M\|v_k - v_{k-1}\|$. Thus, by induction,

$$\|v_k - v_{k-1}\| \le M^{k-1}\|v_1 - v_0\|.$$

Therefore, for any $N > n$,

$$
\begin{aligned}
\|v_N - v_n\| &= \left\| \sum_{k=n+1}^{N} v_k - v_{k-1} \right\| \\
&\le \|v_1 - v_0\| \sum_{k=n+1}^{N} M^{k-1} \\
&\le \frac{M^n}{1-M}\|v_1 - v_0\| \\
&= \frac{M^n}{1-M}\|Tv_0 - v_0\|,
\end{aligned}
$$

(2.7.6)

which shows that $\{v_n\}$ is a Cauchy sequence. Since V is complete, $\{v_n\}$ is convergent. Thus, if $\lim_{n\to\infty} v_n =: v$, we have

$$
\begin{aligned}
v &= \lim_{n\to\infty} v_{n+1} \\
&= \lim_{n\to\infty} Tv_n \\
&= T\left(\lim_{n\to\infty} v_n \right) \quad \text{(because } T \text{ is continuous)} \\
&= Tv;
\end{aligned}
$$

in other words, there exists a fixed point.   □

(2.7.7) Theorem. (Lax-Milgram)   *Given a Hilbert space* $(V, (\cdot, \cdot))$, *a continuous, coercive bilinear form* $a(\cdot, \cdot)$ *and a continuous linear functional* $F \in V'$, *there exists a unique* $u \in V$ *such that*

$$(2.7.8) \qquad\qquad a(u, v) \;=\; F(v) \qquad \forall v \in V.$$

Proof. For any $u \in V$, define a functional Au by $Au(v) = a(u, v) \quad \forall v \in V$. Au is linear since

$$
\begin{aligned}
Au(\alpha v_1 + \beta v_2) &= a(u, \alpha v_1 + \beta v_2) \\
&= \alpha a(u, v_1) + \beta a(u, v_2) \\
&= \alpha Au(v_1) + \beta Au(v_2) \quad \forall v_1, v_2 \in V,\ \alpha, \beta \in \mathbb{R}.
\end{aligned}
$$

Au is also continuous since, for all $v \in V$,

$$|Au(v)| \;=\; |a(u, v)| \le C\|u\|\,\|v\|,$$

where C is the constant from the definition of continuity for $a(\cdot, \cdot)$. Therefore,

$$\|Au\|_{V'} \;=\; \sup_{v \neq 0} \frac{|Au(v)|}{\|v\|} \le C\|u\| < \infty.$$

Thus, $Au \in V'$. Similarly, (see exercise 2.x.8), one can show that the mapping $u \to Au$ is a linear map $V \longrightarrow V'$. Here we also showed that the linear mapping $A : V \longrightarrow V'$ is continuous with $\|A\|_{L(V,V')} \le C$.

Now, by the Riesz Representation Theorem, for any $\phi \in V'$ there exists unique $\tau\phi \in V$ such that $\phi(v) = (\tau\phi, v)$ for any $v \in V$ (by Remark 2.4.6). We must find a unique u such that

$$Au(v) \;=\; F(v) \qquad \forall v \in V.$$

In other words, we want to find a unique u such that

$$Au \;=\; F \qquad (\text{in } V'),$$

or

$$\tau Au \;=\; \tau F \qquad (\text{in } V),$$

since $\tau : V' \longrightarrow V$ is a one-to-one mapping. We solve this last equation by using Lemma 2.7.2. We want to find $\rho \neq 0$ such that the mapping $T : V \longrightarrow V$ is a contraction mapping, where T is defined by

$$(2.7.9) \qquad\qquad Tv \;:=\; v - \rho\big(\tau Av - \tau F\big) \qquad \forall v \in V.$$

If T is a contraction mapping, then by Lemma 2.7.2, there exists a unique $u \in V$ such that

$$(2.7.10) \qquad\qquad Tu \;=\; u - \rho\big(\tau Au - \tau F\big) \;=\; u,$$

that is, $\rho\big(\tau Au - \tau F\big) = 0$, or $\tau Au = \tau F$.

It remains to show that such a $\rho \neq 0$ exists. For any $v_1, v_2 \in V$, let $v = v_1 - v_2$. Then

$$
\begin{aligned}
\|Tv_1 - Tv_2\|^2 &= \|v_1 - v_2 - \rho(\tau A v_1 - \tau A v_2)\|^2 \\
&= \|v - \rho(\tau A v)\|^2 && (\tau, A \text{ are linear}) \\
&= \|v\|^2 - 2\rho(\tau A v, v) + \rho^2\|\tau A v\|^2 \\
&= \|v\|^2 - 2\rho A v(v) + \rho^2 Av(\tau A v) && (\text{definition of } \tau) \\
&= \|v\|^2 - 2\rho\, a(v, v) + \rho^2\, a(v, \tau A v) && (\text{definition of } A) \\
&\leq \|v\|^2 - 2\rho\alpha\|v\|^2 + \rho^2 C\|v\|\,\|\tau A v\| \\
&&& (\text{coercivity and continuity of } A) \\
&\leq (1 - 2\rho\alpha + \rho^2 C^2)\|v\|^2 && (A \text{ bounded}, \tau \text{ isometric}) \\
&= (1 - 2\rho\alpha + \rho^2 C^2)\|v_1 - v_2\|^2 \\
&= M^2\|v_1 - v_2\|^2.
\end{aligned}
$$

Here, α is the constant in the definition of coercivity of $a(\cdot, \cdot)$. Note that $\|\tau A v\| = \|A v\| \leq C\|v\|$ was used in the last inequality. We thus need

$$
1 - 2\rho\alpha + \rho^2 C^2 < 1 \quad \text{for some } \rho, \quad \text{i.e.,}
$$
$$
\rho(\rho C^2 - 2\alpha) < 0.
$$

If we choose $\rho \in (0, 2\alpha/C^2)$ then $M < 1$ and the proof is complete.   □

(2.7.11) Remark. Note that $\|u\|_V \leq (1/\alpha)\|F\|_{V'}$ where α is the coercivity constant (see exercise 2.x.9).

(2.7.12) Corollary. *Under conditions (2.6.1), the variational problem (2.6.2) has a unique solution.*

Proof. Conditions (1) and (2) of (2.6.1) imply that $(V, (\cdot, \cdot))$ is a Hilbert space. Apply the Lax-Milgram Theorem.   □

(2.7.13) Corollary. *Under the conditions (2.6.1), the approximation problem (2.6.3) has a unique solution.*

Proof. Since V_h is a (closed) subspace of V, (2.6.1) holds with V replaced by V_h. Apply the previous corollary.   □

(2.7.14) Remark. Note that V_h need not be finite-dimensional for (2.6.3) to be well-posed.

2.8 Estimates for General Finite Element Approximation

Let u be the solution to the variational problem (2.6.2) and u_h be the solution to the approximation problem (2.6.3). We now want to estimate the error $\|u - u_h\|_V$. We do so by the following theorem.

(2.8.1) Theorem. (Céa)   *Suppose the conditions (2.6.1) hold and that u solves (2.6.2). For the finite element variational problem (2.6.3) we have*

$$(2.8.2) \qquad \|u - u_h\|_V \leq \frac{C}{\alpha} \min_{v \in V_h} \|u - v\|_V \,,$$

where C is the continuity constant and α is the coercivity constant of $a(\cdot, \cdot)$ on V.

Proof. Since $a(u, v) = F(v)$ for all $v \in V$ and $a(u_h, v) = F(v)$ for all $v \in V_h$ we have (by subtracting)

$$(2.8.3) \qquad a(u - u_h, v) = 0 \qquad \forall v \in V_h.$$

For all $v \in V_h$,

$$
\begin{aligned}
\alpha \|u - u_h\|_V^2 &\leq a(u - u_h, u - u_h) &&\text{(by coercivity)} \\
&= a(u - u_h, u - v) + a(u - u_h, v - u_h) \\
&= a(u - u_h, u - v) &&\text{(since } v - u_h \in V_h\text{)} \\
&\leq C \|u - u_h\|_V \|u - v\|_V. &&\text{(by continuity)}
\end{aligned}
$$

Hence,

$$(2.8.4) \qquad \|u - u_h\|_V \leq \frac{C}{\alpha} \|u - v\|_V \qquad \forall v \in V_h.$$

Therefore,

$$
\begin{aligned}
\|u - u_h\|_V &\leq \frac{C}{\alpha} \inf_{v \in V_h} \|u - v\|_V \\
&= \frac{C}{\alpha} \min_{v \in V_h} \|u - v\|_V. &&\text{(since } V_h \text{ is closed)}
\end{aligned}
$$

$\square$

(2.8.5) *Remarks.*

1. Céa's Theorem shows that u_h is *quasi-optimal* in the sense that the error $\|u - u_h\|_V$ is proportional to the best it can be using the subspace V_h.
2. In the symmetric case, we proved
$$\|u - u_h\|_E = \min_{v \in V_h} \|u - v\|_E.$$

Hence,

$$\|u - u_h\|_V \leq \frac{1}{\sqrt{\alpha}} \|u - u_h\|_E$$

$$= \frac{1}{\sqrt{\alpha}} \min_{v \in V_h} \|u - v\|_E$$

$$\leq \sqrt{\frac{C}{\alpha}} \min_{v \in V_h} \|u - v\|_V$$

$$\leq \frac{C}{\alpha} \min_{v \in V_h} \|u - v\|_V,$$

the result of Céa's Theorem. This is really the remark about the relationship between the two formulations, namely, that one can be derived from the other.

2.9 Higher-dimensional Examples

We now show how the theory developed in the previous sections can be applied in some multi-dimensional problems. Consider a variational form defined by

$$a(u, v) = \int_\Omega A(x)\nabla u(x) \cdot \nabla v(x) + \big(\mathbf{B}(x) \cdot \nabla u(x)\big)v(x) + C(x)u(x)v(x) \, dx$$

where A, $\mathbf{B}$ and C are bounded, measurable functions on $\Omega \subset \mathbb{R}^n$. Of course, $\mathbf{B}$ is vector valued. Formally, this variational form corresponds to the differential operator

$$-\nabla \cdot \big(A(x)\nabla u(x)\big) + \mathbf{B}(x) \cdot \nabla u(x) + C(x)u(x),$$

but the variational formulation is well defined even when the differential operator makes no sense in a traditional way. Hölder's inequality implies that $a(\cdot, \cdot)$ is continuous on $H^1(\Omega)$, with the constant c_1 in Definition 2.5.2 depending only on the $L^\infty(\Omega)$ norms of the coefficients. However, it is more complicated to verify coercivity.

To begin with, we consider the symmetric case, $\mathbf{B} \equiv 0$. Next, suppose there is a constant, $\gamma > 0$, such that

(2.9.1) $A(x) \geq \gamma \quad \& \quad C(x) \geq \gamma \quad$ for a.a. $x \in \Omega$.

Then $a(\cdot, \cdot)$ is coercive on all of $H^1(\Omega)$, with the constant c_2 in Definition 2.5.2 equal to γ. Thus, we have the following.

(2.9.2) Theorem. *If* $\mathbf{B} \equiv 0$ *and* (2.9.1) *holds, then there is a unique solution,* u, *to* (2.5.5) *with* $a(\cdot, \cdot)$ *as above and* $V = H^1(\Omega)$. *Moreover, for any* $V_h \subset H^1(\Omega)$, *there is a unique solution,* u_h, *to* (2.5.7) *and the estimate* (2.5.10) *holds for* $u - u_h$.

Note that we have not even assumed continuity of the coefficients A and C; in fact, discontinuous coefficients occur in many important physical models. The condition (2.9.1) of positivity for C is necessary to some degree, since if $C \equiv 0$ then $a(v, v) = 0$ for any constant function, v.

Now consider the general case, when $\mathbf{B}$ is nonzero. We have, by Hölder's inequality and the arithmetic-geometric mean inequality (0.9.5),

$$\left| \int_\Omega \left(\mathbf{B}(x) \cdot \nabla u(x) \right) u(x) \, dx \right| \leq \| \, |\mathbf{B}| \, \|_{L^\infty(\Omega)} \, |u|_{H^1(\Omega)} \, \|u\|_{L^2(\Omega)}$$

$$\leq \| \, |\mathbf{B}| \, \|_{L^\infty(\Omega)} \, \|u\|^2_{H^1(\Omega)} / 2.$$

If (2.9.1) holds and in addition

(2.9.3) $\| \, |\mathbf{B}| \, \|_{L^\infty(\Omega)} < 2\gamma$

then $a(\cdot, \cdot)$ is coercive on $H^1(\Omega)$ and the following holds.

(2.9.4) Theorem. *If* (2.9.1) *and* (2.9.3) *hold, then there is a unique solution,* u, *to* (2.6.2) *with* $a(\cdot, \cdot)$ *as above and* $V = H^1(\Omega)$. *Moreover, for any* $V_h \subset H^1(\Omega)$, *there is a unique solution,* u_h, *to* (2.6.3) *and the estimate* (2.8.2) *holds for* $u - u_h$.

The conditions given here for coercivity are somewhat restrictive, although they are appropriate in the case of Neumann boundary conditions. More complex variational problems will be considered in Chapter 5. In particular, it will be demonstrated that less stringent conditions are necessary in the presence of Dirichlet boundary conditions in order to guarantee coercivity.

2.x Exercises

2.x.1 Prove (2.2.5.ii).

2.x.2 If M is a subspace, prove that $\left(M^\perp \right)^\perp = M$.

2.x.3 Prove parts (1) and (3) of Proposition 2.2.7.

2.x.4 Prove Theorem 2.2.8.

2.x.5 Let $P_M{}^\perp$ be the operator defined in (2.3.4). Prove that $P_M{}^\perp = \Gamma_{M^\perp}$.

2.x.6 Prove the claim in Remark 2.5.11.

2.x.7 Prove that a contraction mapping is always continuous (cf. Lemma 2.7.2).

2.x.8 Prove that the mapping $u \to Au$ in the proof of the Lax-Milgram Theorem 2.7.7 is a linear map $V \longrightarrow V'$.

2.x.9 Prove that the solution u guaranteed by the Lax-Milgram Theorem satisfies

$$\|u\|_V \leq \frac{1}{\alpha} \|F\|_{V'}$$

(cf. Remark 2.7.11).

2.x.10 For the differential equation $-u'' + ku' + u = f$, find a value for k such that $a(v, v) = 0$ but $v \not\equiv 0$ for some $v \in H^1(0, 1)$ (cf. (2.6.6)).

2.x.11 Let $a(\cdot, \cdot)$ be the inner product for a Hilbert space V. Prove that the following two statements are equivalent for $F \in V'$ and an *arbitrary* (closed) subspace U of V:

a) $u \in U$ satisfies $a(u, v) = F(v)$ $\forall v \in U$
b) u minimizes $\frac{1}{2} a(v, v) - F(v)$ over $v \in U$,
i.e. show that existence in one implies existence in the other. (Hint: expand the expression $a(u + \epsilon v, u + \epsilon v) = \ldots.$)

2.x.12 Let $a(u, v) = \int_0^1 (u' \, v' + u' \, v + uv) \, dx$ and $V = \{v \in W_2^1(0, 1) : v(0) = v(1) = 0\}$. Prove that $a(v, v) = \int_0^1 [(v')^2 + v^2] \, dx$ for all $v \in V$ (Hint: write $vv' = \frac{1}{2}(v^2)'$.)

2.x.13 Let $a(\cdot, \cdot)$, V and U be as in exercise 2.x.11. For $g \in V$ define $U_g = \{v + g : v \in U\}$ (note $U_0 = U$). Prove that the following statements are equivalent for arbitrary $g \in V$:
a) $u \in U_g$ satisfies $a(u, v) = 0$ $\forall v \in U_0$
b) u minimizes $a(v, v)$ over all $v \in U_g$.

2.x.14 Show that $\mathcal{D}(\Omega)$ is dense in $L^2(\Omega)$. (Hint: use exercise 1.x.39 to show that $\mathcal{D}(\Omega)^\perp = \{0\}$.)

2.x.15 Let H be a Hilbert space and let $v \in H$ be arbitrary. Prove that

$$\|v\|_H = \sup_{0 \neq w \in H} \frac{(v, w)_H}{\|w\|_H} .$$

(Hint: apply Schwarz' inequality (2.1.5) and also consider $w = v$.)

Chapter 3

The Construction of a Finite Element Space

To approximate the solution of the variational problem,

$$a(u, v) = F(v) \quad \forall\, v \in V,$$

developed in Chapter 0, we need to construct finite-dimensional subspaces $S \subset V$ in a systematic, practical way.

Let us examine the space S defined in Sect. 0.4. To understand fully the functions in the space S, we need to answer the following questions:

1. What does a function look like in a given subinterval?
2. How do we determine the function in a given subinterval?
3. How do the restrictions of a function on two neighboring intervals match at the common boundary?

In this chapter, we will define piecewise function spaces that are similar to S, but which are defined on more general regions. We will develop concepts that will help us answer these questions.

3.1 The Finite Element

We follow Ciarlet's definition of a finite element (Ciarlet 1978).

(3.1.1) Definition. *Let*

    (i) *$K \subseteq \mathrm{I\!R}^n$ be a bounded closed set with nonempty interior and piecewise smooth boundary (the **element domain**),*

    (ii) *$\mathcal{P}$ be a finite-dimensional space of functions on K (the **space of shape functions**) and*

    (iii) *$\mathcal{N} = \{N_1, N_2, \ldots, N_k\}$ be a basis for $\mathcal{P}'$ (the set of **nodal variables**).*

*Then $(K, \mathcal{P}, \mathcal{N})$ is called a **finite element**.*

It is implicitly assumed that the nodal variables, N_i, lie in the dual space of some larger function space, e.g., a Sobolev space.

(3.1.2) Definition. *Let $(K, \mathcal{P}, \mathcal{N})$ be a finite element. The basis $\{\phi_1, \phi_2, \ldots, \phi_k\}$ of $\mathcal{P}$ dual to $\mathcal{N}$ (i.e., $N_i(\phi_j) = \delta_{ij}$) is called the **nodal basis** of $\mathcal{P}$.*

(3.1.3) Example. (the 1-dimensional Lagrange element) Let $K = [0, 1]$, $\mathcal{P} =$ the set of linear polynomials and $\mathcal{N} = \{N_1, N_2\}$, where $N_1(v) = v(0)$ and $N_2(v) = v(1) \quad \forall v \in \mathcal{P}$. Then $(K, \mathcal{P}, \mathcal{N})$ is a finite element and the noda basis consists of $\phi_1(x) = 1 - x$ and $\phi_2(x) = x$.

In general, we can let $K = [a, b]$ and $\mathcal{P}_k =$ the set of all polynomials of degree less than or equal to k. Let $\mathcal{N}_k = \{N_0, N_1, N_2, \ldots, N_k\}$, where $N_i(v) = v(a + (b - a)i/k) \quad \forall v \in \mathcal{P}_k$ and $i = 0, 1, \ldots, k$. Then $(K, \mathcal{P}_k, \mathcal{N}_k)$ is a finite element. The verification of this uses Lemma 3.1.4.

Usually, condition (iii) of Definition 3.1.1 is the only one that requires much work, and the following simplifies its verification.

(3.1.4) Lemma. *Let $\mathcal{P}$ be a d-dimensional vector space and let $\{N_1, N_2, \ldots, N_d\}$ be a subset of the dual space $\mathcal{P}'$. Then the following two statements are equivalent.*

(a) $\{N_1, N_2, \ldots, N_d\}$ *is a basis for $\mathcal{P}'$.*
(b) *Given $v \in \mathcal{P}$ with $N_i v = 0$ for $i = 1, 2, \ldots, d$, then $v \equiv 0$.*

Proof. Let $\{\phi_1, \ldots, \phi_d\}$ be some basis for $\mathcal{P}$. $\{N_1, \ldots, N_d\}$ is a basis for $\mathcal{P}'$ iff given any L in $\mathcal{P}'$,

$$(3.1.5) \qquad\qquad L = \alpha_1 N_1 + \ldots + \alpha_d N_d$$

(because $d = \dim \mathcal{P} = \dim \mathcal{P}'$). The equation (3.1.5) is equivalent to

$$y_i := L(\phi_i) = \alpha_1 N_1(\phi_i) + \ldots + \alpha_d N_d(\phi_i), \quad i = 1, \ldots, d.$$

Let $\mathbf{B} = (N_j(\phi_i))$, $i, j = 1, \ldots, d$. Thus, (a) is equivalent to $\mathbf{B}\alpha = y$ is *always solvable*, which is the same as $\mathbf{B}$ being invertible.

Given any $v \in \mathcal{P}$, we can write $v = \beta_1 \phi_1 + \ldots + \beta_d \phi_d$. $N_i v = 0$ means that $\beta_1 N_i(\phi_1) + \ldots + \beta_d N_i(\phi_d) = 0$. Therefore, (b) is equivalent to

$$(3.1.6) \qquad \begin{array}{c} \beta_1 N_i(\phi_1) + \ldots + \beta_d N_i(\phi_d) = 0 \quad \text{for } i = 1, \ldots, d \\ \implies \beta_1 = \ldots = \beta_d = 0. \end{array}$$

Let $\mathbf{C} = (N_i(\phi_j))$, $i, j = 1, \ldots, d$. Then (b) is equivalent to $\mathbf{C}x = 0$ *only has trivial solutions*, which is the same as $\mathbf{C}$ being invertible. But $\mathbf{C} = \mathbf{B}^T$. Therefore, (a) is equivalent to (b). $\qquad\square$

(3.1.7) Remark. Condition (iii) of Definition 3.1.1 is the same as (a) in Lemma 3.1.4, which can be verified by checking (b) in Lemma 3.1.4. For instance, in Example 3.1.3, $v \in \mathcal{P}_1$ means $v = a + bx$; $N_1(v) = N_2(v) = 0$ means $a = 0$ and $a + b = 0$. Hence, $a = b = 0$, i.e., $v \equiv 0$. More generally, if

$v \in \mathcal{P}_k$ and $0 = N_i(v) = v(a + (b-a)i/k) \ \forall i = 0, 1, \ldots, k$ then v vanishes identically by the fundamental theorem of algebra. Thus, $(K, \mathcal{P}_k, \mathcal{N}_k)$ is a finite element.

We will use the following terminology in subsequent sections.

(3.1.8) Definition. *We say that $\mathcal{N}$ determines $\mathcal{P}$ if $\psi \in \mathcal{P}$ with $N(\psi) = 0 \ \forall N \in \mathcal{N}$ implies that $\psi = 0$.*

(3.1.9) Remark. We will often refer to the hyperplane $\{x : L(x) = 0\}$, where L is a non-degenerate linear function, simply as L.

(3.1.10) Lemma. *Let P be a polynomial of degree $d \geq 1$ that vanishes on a hyperplane L. Then we can write $P = LQ$, where Q is a polynomial of degree $(d-1)$.*

Proof. Make an affine change of coordinates such that $L(\hat{x}, x_n) = x_n$ and the hyperplane $L(\hat{x}, x_n) = 0$ is the $\hat{x}$-axis. Therefore, $P(\hat{x}, 0) \equiv 0$. Since degree$(P) = d$, we have

$$P(\hat{x}, x_n) = \sum_{j=0}^{d} \sum_{|\hat{i}| \leq d-j} c_{\hat{i}j} \hat{x}^{\hat{i}} x_n^j$$

where $\hat{x} = (x_1, \ldots, x_{n-1})$ and $\hat{i} = (i_1, \ldots, i_{n-1})$. Letting $x_n = 0$, we obtain $0 \equiv P(\hat{x}, 0) = \sum_{|\hat{i}| \leq d} c_{\hat{i}0} \hat{x}^{\hat{i}}$, which implies that $c_{\hat{i}0} = 0$ for $|\hat{i}| \leq d$. Therefore,

$$P(\hat{x}, x_n) = \sum_{j=1}^{d} \sum_{|\hat{i}| \leq d-j} c_{\hat{i}j} \hat{x}^{\hat{i}} x_n^j$$

$$= x_n \sum_{j=1}^{d} \sum_{|\hat{i}| \leq d-j} c_{\hat{i}j} \hat{x}^{\hat{i}} x_n^{j-1}$$

$$= x_n Q$$

$$= LQ,$$

where degree $Q = d - 1$. $\qquad\square$

3.2 Triangular Finite Elements

Let K be any triangle. Let $\mathcal{P}_k$ denote the set of all polynomials in two variables of degree $\leq k$. The following table gives the dimension of $\mathcal{P}_k$.

Table 3.1. Dimension of $\mathcal{P}_k$ in two dimensions

k	$\dim \mathcal{P}_k$
1	3
2	6
3	10
⋮	⋮
k	$\frac{1}{2}(k+1)(k+2)$

The Lagrange Element

(3.2.1) Example. (k = 1) Let $\mathcal{P} = \mathcal{P}_1$. Let $\mathcal{N}_1 = \{N_1, N_2, N_3\}$ ($\dim \mathcal{P}_1 = 3$) where $N_i(v) = v(z_i)$ and z_1, z_2, z_3 are the vertices of K. This element is depicted in Fig. 3.1.

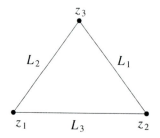

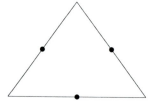

Fig. 3.1. linear Lagrange triangle

Fig. 3.2. Crouzeix-Raviart noncon-forming linear triangle

Note that "•" indicates the nodal variable evaluation at the point where the dot is located.

We verify 3.1.1(iii) using 3.1.4(b), i.e., we prove that $\mathcal{N}_1$ determines $\mathcal{P}_1$. Let L_1, L_2 and L_3 be non-trivial linear functions that define the lines on which lie the edges of the triangle. Suppose that a polynomial $P \in \mathcal{P}$ vanishes at z_1, z_2 and z_3. Since $P|_{L_1}$ is a linear function of one variable that vanishes at two points, $P = 0$ on L_1. By Lemma 3.1.10 we can write $P = c\,L_1$, where c is a constant. But

$$0 = P(z_1) = c\,L_1(z_1) \quad \Longrightarrow \quad c = 0$$

(because $L_1(z_1) \neq 0$). Thus, $P \equiv 0$ and hence $\mathcal{N}_1$ determines $\mathcal{P}_1$. □

(3.2.2) *Remark.* The above choice for $\mathcal{N}$ is not unique. For example, we could have defined

$$N_i(v) = v(\text{midpoint of the } i^{th} \text{ edge}),$$

as shown in Fig. 3.2. By connecting the midpoints, we construct a triangle on which $P \in \mathcal{P}_1$ vanishes at the vertices. An argument similar to the one in Example 3.2.1 shows that $P \equiv 0$ and hence, $\mathcal{N}_1$ determines $\mathcal{P}_1$.

(3.2.3) Example. (k = 2) Let $\mathcal{P} = \mathcal{P}_2$. Let $\mathcal{N}_2 = \{N_1, N_2, \ldots, N_6\}$ (dim $\mathcal{P}_2 = 6$) where

$$N_i(v) = \begin{cases} v(i^{th} \text{ vertex}), & i=1,2,3; \\ v(\text{midpoint of the } (i-3) \text{ edge}), & \\ \quad (\text{or any other point on the } i-3 \text{ edge}) & i=4,5,6. \end{cases}$$

This element is depicted in Fig. 3.3.

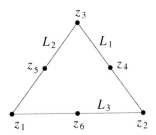

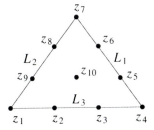

Fig. 3.3. quadratic Lagrange triangle   **Fig. 3.4.** cubic Lagrange triangle

We need to check that $\mathcal{N}_2$ determines $\mathcal{P}_2$. As before, let L_1, L_2 and L_3 be non-trivial linear functions that define the edges of the triangle. Suppose that the polynomial $P \in \mathcal{P}_2$ vanishes at $z_1, z_2, \ldots, z_6$. Since $P|_{L_1}$ is a quadratic function of one variable that vanishes at three points, $P = 0$ on L_1. By Lemma 3.1.10 we can write $P = L_1 Q_1$ where $\deg Q_1 = (\deg P) - 1 = 2 - 1 = 1$. But P also vanishes on L_2. Therefore, $L_1 Q_1|_{L_2} = 0$. Hence, on L_2, either $L_1 = 0$ or $Q_1 = 0$. But L_1 can equal zero only at one point of L_2 since we have a non-degenerate triangle. Therefore, $Q_1 = 0$ on L_2, except possibly at one point. By continuity, we have $Q_1 \equiv 0$ on L_2.

By Lemma 3.1.10, we can write $Q_1 = L_2 Q_2$, where $\deg Q_2 = (\deg L_2) - 1 = 1 - 1 = 0$. Hence, Q_2 is a constant (say c), and we can write $P = c\, L_1 L_2$. But $P(z_6) = 0$ and z_6 does not lie on either L_1 or L_2. Therefore,

$$0 = P(z_6) = c\, L_1(z_6)\, L_2(z_6) \quad \Longrightarrow \quad c = 0,$$

since $L_1(z_6) \neq 0$ and $L_2(z_6) \neq 0$. Thus, $P \equiv 0$.   □

(3.2.4) Example. (k=3) Let $\mathcal{P} = \mathcal{P}_3$. Let $\mathcal{N}_3 = \{N_i : i = 1, 2, \ldots, 10\ (= \dim \mathcal{P}_3)\}$ where

$$N_i(v) = v(z_i), \quad i = 1, 2, \ldots, 9 \ (z_i \text{ distinct points on edges as in Fig. 3.4})$$

and

$$N_{10}(v) = v(\text{any interior point}).$$

We must show that $\mathcal{N}_3$ determines $\mathcal{P}_3$.

Let L_1, L_2 and L_3 be non-trivial linear functions that define the edges of the triangle. Suppose that $P \in \mathcal{P}_3$ vanishes at z_i for $i = 1, 2, \ldots, 10$. Applying Lemma 3.1.10 three times along with the fact that $P(z_i) = 0$ for $i = 1, 2, \ldots, 9$, we can write $P = c \, L_1 \, L_2 \, L_3$. But

$$0 = P(z_{10}) = c \, L_1(z_{10}) \, L_2(z_{10}) \, L_3(z_{10}) \quad \Longrightarrow \quad c = 0$$

since $L_i(z_{10}) \neq 0$ for $i = 1, 2, 3$. Thus, $P \equiv 0$.   □

In general for $k \geq 1$, we let $\mathcal{P} = \mathcal{P}_k$. For $\mathcal{N}_k = \{N_i \; : \; i = 1, 2, \ldots, \frac{1}{2}(k+1)(k+2)\}$, we choose evaluation points at

(3.2.5)

$$3 \text{ vertex nodes,}$$

$$3(k-1) \text{ distinct edge nodes and}$$

$$\frac{1}{2}(k-2)(k-1) \text{ interior points.}$$

(The interior points are chosen, by induction, to determine $\mathcal{P}_{k-3}$.) Note that these choices suffice since

$$3 + 3(k-1) + \frac{1}{2}(k-2)(k-1) = 3k + \frac{1}{2}(k^2 - 3k + 2)$$

$$= \frac{1}{2}(k^2 + 3k + 2)$$

$$= \frac{1}{2}(k+1)(k+2)$$

$$= \dim \mathcal{P}_k.$$

The evaluation points for $k = 4$ and $k = 5$ are depicted in Fig. 3.5.

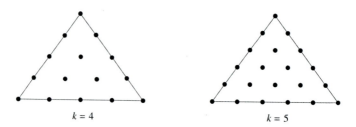

$$k = 4 \qquad\qquad\qquad k = 5$$

Fig. 3.5. quartic and quintic Lagrange triangles

To show that $\mathcal{N}_k$ determines $\mathcal{P}_k$, we suppose that $P \in \mathcal{P}_k$ vanishes at all the nodes. Let L_1, L_2 and L_3 be non-trivial linear functions that define the edges of the triangle. As before, we conclude from the vanishing of P at the edge and vertex nodes that $P = Q \, L_1 \, L_2 \, L_3$ where degree$(Q) \leq k - 3$; Q must vanish at all the interior points, since none of the L_i can be zero there. These points were chosen precisely to determine that $Q \equiv 0$.

The Hermite Element

(3.2.6) Example. (k = 3   Cubic Hermite) Let $\mathcal{P} = \mathcal{P}_3$. Let "•" denote evaluation at the point and "○" denote evaluation of the gradient at the center of the circle. Note that the latter corresponds to two distinct nodal variables, but the particular representation of the gradient is not unique. We claim that $\mathcal{N} = \{N_1, N_2, \ldots, N_{10}\}$, as depicted in Fig. 3.6, determines $\mathcal{P}_3$ (dim $\mathcal{P}_3 = 10$).

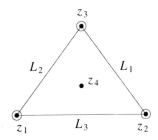

Fig. 3.6. cubic Hermite triangle

Let L_1, L_2 and L_3 again be non-trivial linear functions that define the edges of the triangle. Suppose that for a polynomial $P \in \mathcal{P}_3$, $N_i(P) = 0$ for $i = 1, 2, \ldots, 10$. Restricting P to L_1, we see that z_2 and z_3 are double roots of P since $P(z_2) = 0, P'(z_2) = 0$ and $P(z_3) = 0, P'(z_3) = 0$, where $'$ denotes differentiation along the straight line L_1. But the only third order polynomial in one variable with four roots is the zero polynomial, hence $P \equiv 0$ along L_1. Similarly, $P \equiv 0$ along L_2 and L_3. We can, therefore, write $P = c\,L_1\,L_2\,L_3$. But

$$0 = P(z_4) = c\,L_1(z_4)\,L_2(z_4)\,L_3(z_4) \quad \Longrightarrow \quad c = 0,$$

because $L_i(z_4) \neq 0$ for $i = 1, 2, 3$.   □

(3.2.7) *Remark.* Using directional derivatives, there are various distinct ways to define a finite element using $\mathcal{P}_3$, two of which are shown in Fig. 3.7. Note that arrows represent directional derivatives along the indicated directions at the points. The "global" element to the left has the advantage of ease of computation of directional derivatives in the x or y directions throughout the larger region divided up into triangles. The "local" element to the right holds the advantage in that the nodal parameters of each triangle are invariant with respect to the triangle.

In the general Hermite case, we have

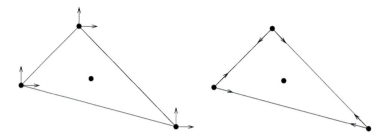

Fig. 3.7. Two different sets of nodal values for cubic Hermite elements.

$$(3.2.9) \quad \begin{cases} 3 \text{ vertex nodes} \\ 6 \text{ directional derivatives (2 for each gradient,} \\ \quad \text{evaluated at each of the 3 vertices)} \\ 3(k-3) \text{ edge nodes} \\ \frac{1}{2}(k-2)(k-1) \text{ interior nodes (as in the Lagrange case).} \end{cases}$$

Note that these sum to $\frac{1}{2}(k+1)(k+2) = \dim \mathcal{P}_k$ as in the Lagrange case.

(3.2.8) Example. (k = 4) We have $(\dim \mathcal{P}_4 = 15)$. Then $\mathcal{N} = \{N_1, N_2, \ldots, N_{15}\}$, as depicted in Fig. 3.8, determines $\mathcal{P}_4$.

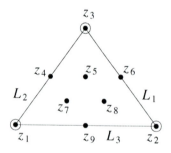

Fig. 3.8. quartic Hermite triangle

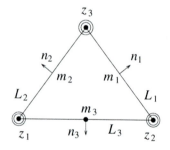

Fig. 3.9. quintic Argyris triangle

The Argyris Element

(3.2.10) Example. (k = 5)   Let $\mathcal{P} = \mathcal{P}_5$. Consider the $21 (= \dim \mathcal{P}_5)$ degrees of freedom shown in Fig. 3.9. As before, let • denote evaluation at the point and the inner circle denote evaluation of the gradient at the center. The outer circle denotes evaluation of the three second derivatives at the center. The arrows represent the evaluation of the normal derivatives at the three midpoints. We claim that $\mathcal{N} = \{N_1, N_2, \ldots, N_{21}\}$ determines $\mathcal{P}_5$.

Suppose that for some $P \in \mathcal{P}_5$, $N_i(P) = 0$ for $i = 1, 2, \ldots, 21$. Let L_i be as before in the Lagrange and Hermite cases. The restriction of P to L_1 is a fifth order polynomial in one variable with triple roots at z_2 and z_3. Hence,

P vanishes identically on L_1. Similarly, P vanishes on L_2 and L_3. Therefore, $P = Q\, L_1\, L_2\, L_3$, where $\deg Q = 2$. Observe that $(\partial_{L_1}\partial_{L_2}P)(z_3) = 0$, where ∂_{L_1} and ∂_{L_2} are the directional derivatives along L_1 and L_2 respectively. Therefore,

$$0 = (\partial_{L_1}\partial_{L_2}P)(z_3) = Q(z_3)\, L_3(z_3)\, \partial_{L_2} L_1\, \partial_{L_1} L_2,$$

since $\partial_{L_i} L_i \equiv 0$ & $L_i(z_3) = 0$, $i = 1, 2$. This implies $Q(z_3) = 0$ because $L_3(z_3) \neq 0$, $\partial_{L_2} L_1 \neq 0$ and $\partial_{L_1} L_2 \neq 0$. Similarly, $Q(z_1) = 0$ and $Q(z_2) = 0$. Also, since $L_1(m_1) = 0$, $\frac{\partial}{\partial n_1}P(m_1) = \left(Q\, \frac{\partial L_1}{\partial n_1}\, L_2\, L_3\right)(m_1)$. Therefore,

$$0 = \frac{\partial}{\partial n_1}P(m_1) \quad \Longrightarrow \quad Q(m_1) = 0$$

because $\frac{\partial L_1}{\partial n_1} \neq 0$, $L_2(m_1) \neq 0$ and $L_3(m_1) \neq 0$. Similarly, $Q(m_2) = 0$ and $Q(m_3) = 0$. So $Q \equiv 0$ by Example 3.2.3. □

We leave to the reader the verification of the following generalization of the Argyris element (exercise 3.x.12).

(3.2.11) Example. Note that $\dim \mathcal{P}_7 = 36$. The nodal variables depicted in Fig. 3.10 determine $\mathcal{P}_7$.

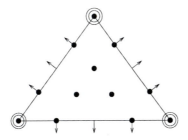

Fig. 3.10. seventh-degree Argyris triangle

3.3 The Interpolant

Now that we have examined a number of finite elements, we wish to piece them together to create subspaces of Sobolev spaces. We begin by defining the (local) interpolant.

(3.3.1) Definition. *Given a finite element* $(K, \mathcal{P}, \mathcal{N})$, *let the set* $\{\phi_i : 1 \leq i \leq k\} \subseteq \mathcal{P}$ *be the basis dual to* $\mathcal{N}$. *If* v *is a function for which all* $N_i \in \mathcal{N}$, $i = 1, \ldots, k$, *are defined, then we define the* **local interpolant** *by*

$$(3.3.2) \qquad \qquad \mathcal{I}_K v := \sum_{i=1}^{k} N_i(v)\, \phi_i.$$

(3.3.3) Example. Let K be the triangle depicted in Fig. 3.11, $\mathcal{P} = \mathcal{P}_1$, $\mathcal{N} = \{N_1, N_2, N_3\}$ as in Example 3.2.1, and $f = e^{xy}$. We want to find $\mathcal{I}_K f$.

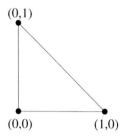

(0,1)

(0,0)   (1,0)

Fig. 3.11. coordinates for linear interpolant

By definition, $\mathcal{I}_K f = N_1(f)\,\phi_1 + N_2(f)\,\phi_2 + N_3(f)\,\phi_3$. We must therefore determine ϕ_1, ϕ_2 and ϕ_3. The line L_1 is given by $y = 1 - x$. We can write $\phi_1 = c\,L_1 = c(1 - x - y)$. But $N_1\,\phi_1 = 1$ implies that $c = \phi_1(z_1) = 1$, hence $\phi_1 = 1 - x - y$. Similarly, $\phi_2 = L_2(x,y)/L_2(z_2) = x$ and $\phi_3 = L_3(x,y)/L_3(z_3) = y$. Therefore,

$$
\begin{aligned}
\mathcal{I}_K f &= N_1(f)\,(1 - x - y) + N_2(f)\,x + N_3(f)\,y \\
&= 1 - x - y + x + y \qquad\qquad \text{(since } f = e^{xy}) \\
&= 1.
\end{aligned}
$$

Properties of the interpolant follow.   □

(3.3.4) Proposition. $\mathcal{I}_K$ is linear.

Proof. See exercise 3.x.2.   □

(3.3.5) Proposition. $N_i\big(\mathcal{I}_K(f)\big) = N_i(f) \quad \forall 1 \le i \le d$.

Proof. We have

$$
\begin{aligned}
N_i\big(\mathcal{I}_K(f)\big) &= N_i\left(\sum_{j=1}^{k} N_j(f)\,\phi_j\right) && \text{(definition of } \mathcal{I}_K(f)) \\
&= \sum_{j=1}^{k} N_j(f)\,N_i(\phi_j) && \text{(linearity of } N_i) \\
&= N_i(f) && (\{\phi_j\} \text{ dual to } \{N_j\}).
\end{aligned}
$$

□

(3.3.6) *Remark.* Proposition 3.3.5 has the interpretation that $\mathcal{I}_K(f)$ is the unique shape function that has the same nodal values as f.

(3.3.7) Proposition. $\mathcal{I}_K(f) = f$ *for* $f \in \mathcal{P}$. *In particular,* $\mathcal{I}_K$ *is idempotent, i.e.,* $\mathcal{I}_K^2 = \mathcal{I}_K$.

Proof. From (3.3.5),

$$N_i(f - \mathcal{I}_K(f)) = 0 \quad \forall i$$

which implies the first assertion. The second is a consequence of the first:

$$\mathcal{I}_K^2 f = \mathcal{I}_K(\mathcal{I}_K f) = \mathcal{I}_K f,$$

since $\mathcal{I}_K f \in \mathcal{P}$. □

We now piece together the elements.

(3.3.8) Definition. *A* **subdivision** *of a domain* Ω *is a finite collection of element domains* $\{K_i\}$ *such that*

(1) *int* $K_i \cap$ *int* $K_j = \emptyset$ *if* $i \neq j$ *and*
(2) $\bigcup K_i = \overline{\Omega}$.

(3.3.9) Definition. *Suppose* Ω *is a domain with a subdivision* $\mathcal{T}$. *Assume each element domain,* K, *in the subdivision is equipped with some type of shape functions,* $\mathcal{P}$, *and nodal variables,* $\mathcal{N}$, *such that* $(K, \mathcal{P}, \mathcal{N})$ *forms a finite element. Let* m *be the order of the highest partial derivatives involved in the nodal variables. For* $f \in C^m(\overline{\Omega})$, *the* **global interpolant** *is defined by*

(3.3.10) $$\mathcal{I}_{\mathcal{T}} f|_{K_i} = \mathcal{I}_{K_i} f$$

for all $K_i \in \mathcal{T}$.

Without further assumptions on a subdivision, no continuity properties can be asserted for the global interpolant. We now describe conditions that yield such continuity. Only the two-dimensional case using triangular elements is considered in detail here; analogous definitions and results can be formulated for higher dimensions and other subdivisions.

(3.3.11) Definition. *A* **triangulation** *of a polygonal domain* Ω *is a subdivision consisting of triangles having the property that*

(3) *no vertex of any triangle lies in the interior of an edge of another triangle.*

(3.3.12) Example. The figure on the left of Fig. 3.12 shows a triangulation of the given domain. The figure on the right is *not* a triangulation.

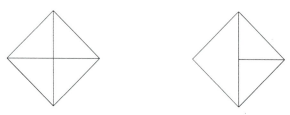

Fig. 3.12. Two subdivisions: the one on the left is a triangulation and the one on the right is not.

(3.3.13) Example. Let Ω be the square depicted in Fig. 3.13. The triangulation $\mathcal{T}$ consists of the two triangles T_1 and T_2, as indicated. The finite element on each triangle is the Lagrange element in Example 3.2.1. The dual basis on T_1 is $\{1 - x - y, x, y\}$ (calculated in Example 3.3.3) and the dual basis on T_2 is (cf. exercise 3.x.3) $\{1 - x, 1 - y, x + y - 1\}$. Let $f = \sin\big(\pi(x+y)/2\big)$. Then

$$\mathcal{I}_\mathcal{T} f = \begin{cases} x + y & \text{on } T_1 \\ 2 - x - y & \text{on } T_2. \end{cases}$$

(3.3.14) *Remark.* For approximating the Dirichlet problem with zero boundary conditions, we use a finite-dimensional space of piecewise polynomial functions satisfying the boundary conditions given by
$$V_\mathcal{T} = \{\mathcal{I}_\mathcal{T} f : f \in C^m(\overline{\Omega}), f|_{\partial\Omega} = 0\}$$
on each triangulation $\mathcal{T}$. This will be discussed further in Chapter 5.

(3.3.15) Definition. *We say that an interpolant has* **continuity order** r *(in short, that it is "C^r") if $\mathcal{I}_\mathcal{T} f \in C^r$ for all $f \in C^m(\overline{\Omega})$. The space, $V_\mathcal{T} = \{\mathcal{I}_\mathcal{T} f : f \in C^m\}$, is said to be a "$C^r$" finite element space.*

(3.3.16) *Remark.* A finite element (or collection of elements) that can be used to form a C^r space as above is often called a "C^r element." Not all choices of nodes will always lead to C^r continuity, however. Some sort of regularity must be imposed. For the elements studied so far, the essential point is that they be placed in a coordinate-free way that is symmetric with respect to the midpoint of the edge.

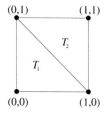

Fig. 3.13. simple triangulation consisting of two triangles

(3.3.17) Proposition. *The Lagrange and Hermite elements are both C^0 elements, and the Argyris element is C^1. More precisely, given a triangulation, $\mathcal{T}$, of Ω, it is possible to choose edge nodes for the corresponding elements $(K, \mathcal{P}, \mathcal{N})$, $K \in \mathcal{T}$, such that the global interpolant satisfies $\mathcal{I}_{\mathcal{T}} f \in C^r$ ($r = 0$ for Lagrange and Hermite, and $r = 1$ for Argyris) for $f \in C^m$ ($m = 0$ for Lagrange, $m = 1$ for Hermite and $m = 2$ for Argyris). In particular, it is sufficient for each edge $\overline{\mathbf{x}\,\mathbf{x}'}$ to have nodes $\xi_i(\mathbf{x}' - \mathbf{x}) + \mathbf{x}$, where $\{\xi_i \, : \, i = 1, \ldots, k - 1 - 2m\}$ is fixed and symmetric around $\xi = 1/2$. Moreover, under these hypotheses, $\mathcal{I}_{\mathcal{T}} f \in W_\infty^{r+1}$.*

Proof. It is sufficient to show that the stated continuity holds across each edge. Let T_i, $i = 1, 2$, denote two triangles sharing an edge, e. Since we assumed that the edge nodes were chosen symmetrically and in a coordinate-free way, we know that the edge nodes on e for the elements on both T_1 and T_2 are at the same location in space. Let $w := \mathcal{I}_{T_1} f - \mathcal{I}_{T_2} f$, where we view both polynomials, $\mathcal{I}_{T_i} f$ to be defined everywhere by extension outside T_i as polynomials. Then w is a polynomial of degree k and its restriction to the edge e has one-dimensional Lagrange, Hermite or Argyris nodes equal to zero. Thus, $w|_e$ must vanish. Hence, the interpolant is continuous across each edge.

Lipschitz continuity of $\mathcal{I}_{\mathcal{T}} f$ follows by showing that it has weak derivatives of order $r + 1$ given by

$$\left(D_{(w)}^\alpha \mathcal{I}_{\mathcal{T}} f\right)|_T = D^\alpha \mathcal{I}_T f \quad \forall T \in \mathcal{T}, \ |\alpha| \le r + 1.$$

The latter is certainly in L^∞. The verification that this is the weak derivative follows from

$$\int_\Omega (D^\alpha \phi)\, (\mathcal{I}_{\mathcal{T}} f)\ dx = \sum_{T \in \mathcal{T}} \int_T (D^\alpha \phi)\, (\mathcal{I}_T f)\ dx$$

$$= \sum_{T \in \mathcal{T}} (-1)^{|\alpha|} \int_T \phi\, (D^\alpha \mathcal{I}_T f)\ dx$$

$$= (-1)^{|\alpha|} \int_\Omega \phi \sum_{T \in \mathcal{T}} \chi_T\, (D^\alpha \mathcal{I}_T f)\ dx,$$

where χ_T denotes the characteristic function of T. The second equality holds because all boundary terms cancel due to the continuity properties of the interpolant. □

3.4 Equivalence of Elements

In the application of the global interpolant, it is essential that we find a uniform bound (independent of $T \in \mathcal{T}$) for the norm of the local interpolation operator $\mathcal{I}_T$. Therefore, we want to compare the local interpolation operators on different elements. The following notions of equivalence are useful for this purpose (cf. Ciarlet & Raviart 1972a).

(3.4.1) Definition. *Let* $(K, \mathcal{P}, \mathcal{N})$ *be a finite element and let* $F(x) = \mathbf{A}x + \mathbf{b}$ *(**A** nonsingular) be an affine map. The finite element* $(\widehat{K}, \widehat{\mathcal{P}}, \widehat{\mathcal{N}})$ *is* **affine equivalent** *to* $(K, \mathcal{P}, \mathcal{N})$ *if*

  (*i*) $F(K) = \widehat{K}$
  (*ii*) $F^*\widehat{\mathcal{P}} = \mathcal{P}$ *and*
  (*iii*) $F_*\mathcal{N} = \widehat{\mathcal{N}}$.

We write $(K, \mathcal{P}, \mathcal{N}) \underset{F}{\cong} (\widehat{K}, \widehat{\mathcal{P}}, \widehat{\mathcal{N}})$ if they are affine equivalent.

(3.4.2) Remark. Recall that the *pull-back* F^* is defined by $F^*(\hat{f}) := \hat{f} \circ F$ and the *push-forward* F_* is defined by $(F_*N)(\hat{f}) := N(F^*(\hat{f}))$.

(3.4.3) Proposition. *Affine equivalence is an equivalence relation.*

Proof. See exercise 3.x.4.   □

(3.4.4) Examples.

  (i) Let K be any triangle, $\mathcal{P} = \mathcal{P}_1$, $\mathcal{N} = \{$evaluation at vertices of $K\}$. All such elements $(K, \mathcal{P}, \mathcal{N})$ are affine equivalent.
  (ii) Let K be any triangle, $\mathcal{P} = \mathcal{P}_2$, $\mathcal{N} = \{$evaluation at vertices and edge midpoints$\}$. All such elements are affine equivalent.
  (iii) Let $\mathcal{P} = \mathcal{P}_2$. In Fig. 3.14, $(T_1, \mathcal{P}, \mathcal{N}_1)$ and $(T_2, \mathcal{P}, \mathcal{N}_2)$ are *not* affine equivalent, but the finite elements $(T_1, \mathcal{P}, \mathcal{N}_1)$ and $(T_3, \mathcal{P}, \mathcal{N}_3)$ are affine equivalent.

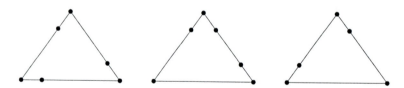

Fig. 3.14. inequivalent quadratic elements: noda placement incompatibility

  (iv) Let $\mathcal{P} = \mathcal{P}_3$. The elements $(T_1, \mathcal{P}, \mathcal{N}_1)$ and $(T_2, \mathcal{P}, \mathcal{N}_2)$ depicted in Fig. 3.15 are *not* affine equivalent since the directional derivatives differ.
  (v) Let $\mathcal{P} = \mathcal{P}_3$. Then the elements $(T_1, \mathcal{P}, \mathcal{N}_1)$ and $(T_2, \mathcal{P}, \mathcal{N}_2)$ depicted in Fig. 3.16 are *not* affine equivalent since the strength of the directional derivatives (indicated by the length of the arrows) differ.

(3.4.5) Proposition. *There exist nodal placements such that all Lagrange elements of a given degree are affine equivalent.*

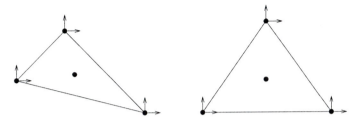

Fig. 3.15. inequivalent cubic Hermite elements: direction incompatibility

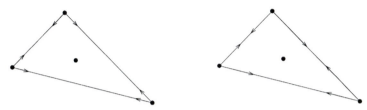

Fig. 3.16. inequivalent cubic Hermite elements: derivative strength incompatibility

Proof. We pick nodes using *barycentric coordinates*, (b_1, b_2, b_3), for each triangle. The i-th barycentric coordinate of a point (x, y) can be defined simply as the value of the i-th linear Lagrange basis function at that point $(b_i(x, y) := \phi_i(x, y))$. Thus, each barycentric coordinate is naturally associated with a given vertex; it is equal to the proportional distance of the point from the opposite edge. Note that the barycentric coordinates sum to one (since this yields the interpolant of the constant, 1). Thus, the mapping $(x, y) \to \mathbf{b}(x, y)$ maps the triangle (invertibly) to a subset of $\{\mathbf{b} \in [0, 1]^3 : b_1 + b_2 + b_3 = 1\}$.

For degree k Lagrange elements, pick nodes at the points whose barycentric coordinates are

$$\left(\frac{i}{k}, \frac{j}{k}, \frac{l}{k}\right) \quad \text{where} \quad 0 \le i, j, l \le k \quad \text{and} \quad i + j + l = k.$$

$\square$

(3.4.6) Definition. *The finite elements* $(K, \mathcal{P}, \mathcal{N})$ *and* $(K, \mathcal{P}, \tilde{\mathcal{N}})$ *are* **interpolation equivalent** *if*

$$\mathcal{I}_\mathcal{N} f = \mathcal{I}_{\tilde{\mathcal{N}}} f \quad \forall f \text{ sufficiently smooth,}$$

where $\mathcal{I}_\mathcal{N}$ *(resp.* $\mathcal{I}_{\tilde{\mathcal{N}}}$*) is defined by the right-hand side of* (3.3.2) *with* $N_i \in \mathcal{N}$ *(resp.* $N_i \in \tilde{\mathcal{N}}$*). We write* $(K, \mathcal{P}, \mathcal{N}) \underset{\mathcal{I}}{\widetilde{=}} (K, \mathcal{P}, \tilde{\mathcal{N}})$.

(3.4.7) Proposition. *Suppose $(K,\mathcal{P},\mathcal{N})$ and $(K,\mathcal{P},\widetilde{\mathcal{N}})$ are finite elements. Every nodal variable in $\mathcal{N}$ is a linear combination of nodal variables in $\widetilde{\mathcal{N}}$ (when viewed as a subset of $C^m(K)'$) if and only if $(K,\mathcal{P},\mathcal{N}) \underset{\mathcal{I}}{\cong} (K,\mathcal{P},\widetilde{\mathcal{N}})$.*

Proof. (only if) We must show that $\mathcal{I}_{\mathcal{N}}f = \mathcal{I}_{\widetilde{\mathcal{N}}}f \quad \forall f \in C^m(K)$. For $N_i \in \mathcal{N}$, we can write $N_i = \sum_{j=1}^{k} c_j \widetilde{N}_j$ since every nodal variable in $\mathcal{N}$ is a linear combination of nodal variables in $\widetilde{\mathcal{N}}$. Therefore,

$$
N_i(\mathcal{I}_{\widetilde{\mathcal{N}}}f) = \left(\sum_{j=1}^{k} c_j \widetilde{N}_j \right)(\mathcal{I}_{\widetilde{\mathcal{N}}}f)
$$

$$
= \sum_{j=1}^{k} c_j \widetilde{N}_j(\mathcal{I}_{\widetilde{\mathcal{N}}}f)
$$

$$
= \sum_{j=1}^{k} c_j \widetilde{N}_j(f)
$$

$$
= N_i(f).
$$

The converse is left to the reader in exercise 3.x.26.   □

(3.4.8) Example. The Hermite elements in Fig. 3.7 (and 3.15–16) are interpolation equivalent (exercise 3.x.29).

(3.4.9) Definition. *If $(K,\mathcal{P},\mathcal{N})$ is a finite element that is affine equivalent to $(\widehat{K},\widehat{\mathcal{P}},\widehat{\mathcal{N}})$ and $(\widehat{K},\widehat{\mathcal{P}},\widehat{\mathcal{N}})$ is interpolation equivalent to $(\widetilde{K},\widetilde{\mathcal{P}},\widetilde{\mathcal{N}})$, then we say that $(K,\mathcal{P},\mathcal{N})$ is* **affine-interpolation equivalent** *to $(\widetilde{K},\widetilde{\mathcal{P}},\widetilde{\mathcal{N}})$.*

(3.4.10) Example.

  (i) All affine equivalent elements (e.g., Lagrange elements with appropriate choices for the edge and interior nodes as described in Proposition 3.4.5) are affine-interpolation equivalent.
  (ii) The Hermite elements with appropriate choices for the edge and interior nodes are affine-interpolation equivalent.
  (iii) The Argyris elements are *not* affine-interpolation equivalent (Ciarlet 1978).

The following is an immediate consequence of the definitions.

(3.4.11) Proposition. *If $(K,\mathcal{P},\mathcal{N})$ is affine-interpolation equivalent to $(\widetilde{K},\widetilde{\mathcal{P}},\widetilde{\mathcal{N}})$ then $\mathcal{I} \circ F^* = F^* \circ \widetilde{\mathcal{I}}$ where F is the affine mapping $K \to \widetilde{K}$.*

3.5 Rectangular Elements

In this section we consider finite elements defined on rectangles. Let $\mathcal{Q}_k = \{\sum_j c_j\, p_j(x)\, q_j(y) : p_j, q_j$ polynomials of degree $\leq k\}$. One can show that

$$(3.5.1) \qquad\qquad \dim \mathcal{Q}_k = (\dim \mathcal{P}_k^1)^2,$$

where $\mathcal{P}_k^1$ denotes the space of polynomials of degree less than or equal to k in one variable (cf. exercise 3.x.6).

Tensor Product Elements

(3.5.2) Example. (k = 1) Let K be any rectangle, $\mathcal{P} = \mathcal{Q}_1$, and $\mathcal{N}$ as depicted in Fig. 3.17.

Suppose that the polynomial $P \in \mathcal{Q}_1$ vanishes at z_1, z_2, z_3 and z_4. The restriction of P to any side of the rectangle is a first-order polynomial of one variable. Therefore, we can write $P = c\, L_1 L_2$ for some constant c. But

$$0 = P(z_4) = c\, L_1(z_4)\, L_2(z_4) \quad \Longrightarrow \quad c = 0,$$

since $L_1(z_4) \neq 0$ and $L_2(z_4) \neq 0$. Thus, $P \equiv 0$.   □

(3.5.3) Example. (k = 2) Let K be any rectangle, $\mathcal{P} = \mathcal{Q}_2$, and $\mathcal{N}$ as depicted in Fig. 3.18. Suppose that a polynomial $P \in \mathcal{Q}_2$ vanishes at z_i, for $i = 1, \ldots, 9$. Then we can write $P = c\, L_1 L_2 L_3 L_4$ for some constant c. But

$$0 = P(z_9) = c\, L_1(z_9)\, L_2(z_9)\, L_3(z_9)\, L_4(z_9) \quad \Longrightarrow \quad c = 0,$$

since $L_i(z_9) \neq 0$ for $i = 1, 2, 3, 4$.   □

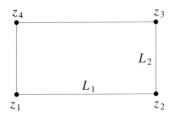

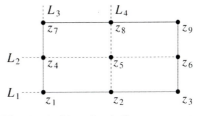

Fig. 3.17. bilinear Lagrange rectangle

Fig. 3.18. biquadratic Lagrange rectangle

(3.5.4) Example. (arbitrary k) Let K be any rectangle, $\mathcal{P} = \mathcal{Q}_k$, and $\mathcal{N}$ denote point evaluations at $\{(t_i, t_j) : i, j = 0, 1, \ldots k\}$ where $\{0 = t_0 < t_1 < \ldots < t_k = 1\}$. (The case $k = 3$ is depicted in Fig. 3.19.) Then $(K, \mathcal{P}, \mathcal{N})$ is a finite element (cf. exercise 3.x.7).

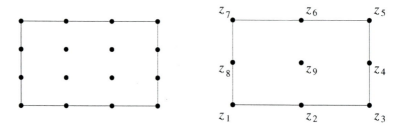

Fig. 3.19. bicubic Lagrange rectangle   **Fig. 3.20.** notation for Lemma 3.5.6

The Serendipity Element

(3.5.5) Example. (Quadratic Case) To define the shape functions for this case, we need the following lemma (see Fig. 3.20 for the notation).

(3.5.6) Lemma. *There exist constants* $c_1, \ldots, c_8$ *such that*

$$\phi(z_9) = \sum_{i=1}^{8} c_i \, \phi(z_i) \quad \text{for } \phi \in \mathcal{P}_2.$$

Proof. Note that evaluation at $z_1, \ldots, z_6$ forms a nodal basis for $\mathcal{P}_2$. Let $\{\phi_1, \ldots, \phi_6\}$ be the dual basis of $\mathcal{P}_2$, i.e., $N_i \phi_j = \delta_{ij}$ for $i, j = 1, \ldots, 6$. If $\phi \in \mathcal{P}_2$, then

$$\phi = N_1(\phi) \, \phi_1 + N_2(\phi) \, \phi_2 + \ldots + N_6(\phi) \, \phi_6.$$

Therefore,

$$\phi(z_9) = \phi(z_1) \, \phi_1(z_9) + \phi(z_2) \, \phi_2(z_9) + \ldots + \phi(z_6) \, \phi_6(z_9)$$
$$= c_1 \, \phi(z_1) + c_2 \, \phi(z_2) + \ldots + c_8 \, \phi(z_8)$$

(let $c_7 = c_8 = 0$).   □

Let K be any rectangle, $\mathcal{P} = \{\phi \in \mathcal{Q}_2 : \sum_{i=1}^{8} c_i \, \phi(z_i) - \phi(z_9) = 0\}$, and $\mathcal{N}$ as depicted in Fig. 3.21. Then $(K, \mathcal{P}, \mathcal{N})$ is a finite element because if $\phi \in \mathcal{P}$ vanishes at $z_1, \ldots, z_8$ we can write $\phi = c \, L_1 L_2 L_3 L_4$ for some constant c. But

$$0 = \sum_{i=1}^{8} c_i \, \phi(z_i) = \phi(z_9) \quad \Longrightarrow \quad c = 0$$

(since $L_i(z_9) \neq 0$, $i = 1, \ldots, 4$). Therefore, $\phi \equiv 0$.   □

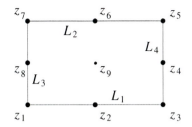

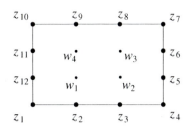

Fig. 3.21. quadratic serendipity element

Fig. 3.22. cubic serendipity element

(3.5.7) Example. (Cubic Case) Let K be any rectangle. There exist constants c_j^i such that for $\phi \in \mathcal{P}_3$, $\phi(w_i) = \sum_{j=1}^{12} c_j^i \phi(z_j)$ $i = 1, 2, 3, 4$ (for w_i and z_j as depicted in Fig. 3.22), then let $\mathcal{P} = \{\phi \in \mathcal{Q}_3 : \phi(w_i) - \sum_{j=1}^{12} c_j^i \phi(z_j) = 0$ for $i = 1, 2, 3, 4\}$ and $\mathcal{N}$ as depicted. Then $(K, \mathcal{P}, \mathcal{N})$ is a finite element (cf. exercise 3.x.8).

(3.5.8) *Remark.* The notion of a C^r rectangular element can be defined similarly to Definition 3.3.15. Following the proof of Proposition 3.3.17, we can see that all the rectangular elements defined in this section are C^0. An example of a C^1 rectangular element is in exercise 3.x.16.

(3.5.9) *Remark.* The space, $\mathcal{P}$, of shape functions for serendipity elements is not uniquely defined by the choice of nodal variables. Another way to choose them is described in Sect. 4.6.

3.6 Higher-dimensional Elements

Higher-dimensional elements can be constructed inductively just the way we constructed two-dimensional elements using properties of one-dimensional elements as building blocks. As an illustration, we describe tetrahedral elements in three dimensions. The Lagrange elements can be defined as before, inductively in the degree, k, as follows.

We pick nodal variables at the vertices (of which there are four), at $k-1$ points on the interior of each edge (there are six edges) and at $(k-2)(k-1)/2$ points in the interior of each face (again four of these). The face points, which exist only for $k \geq 3$, should be chosen so as to determine polynomials in two variables (in the plane of the face) of degree $k-3$. For $k \geq 4$, we also pick points in the interior of the tetrahedron so as to determine polynomials in three variables of degree $k-4$. The existence of the latter will be demonstrated by induction on k, as in the two-dimensional case.

Suppose these nodal values vanish for $v \in \mathcal{P}_k$. The restriction of v to each face, F_i, of the tetrahedron is a polynomial in two variables (the coordinates for the plane of F_i), and the nodal variables have been chosen to determine this restriction. Thus, $v|_{F_i}$ is identically zero. Let L_i denote a nontrivial linear function vanishing on F_i. By applying (3.1.10) four times,

$$v = L_1 L_2 L_3 L_4 R$$

where the remainder is a polynomial of degree $k - 4$. For $k \leq 3$ this implies that $R = 0$, so that $v = 0$. In the general case, we use the interior nodes to determine that $R = 0$. It simply remains to count the number of nodes and check that it equals $\dim \mathcal{P}_k$.

We have enumerated

(3.6.1) $C(k) := 4 + 6(k - 1) + 2(k - 2)(k - 1) + \dim \mathcal{P}_{k-4}$

nodes above. The dimension of $\mathcal{P}_k$ can be computed as follows. We can decompose an arbitrary polynomial, P, of degree k in three variables uniquely as

(3.6.2) $P(x, y, z) = p(x, y) + zq(x, y, z)$

where the degree of p is k and the degree of q is $k - 1$. Simply let $p(x, y) := P(x, y, 0)$ and apply (3.1.10) to $P - p$ with $L(x, y, z) = z$. Therefore,

(3.6.3) $\dim \mathcal{P}_k = (k + 1)(k + 2)/2 + \dim \mathcal{P}_{k-1} = \sum_{j=0}^{k} (j + 1)(j + 2)/2,$

where the second equality follows from the first by induction. The first few of these are given in the following table, and it is easily checked that they agree with (3.6.1).

Table 3.2. dimension of polynomials of degree k, $\mathcal{P}_k$, in three dimensions

k	$\dim \mathcal{P}_k$
1	4
2	10
3	20
4	35

Since (3.6.3) implies $\dim \mathcal{P}_k$ is a cubic polynomial in k (with leading coefficient $1/6$), we conclude that $C(k)$ is also a cubic polynomial in k. Since these cubics agree for $k = 1, 2, 3, 4$, they must be identical.

The above arguments also show that the nodes can be arranged so as to insure that the Lagrange elements are C^0. As in the proof of Proposition 3.3.17, it suffices to see that the restrictions of the global interpolant to neighboring tetrahedra agree on the common face. This is possible because of our choice of facial nodes to determine polynomials in two variables on that face. One must again choose the facial nodes in a symmetric and coordinate free way. In particular, it is sufficient to let the nodes be located at points whose barycentric coordinates (see (3.4.5)) B on each face satisfy

$$(b_1, b_2, b_3) \in B \implies (b_{\sigma(1)}, b_{\sigma(2)}, b_{\sigma(3)}) \in B$$

for any permutation σ of the indices.

(3.6.4) *Remark.* The notion of a C^r tetrahedral element can be defined similarly to Definition 3.3.15. Following the proof of Proposition 3.3.17, we see that the tetrahedral elements defined in this section are all C^0.

3.7 Exotic Elements

All the elements $(K, \mathcal{P}, \mathcal{N})$ studied so far have shape functions consisting of polynomials. However, this is not at all necessary. We consider some of the possibilities briefly here. We restrict our discussion to the class of *macro-finite-elements*, for which the shape functions, $\mathcal{P}$, are themselves piecewise polynomials. Other types of shape functions have been proposed, e.g., rational functions (Wachspress 1975).

Let K denote a triangle, and let it be divided into four subtriangles by connecting edge midpoints as shown in Fig. 3.23. Define $\mathcal{P}$ to be the set of continuous piecewise linear functions on this subtriangulation. If $\mathcal{N}$ consists of point-evaluations at the vertices and edge midpoints of K, we clearly have a well defined C^0 finite element.

A more complex element is that of Clough and Tocher (Ciarlet 1978). Let K denote a triangle, and let it be divided into three subtriangles as shown in Fig. 3.24. Let $\mathcal{P}$ be the set of C^1 piecewise cubic functions on this subtriangulation. Let $\mathcal{N}$ consist of point- and gradient-evaluations at the vertices and normal-derivative-evaluations at the edge midpoints of K. Then $(K, \mathcal{P}, \mathcal{N})$ is a well defined, C^1 finite element (Ciarlet 1978).

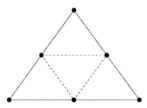

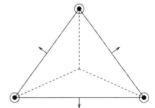

Fig. 3.23. macro-piecewise-linear triangle

Fig. 3.24. Clough-Tocher C^1 macro-piecewise-cubic triangle

3.x Exercises

3.x.1 Let m and k be nonnegative integers, and let P be a polynomial in one variable of degree $2m + k + 1$. Suppose that $P^{(j)}(a) = 0$ for $a = 0, 1$ and $j = 0, \ldots, m$, and further that $P(\xi_j) = 0$ for $0 < \xi_1 < \ldots < \xi_k < 1$. Prove that $P \equiv 0$.

3.x.2 Prove that the local interpolant is linear (cf. Proposition 3.3.4).

3.x.3 Find the dual basis for triangle T_2 in Example 3.3.13.

3.x.4 Show that affine equivalence is an equivalence relation (cf. Proposition 3.4.3).

3.x.5 Show that interpolation equivalence is an equivalence relation.

3.x.6 Show that $\dim \mathcal{Q}_k = (\dim \mathcal{P}_k^1)^2$, where $\mathcal{P}_k^1 = \{$polynomials in one variable of degree less than or equal to $k\}$ and $\{x^i y^j : i, j = 0, \ldots, k\}$ is a basis of $\mathcal{Q}_k$.

3.x.7 Prove that $(K, \mathcal{P}, \mathcal{N})$ in Example 3.5.4 is a finite element.

3.x.8 Prove that $(K, \mathcal{P}, \mathcal{N})$ in Example 3.5.7 is a finite element.

3.x.9 Construct nodal basis functions for $K = $ the rectangle with vertices $(-1, 0)$, $(1, 0)$, $(1, 1)$ and $(-1, 1)$, $\mathcal{P} = \mathcal{Q}_1$, and $\mathcal{N} = $ evaluation at the vertices.

3.x.10 Construct nodal basis functions for $K = $ the triangle with vertices $(0, 0)$, $(1, 0)$ and $(0, 1)$, $\mathcal{P} = \mathcal{P}_2$, and $\mathcal{N} = $ evaluation at the vertices and at the midpoints of the edges.

3.x.11 Prove that the set of nodal variables

$$\Sigma_n = \{P(a), P'(a), P^{(3)}(a), \ldots, P^{(2n-1)}(a) : a = 0, 1\}$$

determine unique polynomials (in one variable) of degree $2n + 1$. (For $n = 1$, this is just Hermite interpolation, as in exercise 3.x.1.)

3.x.12 Show that the nodal variables for the Argyris element described in Example 3.2.11 determine $\mathcal{P}_7$. Give a general description of the Argyris element for arbitrary degree $k \geq 5$.

3.x.13 Show that if $\mathcal{P} = \mathcal{Q}_1$, then the nodal variables depicted in Fig. 3.25 do *not* determine $\mathcal{P}$.

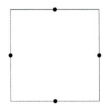

Fig. 3.25. a non-element

3.x.14 **Nonconforming piecewise linear element** Show that the edge mid-points in a triangulation can be used to parametrize the space of piecewise linear functions (in general discontinuous) that are continuous at each edge midpoint. Can you generalize this to quadratics (i.e., find a nodal basis for piecewise quadratics that are continuous at two points on each edge)?

3.x.15 **Rotated nonconforming bilinear element** Let K be the square $[-1,1] \times [-1,1]$, $\mathcal{P}$ be the space of shape functions spanned by 1, x, y and $x^2 - y^2$, and $\mathcal{N}$ consist of the evaluations of the shape functions at the four midpoints (cf. Fig. 3.25). Show that $(K, \mathcal{P}, \mathcal{N})$ is a finite element. Let $a_j(v) = (1/|e_j|) \int_{e_j} v \, ds$ be the mean value of the function v on the edge e_j of K, and $\mathcal{N}_* = \{a_1, \ldots, a_4\}$. Show that $(K, \mathcal{P}, \mathcal{N}_*)$ is also a finite element. Are these two elements interpolant equivalent?

3.x.16 **Bicubic Hermite elements** Prove that a tensor product cubic in two variables is uniquely determined by

$$\Sigma = \{P(a_i), \frac{\partial P}{\partial x_1}(a_i), \frac{\partial P}{\partial x_2}(a_i), \frac{\partial^2 P}{\partial x_1 \partial x_2}(a_i) : i = 1, \ldots, 4\}$$

where a_i are the rectangle vertices. Will this generate a C^1 piecewise cubic on a rectangular subdivision?

3.x.17 Let $\mathcal{I}$ be the interpolation operator associated with continuous, piecewise linears on triangles, i.e., $\mathcal{I}u = u$ at vertices. Prove that $\|\mathcal{I}\|_{C^0 \to C^0} = 1$, i.e., for any continuous function u, $\|\mathcal{I}u\|_{L^\infty} \le \|u\|_{L^\infty}$. (Hint: where does the maximum of $|\mathcal{I}u|$ occur on a triangle?) Is this true for piecewise quadratics?

3.x.18 Let "∠" denote the second derivative that is the concatenation of the directional derivatives in the two directions indicated by the line segments. Show that $\mathcal{P}_4$ is determined by (i) the value, gradient and "∠" second derivative at each vertex (the directions used for "∠" at each vertex are given by the edges meeting there, as shown in Fig. 3.26) and (ii) the value at each edge midpoint.

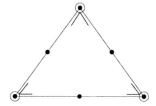

Fig. 3.26. a quartic finite element

3.x.19 Suppose that the nodes for the Lagrange element are chosen at the barycentric lattice points introduced in the proof of Proposition

3.4.5. Show that the corresponding nodal basis functions for $\mathcal{P}_k$ can be written as a product of k linear functions. (Hint: for each node determine k lines that contain all other nodes.)

3.x.20 Generalize Proposition 3.4.5 and exercise 3.x.19 to three dimensions.

3.x.21 Generalize 3.4.5 and exercise 3.x.19 to n dimensions.

3.x.22 Prove that no analog of the serendipity element exists for biquartic polynomials. (Hint: show that one can not remove all interior points.)

3.x.23 Show that the decomposition (3.6.2) is unique.

3.x.24 Develop three-dimensional Hermite and Argyris elements. Are the latter C^1?

3.x.25 Develop four-dimensional Lagrange elements.

3.x.26 Prove the "if" part of Proposition 3.4.7.

3.x.27 Show that the Hermite element is not C^1.

3.x.28 Can the derivative nodes for the Hermite elements be chosen to give an affine-equivalent family for arbitrary triangles?

3.x.29 Use (3.4.7) to prove (3.4.8).

3.x.30 Let $\mathcal{P}_k^n$ denote the space of polynomials of degree k in n variables. Prove that $\dim \mathcal{P}_k^n = \begin{pmatrix} n+k \\ k \end{pmatrix}$, where the latter is the binomial coefficient. (Hint: show that (3.6.2) holds in n-dimensions and use this to prove that the numbers $\dim \mathcal{P}_k^n$ form Pascal's triangle.)

3.x.31 Develop three-dimensional tensor-product and serendipity elements.

3.x.32 Give conditions on rectangular subdivisions that allow the tensor-product elements to be C^0.

3.x.33 Give conditions on rectangular subdivisions that allow the bicubic Hermite elements to be C^1 (see exercise 3.x.16). Are the conditions the same as in exercise 3.x.32?

3.x.34 What conditions on simplicial subdivisions allow three-dimensional Lagrange elements to be C^0?

Chapter 4

Polynomial Approximation Theory in Sobolev Spaces

We will now develop the approximation theory appropriate for the finite elements developed in Chapter 3. We take a constructive approach, defining an averaged version of the Taylor polynomial familiar from calculus. The key estimates are provided by some simple lemmas from the theory of Riesz potentials, which we derive. As a corollary, we provide a proof of Sobolev's inequality, much in the spirit given originally by Sobolev.

Initially, we derive estimates appropriate on individual element domains. Later, we show how these can be combined to provide error estimates for interpolants that are globally defined on a collection of element domains that subdivide a larger domain. This will concentrate primarily on the case of polyhedral domains, but generalizations to "isoparametric" elements are presented in the last section. These elements are extremely convenient and effective for approximating problems on domains with curved boundaries.

4.1 Averaged Taylor Polynomials

We turn our attention to finding a polynomial approximation of order m for a function in a Sobolev space. Let $B = \{x \in \mathbb{R}^n : |x - x_0| < \rho\}$. A function $\phi \in C_0^\infty(\mathbb{R}^n)$ with the properties (i) supp $\phi = \overline{B}$ and (ii) $\int_{\mathbb{R}^n} \phi(x) \, dx = 1$ will be called a *cut-off function*. For example, let

$$\psi(x) = \begin{cases} e^{-\left(1-(|x-x_0|/\rho)^2\right)^{-1}} & \text{if } |x - x_0| < \rho \\ 0 & \text{if } |x - x_0| \geq \rho. \end{cases}$$

Let $c = \int_{\mathbb{R}^n} \psi(x) \, dx$ $(c > 0)$, then $\phi(x) = (1/c)\psi(x)$ satisfies (i) and (ii), and $\max |\phi| \leq$ constant $\cdot \rho^{-n}$.

Let us assume that $u \in C^{m-1}(\mathbb{R}^n)$.

(4.1.1) Definition. *The **Taylor polynomial of order** m **evaluated at** y *is given by*

$$(4.1.2) \qquad T_y^m u(x) = \sum_{|\alpha| < m} \frac{1}{\alpha!} D^\alpha u(y)(x - y)^\alpha,$$

where $\alpha = (\alpha_1, \alpha_2, \dots, \alpha_n)$ is an n-tuple of nonnegative integers, $x \in \mathbb{R}^n$, $x^\alpha = \prod_{i=1}^n x_i^{\alpha_i}$, $\alpha! = \prod_{i=1}^n \alpha_i!$ and $|\alpha| = \sum_{i=1}^n \alpha_i$.

In general, if u is in a Sobolev space, $D^\alpha u$ may not exist in the usual (pointwise) sense. How do we then define a Taylor polynomial for such a function? We accomplish this by taking an "average" of $T_y^m u(x)$ over y, as given in the following definition.

(4.1.3) Definition. *Suppose u has weak derivatives of order strictly less than m in a region Ω such that $B \subset\subset \Omega$. The corresponding Taylor polynomial of order m of u averaged over B is defined as*

$$(4.1.4) \qquad Q^m u(x) = \int_B T_y^m u(x) \phi(y) \, dy,$$

where $T_y^m u(x)$ is defined as in (4.1.2), B is the ball centered at x_0 with radius ρ and ϕ is the cut-off function supported in $\overline{B}$.

Note that (4.1.4) does indeed make sense for $u \in W_p^{m-1}(\Omega)$ since a typical term takes the form

$$(4.1.5) \qquad \int_B \frac{1}{\alpha!} D^\alpha u(y)(x-y)^\alpha \phi(y) \, dy,$$

which exists because $D^\alpha u$ is in $L_{loc}^1(\Omega)$. If we write

$$(4.1.6) \qquad (x-y)^\alpha = \prod_{i=1}^n (x_i - y_i)^{\alpha_i} = \sum_{\gamma + \beta = \alpha} a_{(\gamma, \beta)} x^\gamma y^\beta,$$

where $\gamma = (\gamma_1, \gamma_2, \dots, \gamma_n)$ and $\beta = (\beta_1, \beta_2, \dots, \beta_n)$ are n-tuples of nonnegative integers and $a_{(\gamma, \beta)}$ are constants, we find that (4.1.5) may be written as

$$(4.1.7) \qquad \sum_{\gamma + \beta = \alpha} \frac{1}{\alpha!} a_{(\gamma, \beta)} x^\gamma \int_B D^\alpha u(y) y^\beta \phi(y) \, dy.$$

Therefore,

$$(4.1.8) \qquad Q^m u(x) = \sum_{|\alpha| < m} \sum_{\gamma + \beta = \alpha} \frac{1}{\alpha!} a_{(\gamma, \beta)} x^\gamma \int_B D^\alpha u(y) y^\beta \phi(y) \, dy.$$

Thus, the next proposition follows directly.

(4.1.9) Proposition. *$Q^m u$ is a polynomial of degree less than m in x.*

In fact, $Q^m u$ can be defined for functions in $L^1(B)$. We only need to rewrite (4.1.8) by integrating by parts:

$$Q^m u(x) =$$

(4.1.10)
$$\sum_{|\alpha| < m} \sum_{\gamma + \beta = \alpha} \frac{(-1)^{|\alpha|}}{\alpha!} a_{(\gamma,\beta)} x^\gamma \int_B u(y) D^\alpha \left(y^\beta \phi(y)\right) dy.$$

(4.1.11) Remark. Note that if u has weak derivatives of all orders less than m in Ω, then (4.1.8) is equivalent to (4.1.10) by using the definition of weak derivative (1.2.4).

(4.1.12) Proposition. $Q^m u$ is defined for all $u \in L^1(B)$ and

(4.1.13)
$$Q^m u(x) = \sum_{|\lambda| < m} x^\lambda \int_B \psi_\lambda(y) u(y) \, dy,$$

where $\psi_\lambda \in C_0^\infty(\mathbb{R}^n)$ and $\operatorname{supp} \psi_\lambda \subset \overline{B}$.

Proof. This follows from (4.1.10) if we define

(4.1.14)
$$\psi_\lambda(y) = \sum_{\lambda \leq \alpha, |\alpha| < m} \frac{(-1)^{|\alpha|}}{\alpha!} a_{(\lambda, \alpha - \lambda)} D^\alpha \left(y^{\alpha - \lambda} \phi(y)\right). \quad \square$$

(4.1.15) Corollary. If Ω is a bounded domain in $\mathbb{R}^n$, then for any k

(4.1.16)
$$\|Q^m u\|_{W_\infty^k(\Omega)} \leq C_{m,n,\rho,\Omega} \|u\|_{L^1(B)}.$$

Proof. This follows directly from equality (4.1.13) and the fact that both $\sup_{y \in B} |\psi_\lambda(y)|$ and $\sup_{x \in \Omega} |D^\alpha x^\lambda|$ are bounded. $\quad \square$

As a result of Corollary 4.1.15, we note that Q^m is a bounded map of L^1 into W_∞^k.

(4.1.17) Proposition. For any α such that $|\alpha| \leq m - 1$,

(4.1.18)
$$D^\alpha Q^m u = Q^{m - |\alpha|} D^\alpha u \quad \text{for all} \quad u \in W_1^{|\alpha|}(B).$$

Proof. If $u \in C^\infty(\Omega)$, then

$$D_x^\alpha Q^m u(x) = \int_B D_x^\alpha T_y^m u(x) \phi(y) \, dy$$

$$= \int_B T_y^{m - |\alpha|} D_x^\alpha u(x) \phi(y) \, dy \quad \text{(exercise 4.x.1)}$$

$$= Q^{m - |\alpha|} D^\alpha u(x).$$

The proof of the proposition is completed via a density argument. $\quad \square$

(4.1.19) *Remark.* The polynomial Q^m could be called a Sobolev polynomial, as the construction of a similar polynomial was given and used by Sobolev (Sobolev 1963 & 1991) in the study of the spaces that bear his name. Such a polynomial is not unique, due to the choice of cut-off function ϕ. Moreover, the actual construction in (Sobolev 1963 & 1991) apparently does not satisfy (4.1.18), an identity that leads to certain simplifications later.

4.2 Error Representation

We need the following form of Taylor's Theorem (cf. exercise 4.x.2). For $f \in C^m([0,1])$, we have

$$(4.2.1) \qquad f(1) = \sum_{k=0}^{m-1} \frac{1}{k!} f^{(k)}(0) + m \int_0^1 \frac{1}{m!} s^{m-1} f^{(m)}(1-s) \, ds.$$

(4.2.2) Definition. *Ω is* **star-shaped with respect to B** *if, for all $x \in \Omega$, the closed convex hull of $\{x\} \cup B$ is a subset of Ω.*

(4.2.3) Example. Fig. 4.1 is star-shaped with respect to ball B, but not with respect to ball B'. Fig. 4.2 is an example of a domain that is not star-shaped with respect to any ball.

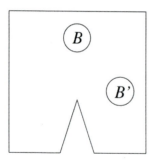

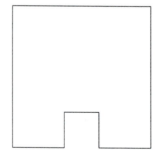

Fig. 4.1. domain star-shaped with respect to B but not B'

Fig. 4.2. domain which is not star-shaped with respect to any ball

From now on, we assume that Ω is star-shaped with respect to B. Let u be a C^m function on Ω. For $x \in \Omega$ and $y \in B$, define $f(s) = u(y+s(x-y))$. Then, by using the chain rule, we obtain

$$(4.2.4) \qquad \frac{1}{k!} f^{(k)}(s) = \sum_{|\alpha|=k} \frac{1}{\alpha!} D^\alpha u(y+s(x-y)) (x-y)^\alpha.$$

Combining (4.2.1) and (4.2.4), we obtain

$$
\begin{aligned}
u(x) = & \sum_{|\alpha|<m} \frac{1}{\alpha!} D^\alpha u(y)(x-y)^\alpha \\
& + \sum_{|\alpha|=m} (x-y)^\alpha \int_0^1 \frac{m}{\alpha!} s^{m-1} D^\alpha u\big(x+s(y-x)\big)\, ds \\
= & \; T_y^m u(x) + m \sum_{|\alpha|=m} (x-y)^\alpha \int_0^1 \frac{1}{\alpha!} s^{m-1} D^\alpha u\big(x+s(y-x)\big)\, ds.
\end{aligned}
$$

(4.2.5)

(4.2.6) Definition. *The mth-order remainder term is given by*

$$
R^m u(x) = u(x) - Q^m u(x).
$$

Using Definition 4.2.6, part (ii) of the definition of a cut-off function, (4.1.4) and (4.2.5) we obtain

$$
\begin{aligned}
R^m u(x) = & \int_B u(x)\phi(y)dy - \int_B T_y^m u(x)\phi(y)\, dy \\
= & \int_B \big[u(x) - T_y^m u(x)\big]\, \phi(y)\, dy \\
= & \int_B \phi(y) m \Bigg(\sum_{|\alpha|=m} (x-y)^\alpha \\
& \times \int_0^1 \frac{s^{m-1}}{\alpha!} D^\alpha u\big(x+s(y-x)\big)\, ds \Bigg) dy.
\end{aligned}
$$

(4.2.7)

Let C_x denote the convex hull of $\{x\} \cup B$.

(4.2.8) Proposition. *The remainder $R^m u := u - Q^m u$ satisfies*

$$
R^m u(x) = m \sum_{|\alpha|=m} \int_{C_x} k_\alpha(x,z) D^\alpha u(z)\, dz,
$$

(4.2.9)

where $z = x + s(y-x)$, $k_\alpha(x,z) = (1/\alpha!)(x-z)^\alpha k(x,z)$ and

$$
|k(x,z)| \le C \left(1 + \frac{1}{\rho}|x - x_0| \right)^n |z - x|^{-n}.
$$

(4.2.10)

Proof. We first make a change of variables from the (y,s)-space to the (z,s)-space, where

$$
z = x + s(y-x).
$$

(4.2.11)

By the change of variable formula,

$$(4.2.12) \qquad\qquad ds\, dy = s^{-n} ds\, dz.$$

The domain of integration in the (y, s)-space is $B \times (0, 1]$ and the corresponding domain in the (z, s)-space is the set

$$A = \{(z, s) : s \in (0, 1], \quad |(1/s)(z - x) + x - x_0| < \rho\}.$$

Note that

$$(4.2.13) \qquad (z, s) \in A \text{ implies that } \frac{|z - x|}{|x - x_0| + \rho} < s.$$

Also,

$$(4.2.14) \qquad (x - y)^\alpha = s^{-m}(x - z)^\alpha \quad \text{if} \quad |\alpha| = m.$$

Letting χ_A be the characteristic function of A, from (4.2.7) and (4.2.14) we obtain

$$(4.2.15) \qquad
\begin{aligned}
R^m u(x) &= \sum_{|\alpha|=m} \iint \chi_A(z, s) \phi\left(x + \frac{(z - x)}{s}\right) \\
&\qquad \times \frac{m}{\alpha!} s^{-n-1}(x - z)^\alpha D^\alpha u(z)\, ds\, dz.
\end{aligned}$$

The projection of A onto the z-space is C_x. Therefore, by Fubini's Theorem,

$$
\begin{aligned}
R^m u(x) &= m \sum_{|\alpha|=m} \int_{C_x} \frac{1}{\alpha!} D^\alpha u(z)(x - z)^\alpha \\
&\qquad \times \left[\int_0^1 \phi(x + (1/s)(z - x))\chi_A(z, s)s^{-n-1}\, ds\right] dz \\
&= m \sum_{|\alpha|=m} \int_{C_x} k_\alpha(x, z) D^\alpha u(z)\, dz,
\end{aligned}
$$

if we define $k(x, z) = \int_0^1 \phi(x + (1/s)(z - x))\chi_A(z, s)\, s^{-n-1}\, ds$ and $k_\alpha(x, z) = (1/\alpha!)(x - z)^\alpha k(x, z)$. It remains to prove estimate (4.2.10) for $k(x, z)$.

Let $t = |z - x|/(|x - x_0| + \rho)$. Then

$$
\begin{aligned}
|k(x, z)| &= \left| \int_0^1 \chi_A(z, s)\, \phi(x + (1/s)(z - x)) s^{-n-1}\, ds \right| \\
&\leq \int_t^1 |\phi(x + (1/s)(z - x))|\, s^{-n-1}\, ds \qquad \text{(by 4.2.13)} \\
&\leq \|\phi\|_{L^\infty(B)} \left. \frac{s^{-n}}{n} \right|_1^t \\
&\leq (1/n) \|\phi\|_{L^\infty(B)}\, t^{-n}
\end{aligned}
$$

$$= (1/n) \, \|\phi\|_{L^\infty(B)} \left(\rho + |x - x_0|\right)^n |z - x|^{-n}$$

$$\leq C \rho^{-n} \left(\rho + |x - x_0|\right)^n |z - x|^{-n} \quad \text{(example in Sect. 4.1)}$$

$$= C \left(1 + \frac{1}{\rho}|x - x_0|\right)^n |z - x|^{-n}.$$

Note that the use of Fubini's Theorem above is justified by the following calculation:

$$\int_{C_x} \int_0^1 \left|\phi\big(x + (1/s)(z - x)\big)\right| \chi_A(z, s) \, s^{-n-1} \, \frac{|x - z|^m}{\alpha!} \left|D^\alpha u(z)\right| ds \, dz$$

$$\leq \int_{C_x} |D^\alpha u(z)| (1/\alpha!) |x - z|^{m-n} C \left(1 + \frac{1}{\rho}|x - x_0|\right)^n dz$$

$$< \infty.$$

$\square$

(4.2.16) Definition. *Suppose Ω has diameter d and is star-shaped with respect to a ball B. Let $\rho_{\max} = \sup\{\rho : \Omega \text{ is star-shaped with respect to a ball of radius } \rho\}$. Then the* **chunkiness parameter** *of Ω is defined by*

(4.2.17)
$$\gamma = \frac{d}{\rho_{\max}}.$$

(4.2.18) Corollary. *The ball B can be chosen so that the function $k(x, z)$ in Proposition 4.2.8 satisfies the following estimate:*

(4.2.19)
$$|k(x, z)| \leq C \left(\gamma + 1\right)^n |z - x|^{-n}, \qquad \forall x \in \Omega,$$

where γ is the chunkiness parameter of Ω.

Proof. Choose a ball B such that Ω is star-shaped with respect to B and such that its radius $\rho > (1/2)\rho_{\max}$. Then

$$|k(x, z)| \leq C \left(1 + \frac{1}{\rho}|x - x_0|\right)^n |z - x|^{-n} \qquad \text{(from 4.2.10)}$$

$$\leq C \left(1 + \frac{2d}{\rho_{\max}}\right)^n |z - x|^{-n}$$

$$\leq C 2^n \left(1 + \frac{d}{\rho_{\max}}\right)^n |z - x|^{-n}$$

$$= C(1 + \gamma)^n |z - x|^{-n}. \qquad \text{(from 4.2.17)}$$

$\square$

4.3 Bounds for Riesz Potentials

We have derived bounds for the remainder R^m in terms of a "Riesz potential." In this section, we derive various bounds for such potentials.

(4.3.1) Lemma. *If* $f \in L^p(\Omega)$ *for* $1 < p < \infty$ *and* $m > n/p$, *then*

$$\int_\Omega |x - z|^{-n+m} |f(z)| \, dz \le C_p \, d^{m-n/p} \|f\|_{L^p(\Omega)} \quad \forall \, x \in \Omega.$$

This inequality also holds for $p = 1$ *if* $m \ge n$.

Proof. First assume that $1 < p < \infty$ and $m > n/p$. Let $1/p + 1/q = 1$.

$$\int_\Omega |x - z|^{-n+m} |f(z)| \, dz$$

$$\le \left(\int_\Omega |x - z|^{(-n+m)q} \right)^{1/q} \|f\|_{L^p(\Omega)} \qquad \text{(Hölder's inequality)}$$

$$\le C \left(\int_0^d r^{(-n+m)q+n-1} \, dr \right)^{1/q} \|f\|_{L^p(\Omega)} \quad \text{(using polar coordinates)}$$

$$= C \left(d^{(-n+m)q+n} \right)^{1/q} \|f\|_{L^p(\Omega)} \qquad \text{(note that C is different)}$$

$$= C d^{m-(n/p)} \|f\|_{L^p(\Omega)} \qquad \text{(using } 1/q - 1 = -1/p\text{).}$$

Next, we assume $p = 1$ and $m \ge n$. Then

$$\int_\Omega |x - z|^{-n+m} |f(z)| \, dz \le \|(x - z)^{-n+m}\|_{L^\infty(\Omega)} \|f\|_{L^1(\Omega)}$$

$$\le d^{-n+m} \|f\|_{L^1(\Omega)} \,.$$

$\square$

(4.3.2) Proposition. *For* $u \in W_p^m(\Omega)$,

$$(4.3.3) \qquad \|R^m u\|_{L^\infty(\Omega)} \le C_{m,n,\gamma,p} \, d^{m-n/p} \, |u|_{W_p^m(\Omega)},$$

provided that $1 < p < \infty$ *and* $m > n/p$, *or* $p = 1$ *and* $m \ge n$.

Proof. First assume that $u \in C^m(\Omega) \cap W_p^m(\Omega)$ so that we can use the pointwise representation of $R^m u(x)$ in Proposition 4.2.8:

$$|R^m u(x)| = m \left| \sum_{|\alpha|=m} \int_{C_x} k_\alpha(x, z) \, D^\alpha u(z) \, dz \right| \qquad \text{(by 4.2.9)}$$

$$\le C'_{m,n,\gamma} \sum_{|\alpha|=m} \int_\Omega |x - z|^{m-n} |D^\alpha u(z)| \, dz \qquad \text{(using 4.2.19)}$$

$$\le C_{m,n,\gamma,p} \, d^{m-n/p} \, |u|_{W_p^m(\Omega)} \qquad \text{(by Lemma 4.3.1).}$$

The proof is now completed via a density argument (exercise 4.x.16). $\square$

As a corollary to Proposition 4.3.2, we obtain the following special case of Sobolev's Lemma.

(4.3.4) Lemma. (Sobolev's Inequality) *Suppose Ω has diameter d and is star-shaped with respect to a ball B. If u is in $W_p^m(\Omega)$ where either (i) $1 < p < \infty$ and $m > n/p$ or (ii) $p = 1$ and $m \geq n$, then u is continuous on Ω and*

$$\|u\|_{L^\infty(\Omega)} \leq C_{m,n,\gamma,d} \|u\|_{W_p^m(\Omega)}.$$

Proof. First we show that the inequality holds:

$$
\begin{aligned}
\|u\|_{L^\infty(\Omega)} &\leq \|u - Q^m u\|_{L^\infty(\Omega)} + \|Q^m u\|_{L^\infty(\Omega)} && \text{(triangle inequality)}\\
&\leq C_{m,n,\gamma} |u|_{W_p^m(\Omega)} + C_{m,n,\rho} \|u\|_{L^1(\Omega)} && \text{(using 4.3.3 and 4.1.16)}\\
&\leq C_{m,n,\gamma,d} \|u\|_{W_p^m(\Omega)}.
\end{aligned}
$$

It remains to show that u is continuous on Ω.

Let $u_j \in C^\infty(\Omega) \cap W_p^m(\Omega)$ such that $\|u_j - u\|_{W_p^m(\Omega)} \to 0$ as $j \to \infty$. Then the above inequality implies that $\|u - u_j\|_{L^\infty(\Omega)} \to 0$ as $j \to \infty$. In other words, $u_j \to u$ uniformly. Therefore, u is continuous on Ω. □

If we let $p \to \infty$ in (4.3.3), then we have

$$(4.3.5) \qquad \|R^m u\|_{L^\infty(\Omega)} \leq C_{m,n,\gamma}\, d^m\, |u|_{W_\infty^m(\Omega)}.$$

It turns out that the same inequality is true for general L^p spaces. This result is the Bramble-Hilbert Lemma, which we will now prove. First, however, we need another result regarding Riesz potentials.

(4.3.6) Lemma. *Let $f \in L^p(\Omega)$ for $p \geq 1$ and $m \geq 1$ and let*

$$g(x) = \int_\Omega |x - z|^{m-n} |f(z)|\, dz.$$

Then

$$(4.3.7) \qquad \|g\|_{L^p(\Omega)} \leq C_{m,n}\, d^m\, \|f\|_{L^p(\Omega)}.$$

Proof. First assume $1 < p < \infty$. Then

$$
\begin{aligned}
\|g\|_{L^p(\Omega)}^p &= \int_\Omega |g(x)|^p\, dx\\
&= \int_\Omega \left(\int_\Omega |x - z|^{m-n} |f(z)|\, dz \right)^p dx\\
&\leq \int_\Omega \left[\left(\int_\Omega |f(z)|^p\, |x - z|^{m-n}\, dz \right)^{1/p} \left(\int_\Omega |x - z|^{m-n}\, dz \right)^{1/q} \right]^p dx\\
&\qquad\qquad\qquad \text{(by Hölder's inequality with } 1/p + 1/q = 1)
\end{aligned}
$$

$$\leq C_{m,n}\, d^{mp/q} \int_\Omega \int_\Omega |f(z)|^p |x-z|^{m-n}\, dz\, dx$$

$$\leq C_{m,n}\, d^{mp/q} \int_\Omega \left(\int_\Omega |x-z|^{m-n}\, dx \right) |f(z)|^p\, dz$$

(using Fubini's Theorem)

$$\leq C_{m,n} d^{(1+p/q)m}\, \|f\|_{L^p(\Omega)}^p$$

$$= C_{m,n}\, d^{mp}\, \|f\|_{L^p(\Omega)}^p\,.$$

Therefore,

$$\|g\|_{L^p(\Omega)} \leq C_{m,n}\, d^m\, \|f\|_{L^p(\Omega)}\,.$$

In the case $p=1$,

$$\|g\|_{L^1(\Omega)} = \int_\Omega |g(x)|\, dx$$

$$= \int_\Omega \left(\int_\Omega |x-z|^{m-n}\, |f(z)|\, dz \right) dx$$

$$= \int_\Omega \left(\int_\Omega |x-z|^{m-n}\, dx \right) |f(z)|\, dz \quad \text{(using Fubini's Theorem)}$$

$$\leq C_{m,n}\, d^m\, \|f\|_{L^1(\Omega)}\,.$$

In the case $p=\infty$,

$$|g(x)| = \int_\Omega |x-z|^{m-n}\, |f(z)|\, dz$$

$$\leq \|f\|_{L^\infty(\Omega)} \int_\Omega |x-z|^{m-n}\, dz$$

$$\leq C_{m,n}\, d^m\, \|f\|_{L^\infty(\Omega)}\,.$$

Therefore, $\|g\|_{L^\infty(\Omega)} \leq C_{m,n}\, d^m\, \|f\|_{L^\infty(\Omega)}$. $\qquad\qquad\qquad\square$

(4.3.8) Lemma. (Bramble-Hilbert) *Let B be a ball in Ω such that Ω is star-shaped with respect to B and such that its radius $\rho > (1/2)\rho_{\max}$. Let $Q^m u$ be the Taylor polynomial of order m of u averaged over B where $u \in W_p^m(\Omega)$ and $p \geq 1$. Then*

$$(4.3.9) \qquad |u - Q^m u|_{W_p^k(\Omega)} \leq C_{m,n,\gamma}\, d^{m-k}\, |u|_{W_p^m(\Omega)} \qquad k = 0, 1, \ldots, m,$$

where $d = \operatorname{diam}(\Omega)$.

Proof. It suffices to assume that $\operatorname{diam}(\Omega) = 1$ and to prove that for $u \in C^m(\Omega) \cap W_p^m(\Omega)$ we have

$$(4.3.10) \qquad |u - Q^m u|_{W_p^k(\Omega)} \leq C_{m,n,\gamma}\, |u|_{W_p^m(\Omega)} \qquad \forall\, k = 0, 1, \ldots, m.$$

The general case then follows from a standard homogeneity argument.

For $k = m$,
$$|u - Q^m u|_{W_p^m(\Omega)} = |u|_{W_p^m(\Omega)}.$$

For $k = 0$,
$$\|u - Q^m u\|_{L^p(\Omega)} = \|R^m u\|_{L^p(\Omega)}$$

$$\leq m \sum_{|\alpha|=m} \left\| \int_\Omega k_\alpha(x,z) D^\alpha u(z)\, dz \right\|_{L^p(\Omega)}$$

$$\leq C_{m,n} (1 + 1/\rho)^n \sum_{|\alpha|=m} \left\| \int_\Omega |x - z|^{m-n} \, |D^\alpha u(z)|\, dz \right\|_{L^p(\Omega)}$$

$$\text{(using 4.2.10)}$$

$$\leq C_{m,n,\gamma} |u|_{W_p^m(\Omega)}. \qquad \text{(Lemma 4.3.6 and the fact that } 1/\rho < 2\gamma)$$

For $0 < k < m$,
$$|u - Q^m u|_{W_p^k(\Omega)} = |R^m u|_{W_p^k(\Omega)}$$

$$= \left(\sum_{|\alpha|=k} \|R^{m-k} D^\alpha u\|_{L^p(\Omega)}^p \right)^{1/p} \qquad \text{(using 4.1.18)}$$

$$\leq C_{m,n,\gamma} \left(\sum_{|\alpha|=k} |D^\alpha u|_{W_p^{m-k}(\Omega)}^p \right)^{1/p} \qquad \text{(case } k = 0 \text{ above)}$$

$$\leq C_{m,n,\gamma} |u|_{W_p^m(\Omega)}.$$

For a general domain Ω, define

(4.3.11)
$$\widehat{\Omega} = \{(1/d)x : x \in \Omega\}.$$

If $u \in W_p^m(\Omega)$, let $\hat{u}(y) := u(dy)$. It is clear that $\hat{u} \in W_p^m(\widehat{\Omega})$. By a change of variables, we have

(4.3.12)
$$|\hat{u}|_{W_p^k(\widehat{\Omega})} = d^{k-n/p} |u|_{W_p^k(\Omega)} \quad \text{for } 0 \leq k \leq m.$$

The definition of $Q^m u(x)$ (cf. (4.1.4)) leads to

(4.3.13)
$$\widehat{Q^m} \hat{u} = \widehat{Q^m u}$$

(cf. exercise 4.x.3). Therefore,

$$|\hat{u} - \widehat{Q^m} \hat{u}|_{W_p^k(\widehat{\Omega})} \leq C_{m,n,\gamma} |\hat{u}|_{W_p^m(\widehat{\Omega})} \qquad \text{(by 4.3.10)}$$

$$= C_{m,n,\gamma} d^{m-n/p} |u|_{W_p^m(\Omega)}. \qquad \text{(by 4.3.12)}$$

On the other hand,

$$|\hat{u} - \widehat{Q}^m\hat{u}|_{W_p^k(\widehat{\Omega})} = |\hat{u} - \widehat{Q^m u}|_{W_p^k(\widehat{\Omega})} \qquad \text{(by 4.3.13)}$$

$$= d^{k-n/p} |u - Q^m u|_{W_p^k(\Omega)}.$$

Putting these together, we have proved that for $0 \le k \le m$,

$$|u - Q^m u|_{W_p^k(\Omega)} \le C_{m,n,\gamma} \, d^{m-k} \, |u|_{W_p^m(\Omega)}.$$

$$\square$$

As an application of the Bramble-Hilbert Lemma, we prove Friedrichs' inequality.

(4.3.14) Lemma. (Friedrichs' Inequality) *Suppose Ω is star-shaped with respect to a ball B. Then for all $u \in W_p^1(\Omega)$,*

$$(4.3.15) \qquad \|u - \overline{u}\|_{W_p^1(\Omega)} \le C_\Omega \, |u|_{W_p^1(\Omega)}$$

where $\overline{u} = \frac{1}{|\Omega|} \int_\Omega u(x) \, dx$.

Proof. Since $T_y^1 u(x) = u(y)$, $Q^1 u = \int_B u(y)\phi(y) \, dy$ is a constant. Suppose there does not exist a constant C_Ω such that (4.3.15) holds. Then there exists a sequence $\{v_j\} \subset W_p^1(\Omega)$ such that (a) $\|v_j - \overline{v}_j\|_{W_p^1(\Omega)} = 1$ and (b)$|v_j|_{W_p^1(\Omega)} < 1/j$. Let $w_j = v_j - \overline{v}_j$. Then (i) $\|w_j\|_{W_p^1(\Omega)} = 1$, (ii) $|w_j|_{W_p^1(\Omega)} = |v_j|_{W_p^1(\Omega)} < 1/j$, and (iii) $\overline{w}_j = 0$.

By the Bramble-Hilbert Lemma and (ii),

$$(4.3.16) \qquad \|w_j - Q^1 w_j\|_{W_p^1(\Omega)} \le C_1 \cdot (1/j).$$

Also,

$$|Q^1 w_j| = \Big| \int_B \phi(y) \, w_j(y) \, dy \Big|$$

$$\le \|\phi\|_{L^q(\Omega)} \|w_j\|_{L^p(\Omega)}$$

$$\le C_2. \qquad \text{(by (i))}$$

Hence, by the Bolzano-Weierstrass Theorem, there exists a subsequence $\{Q^1 w_{j_k}\}$ such that

$$(4.3.17) \qquad \lim_{k \to \infty} Q^1 w_{j_k} = \tilde{w} \in [-C_2, C_2].$$

It then follows from (4.3.16), (4.3.17) and the triangle inequality that $w_{j_k} \longrightarrow \tilde{w}$ in W_p^1.

On the other hand,

$$\tilde{w} \, |\Omega| = \int_\Omega \tilde{w} \, dx$$

$$= \lim_{k \to \infty} \int_\Omega w_{j_k} \, dx$$

$$= \lim_{k \to \infty} |\Omega| \, \overline{w}_{j_k}$$

$$= 0. \qquad \qquad \text{(by (iii))}$$

Thus, $\tilde{w} = 0$ and $w_{j_k} \to 0$ in W_p^1, which contradicts $\|w_j\|_{W_p^1(\Omega)} = 1$. □

(4.3.18) Remark. Friedrichs' inequality will be used in Chapter 5 to prove coercivity for the Neumann problem for the Laplace operator.

In (Dupont and Scott 1980) it is shown how to extend the above results to domains that are a finite union of star-shaped domains. For example, the domain in Fig. 4.2 satisfies this condition. More general results have been established by (Dechevski & Quak 1990).

4.4 Bounds for the Interpolation Error

So far we have estimated $R^m u = u - Q^m u$, which is a local estimate. We will now estimate the interpolation error. We begin by estimating the norm of the interpolation operator.

(4.4.1) Lemma. *Let* $(K, \mathcal{P}, \mathcal{N})$ *be a finite element such that* diam $K = 1$, $\mathcal{P} \subseteq W_\infty^m(K)$ *and* $\mathcal{N} \subseteq \left(C^l \left(\overline{K} \right) \right)'$ *(i.e. the nodal variables in* $\mathcal{N}$ *involve derivatives up to order* l*). Then the interpolation operator is bounded from* $C^l \left(\overline{K} \right)$ *into* $W_p^m(K)$ *for* $1 \le p \le \infty$.

Proof. Let $\mathcal{N} = \{N_1, \dots, N_k\}$, and let $\{\phi_1, \dots, \phi_k\} \subseteq \mathcal{P}$ be the dual basis. The interpolant is defined by $\mathcal{I}u = \sum_{i=1}^k N_i(u)\phi_i$, and each $\phi_i \in W_\infty^m(K) \subseteq W_p^m(K)$ by assumption. Then

$$\|\mathcal{I}u\|_{W_p^m(K)} \le \sum_{i=1}^k |N_i(u)| \, \|\phi_i\|_{W_p^m(K)} \qquad \text{(triangle inequality)}$$

$$\le \left(\sum_{1 \le i \le k} \|N_i\|_{C^l(\overline{K})'} \, \|\phi_i\|_{W_p^m(K)} \right) \|u\|_{C^l(\overline{K})}$$

$$= C \, \|u\|_{C^l(\overline{K})}.$$

□

(4.4.2) Definition. *If* $(K, \mathcal{P}, \mathcal{N})$ *satisfies the conditions in Lemma 4.4.1, then* $\sigma(K)$ *is defined to be the operator norm of* $\mathcal{I} : C^l \left(\overline{K} \right) \longrightarrow W_p^m(K)$.

(4.4.3) Definition. *For any bounded region K, we define*

$$\widehat{K} := \{(1/\text{diam } K)x : x \in K\}.$$

Let $\mathcal{P}_k$ be the set of polynomials in n variables of degree less than or equal to k.

(4.4.4) Theorem. *Let $(K, \mathcal{P}, \mathcal{N})$ be a finite element satisfying*
    (i) K is star-shaped with respect to some ball,
    (ii) $\mathcal{P}_{m-1} \subseteq \mathcal{P} \subseteq W_\infty^m(K)$ and
    (iii) $\mathcal{N} \subseteq \left(C^l\left(\overline{K}\right)\right)'.$
Suppose $1 \leq p \leq \infty$ and either $m - l - n/p > 0$ when $p > 1$ or $m - l - n \geq 0$ when $p = 1$. Then for $0 \leq i \leq m$ and $v \in W_p^m(K)$ we have

$$(4.4.5) \qquad |v - \mathcal{I}v|_{W_p^i(K)} \leq C_{m,n,\gamma,\sigma(\widehat{K})}(\text{diam } K)^{m-i} |v|_{W_p^m(K)},$$

where $\widehat{K} = \{(1/\text{diam } K)\, x : x \in K\}$ and γ is the chunkiness parameter for K.

Proof. It suffices to take diam $K = 1$ (in which case $K = \widehat{K}$). The general case then follows by a homogeneity argument (cf. exercise 4.x.4). Also, the interpolation operator is well defined on $W_p^m(K)$ by the Sobolev Lemma 4.3.4.

Let B be a ball in K such that K is star-shaped with respect to B and such that its radius $\rho > (1/2)\rho_{\max}$. Let $Q^m v$ be the Taylor polynomial of order m of v averaged over B. Since $\mathcal{I}f = f$ for $f \in \mathcal{P}$,

$$(4.4.6) \qquad \mathcal{I}Q^m v = Q^m v \qquad \text{because} \quad Q^m v \in \mathcal{P}_{m-1} \subseteq \mathcal{P}.$$

Now

$$
\begin{aligned}
\|v - \mathcal{I}v\|_{W_p^m(K)} &\leq \|v - Q^m v\|_{W_p^m(K)} + \|Q^m v - \mathcal{I}v\|_{W_p^m(K)} \\
&= \|v - Q^m v\|_{W_p^m(K)} + \|\mathcal{I}(Q^m v - v)\|_{W_p^m(K)} & (4.4.6) \\
&\leq \|v - Q^m v\|_{W_p^m(K)} + \sigma(K)\|Q^m v - v\|_{C^l(\overline{K})} \\
&\leq (1 + \sigma(K)\, C_{m,n,\gamma})\, \|v - Q^m v\|_{W_p^m(K)} & (4.3.4) \\
&\leq (1 + \sigma(K)\, C_{m,n,\gamma})\, C'_{m,n,\gamma} |v|_{W_p^m(K)} & \text{(Bramble-Hilbert)} \\
&= C_{m,n,\gamma,\sigma(K)} |v|_{W_p^m(K)}.
\end{aligned}
$$

$\square$

(4.4.7) Corollary. *Under the same hypotheses except $i \leq l$,*

$$(4.4.8) \qquad |v - \mathcal{I}v|_{W_\infty^i(K)} \leq C_{m,n,\gamma,\sigma(\widehat{K})}\, (\text{diam } K)^{m-i-n/p} |v|_{W_p^m(K)}.$$

Proof. Take diam $K = 1$. Then

$$|v - \mathcal{I}v|_{W^l_\infty(K)} \leq C_{m,n,\gamma} \|v - \mathcal{I}v\|_{W^m_p(K)} \qquad \text{(Sobolev Lemma 4.3.4)}$$

$$\leq C_{m,n,\gamma,\sigma(K)} |v|_{W^m_p(K)} \qquad \text{(Theorem 4.4.4)}$$

The general case then follows by a homogeneity argument. □

Our goal is to find a uniform bound for $C_{m,n,\gamma,\sigma(\widehat{K})}$, where K ranges over a collection of elements. Thus, we must study the dependence of $\sigma(\widehat{K})$ on affine transformations.

Let the reference element $(K, \mathcal{P}, \mathcal{N})$ be affine equivalent (cf. (3.4.1)) to $(\widetilde{K}, \widetilde{\mathcal{P}}, \widetilde{\mathcal{N}})$ through the transformation $Ax = ax + b$, $a = (a_{ij})$, with the coefficients of the inverse of a denoted by $(a^{-1})_{ij}$. The definition of affine-equivalence yields

$$(4.4.9) \qquad \widetilde{\mathcal{I}}\tilde{v}(\tilde{x}) = \sum_{N \in \mathcal{N}} (A_* N)\tilde{v} \cdot (A^{-1})^* \phi_N(\tilde{x}).$$

Recall that $(A_* N)\tilde{v} = N(A^*\tilde{v})$, where $(A^*\tilde{v})x = \tilde{v}(Ax)$. Therefore,

$$|(A_* N)\tilde{v}| = |N(A^*\tilde{v})|$$

$$\leq C_N \|A^*\tilde{v}\|_{C^l(\overline{K})}$$

$$\leq C_{N,n,l} \left(1 + \max_{1 \leq i,j \leq n} |a_{ij}|\right)^l \|\tilde{v}\|_{C^l(\overline{\widetilde{K}})}.$$

Also,

$$\|(A^{-1})^* \phi_N\|_{W^m_p(\widetilde{K})} \leq C'_{N,n,m} \left(1 + \max_{1 \leq i,j \leq n} \left|(a^{-1})_{ij}\right|\right)^m \times$$

$$\|\phi_N\|_{W^m_p(K)} |\det a|^{1/p}.$$

Since $\|\phi_N\|_{W^m_p(K)}$ is bounded on the reference element, we have

$$\|\widetilde{\mathcal{I}}\tilde{v}\|_{W^m_p(\widetilde{K})} \leq C_{\text{ref}} \left(1 + \max_{1 \leq i,j \leq n} |a_{ij}|\right)^l \times$$

$$\left(1 + \max_{1 \leq i,j \leq n} \left|(a^{-1})_{ij}\right|\right)^m |\det a|^{1/p} \|\tilde{v}\|_{C^l(\overline{\widetilde{K}})},$$

where

$$C_{\text{ref}} = |\mathcal{N}| \cdot \max_{N \in \mathcal{N}} \{C_{N,n,l}\} \cdot \max_{N \in \mathcal{N}} \{C'_{N,n,m}\} \cdot \max_{N \in \mathcal{N}} \{\|\phi_N\|_{W^m_p(K)}\}$$

and $|\mathcal{N}|$ denotes the number of nodal variables ($= \dim \mathcal{P}$). Therefore,

$$(4.4.10) \qquad \sigma(\widetilde{K}) \leq C_{\text{ref}} \left(1 + \max_{1 \leq i,j \leq n} |a_{ij}|\right)^l \times$$

$$\left(1 + \max_{1 \leq i,j \leq n} \left|(a^{-1})_{ij}\right|\right)^m |\det a|^{1/p}.$$

We have proved the following.

(4.4.11) Proposition. *Given a reference element $(K, \mathcal{P}, \mathcal{N})$ and an affine-equivalent element $(\widetilde{K}, \widetilde{\mathcal{P}}, \widetilde{\mathcal{N}})$ with the affine map $Ax = ax + b$, we have*

$$(4.4.12) \qquad\qquad \sigma(\widetilde{K}) \leq C_{\text{ref}}\, \chi(a),$$

where χ is a continuous function on $GL(\mathrm{I\!R}^n)$. For example, we can take
$$\chi(a) := \left(1 + \max_{1 \leq i,j \leq n} |a_{ij}|\right)^l \left(1 + \max_{1 \leq i,j \leq n} \left|\left(a^{-1}\right)_{ij}\right|\right)^m |\det a|^{1/p}.$$

(4.4.13) Definition. *Let Ω be a given domain and let $\{\mathcal{T}^h\}$, $0 < h \leq 1$, be a family of subdivisions such that*

$$(4.4.14) \qquad\qquad \max\{\operatorname{diam} T : T \in \mathcal{T}^h\} \leq h \operatorname{diam} \Omega.$$

The family is said to be **quasi-uniform** *if there exists $\rho > 0$ such that*

$$(4.4.15) \qquad\qquad \min\{\operatorname{diam} B_T : T \in \mathcal{T}^h\} \geq \rho h \operatorname{diam} \Omega$$

for all $h \in (0,1]$, where B_T is the largest ball contained in T such that T is star-shaped with respect to B_T. The family is said to be **non-degenerate** *if there exists $\rho > 0$ such that for all $T \in \mathcal{T}^h$ and for all $h \in (0,1]$,*

$$(4.4.16) \qquad\qquad \operatorname{diam} B_T \geq \rho \operatorname{diam} T.$$

(4.4.17) Remarks.

(i) $\{\mathcal{T}^h\}$ is non-degenerate if and only if the chunkiness parameter is uniformly bounded for all $T \in \mathcal{T}^h$ and for all $h \in (0,1]$.

(ii) If a family is quasi-uniform, then it is non-degenerate, but not conversely (cf. exercise 4.x.5).

(iii) If we start with an arbitrary triangulation in two dimensions and subdivide by connecting edge midpoints, we obtain a quasi-uniform family of triangulations. A similar, but more complicated, construction can be made in three dimensions (Zhang 1995).

Recall (from Definition 3.3.15) that a reference element $(K, \mathcal{P}, \mathcal{N})$ is said to be a C^r *element* if r is the largest non-negative integer for which

$$(4.4.18) \qquad\qquad V^h = \mathcal{I}^h C^l\left(\overline{\Omega}\right) \subseteq C^r(\Omega) \cap W_\infty^{r+1}(\Omega).$$

Here $\mathcal{I}^h : C^l\left(\overline{\Omega}\right) \longrightarrow L^1(\Omega)$ is the global interpolation operator defined by

$$(4.4.19) \qquad\qquad \mathcal{I}^h u|_T := \mathcal{I}_T^h u \quad \text{for } T \in \mathcal{T}^h, h \in (0,1],$$

where $\mathcal{I}_T^h$ is the interpolation operator for the affine-equivalent element $(T, \mathcal{P}_T, \mathcal{N}_T)$.

(4.4.20) Theorem. *Let* $\{\mathcal{T}^h\}$, $0 < h \leq 1$, *be a non-degenerate family of subdivisions of a polyhedral domain* Ω *in* $\mathbb{R}^n$. *Let* $(K, \mathcal{P}, \mathcal{N})$ *be a reference element, satisfying the conditions of Theorem 4.4.4 for some l, m and p. For all* $T \in \mathcal{T}^h, 0 < h \leq 1$, *let* $(T, \mathcal{P}_T, \mathcal{N}_T)$ *be the affine-equivalent element. Then there exists a positive constant C depending on the reference element, n, m, p and the number* ρ *in (4.4.16) such that for* $0 \leq s \leq m$,

$$(4.4.21) \qquad \left(\sum_{T \in \mathcal{T}^h} \|v - \mathcal{I}^h v\|^p_{W^s_p(T)} \right)^{1/p} \leq C\, h^{m-s} \, |v|_{W^m_p(\Omega)}$$

for all $v \in W^m_p(\Omega)$, *where the left-hand side should be interpreted, in the case* $p = \infty$, *as* $\max_{T \in \mathcal{T}^h} \|v - \mathcal{I}^h v\|_{W^s_\infty(T)}$. *For* $0 \leq s \leq l$,

$$(4.4.22) \qquad \max_{T \in \mathcal{T}^h} \|v - \mathcal{I}^h v\|_{W^s_\infty(T)} \leq C\, h^{m-s-n/p} \, |v|_{W^m_p(\Omega)} \quad \forall v \in W^m_p(\Omega).$$

Proof. We will first prove

$$(4.4.23) \qquad \sup\{\sigma(\widehat{T}) : T \in \mathcal{T}^h, 0 < h \leq 1\} = C(\rho, m, n, p, K) < \infty.$$

Since $(\widehat{T}, \widehat{\mathcal{P}}_T, \widehat{\mathcal{N}}_T)$ is also affine equivalent to $(K, \mathcal{P}, \mathcal{N})$ with some affine map, say $Ax = ax + b$, we have $\sigma(\widehat{T}) \leq C_{\mathrm{ref}}\, \chi(a)$ where χ is continuous on $GL(\mathbb{R}^n)$ by Proposition 4.4.11. We need to show that nondegeneracy implies that a is an element of a compact subset of $GL(\mathbb{R}^n)$, which then implies (4.4.23). Since the family is non-degenerate, there exists $B \subset \widehat{T}$ such that diam $B \geq \rho > 0$ (by (4.4.16)). We have

$$\mathrm{meas}\, B \leq \mathrm{meas}\, \widehat{T}$$
$$= \int_{\widehat{T}} d\hat{x}$$
$$= |\det a| \int_K dx$$
$$\leq |\det a|\, (\mathrm{meas}\, K).$$

On the other hand, since diam $B \geq \rho$, we have meas $B \geq C_n \rho^n$. Therefore, $0 < C_n \rho^n \leq |\det a|\, (\mathrm{meas}\, K)$; in other words, $|\det a| \geq \epsilon > 0$, where ϵ depends on ρ, K and n. Thus, $a \in \{b : |\det b| \geq \epsilon > 0\}$, which is a closed set. Without loss of generality, we may assume that $\{(x_1, \ldots, x_n) : \sum_{i=1}^n x_i \leq t_0, x_i > 0 \text{ for } i = 1, \ldots, n\} \subset K$, where t_0 depends only on K. Let e^i be the i-th unit vector. Then $A(te^i) = tae^i + b \in$ the closure of $\widehat{T}$ for $0 \leq t \leq t_0$. Therefore, $\|ae^i\| \leq \mathrm{diam}\, \widehat{T}/t_0 = 1/t_0$, for $1 \leq i \leq n$, which implies that $|a_{ij}| \leq 1/t_0$, $i, j = 1, \ldots, n$. Therefore, a is an element of the compact set $\{b : |\det b| \geq \epsilon > 0, |b_{ij}| \leq 1/t_0\}$. Thus, (4.4.23) holds, and in view of the form of $\chi(a)$, it follows that $\sigma(\widehat{T}) \leq C$, for all

$T \in \mathcal{T}^h$ and $0 < h \le 1$, where C depends on the reference element, n, m, l and ρ.

To prove (4.4.21), observe that

$$\sum_{T \in \mathcal{T}^h} \|v - \mathcal{I}_T^h v\|_{W_p^s(T)}^p$$

$$\le \sum_{T \in \mathcal{T}^h} C_{m,n,\gamma,\sigma(\widehat{T})}^p \sum_{i=0}^s (\operatorname{diam} T)^{p(m-i)} |v|_{W_p^m(T)}^p \qquad \text{(by 4.4.5)}$$

$$\le \sum_{T \in \mathcal{T}^h} C_{m,n,\gamma,\sigma(\widehat{T})}^p \sum_{i=0}^s (h \operatorname{diam} \Omega)^{p(m-i)} |v|_{W_p^m(T)}^p \qquad \text{(by 4.4.14)}$$

$$\le C h^{p(m-s)} \sum_{T \in \mathcal{T}^h} |v|_{W_p^m(T)}^p \qquad \text{(by 4.4.23 and since } h \le 1)$$

$$= C h^{p(m-s)} |v|_{W_p^m(\Omega)}^p,$$

where C depends on m, n, ρ and the reference element.

The inequality (4.4.22) is similarly proved by using (4.4.8) (cf. exercise 4.x.6). $\qquad\square$

(4.4.24) Corollary. *Let* $\{\mathcal{T}^h\}$, $0 < h \le 1$, *be a non-degenerate family of subdivisions of a polyhedral domain* Ω. *Let* $(K, \mathcal{P}, \mathcal{N})$ *be a reference element, satisfying the conditions of Theorem 4.4.4 for some* l, m *and* p. *Suppose that all* $(T, \mathcal{P}_T, \mathcal{N}_T)$, *for all* $T \in \mathcal{T}^h, 0 < h \le 1$, *are affine-interpolation equivalent to* $(K, \mathcal{P}, \mathcal{N})$. *Then there exists a positive constant* C *depending on the reference element,* l, m, n, p *and the number* ρ *in (4.4.16) such that for* $0 \le s \le m$,

$$(4.4.25) \qquad \left(\sum_{T \in \mathcal{T}^h} \|v - \mathcal{I}^h v\|_{W_p^s(T)}^p \right)^{1/p} \le C h^{m-s} |v|_{W_p^m(\Omega)}$$

for all $v \in W_p^m(\Omega)$, *where the left-hand side should be interpreted, in the case* $p = \infty$, *as* $\max_{T \in \mathcal{T}^h} \|v - \mathcal{I}^h v\|_{W_\infty^s(T)}$. *For* $0 \le s \le l$,

$$(4.4.26) \quad \max_{T \in \mathcal{T}^h} \|v - \mathcal{I}^h v\|_{W_\infty^s(T)} \le C h^{m-s-n/p} |v|_{W_p^m(\Omega)} \quad \forall v \in W_p^m(\Omega).$$

(4.4.27) *Remark.* In the event that the elements in the previous results form C^r elements for some $r \ge 0$, then for $0 \le s \le r + 1$ we have

$$\sum_{T \in \mathcal{T}^h} \|v - \mathcal{I}_T^h v\|_{W_p^s(T)}^p = \|v - \mathcal{I}^h v\|_{W_p^s(\Omega)}^p$$

and

$$\max_{T \in \mathcal{T}^h} \|v - \mathcal{I}_T^h v\|_{W_\infty^s(T)} = \|v - \mathcal{I}^h v\|_{W_\infty^s(\Omega)}.$$

Substituting these expressions in the left-hand side leads to estimates of the form

$$(4.4.28) \qquad \|v - \mathcal{I}^h v\|_{W_p^s(\Omega)} \leq C\, h^{m-s}\, |v|_{W_p^m(\Omega)}$$

for all $v \in W_p^m(\Omega)$ and $0 \leq s \leq \min\{m, r+1\}$ and

$$(4.4.29) \qquad \|v - \mathcal{I}^h v\|_{W_\infty^s(\Omega)} \leq C\, h^{m-s-n/p}\, |v|_{W_p^m(\Omega)}$$

for all $v \in W_p^m(\Omega)$ and $0 \leq s \leq \min\{l, r+1\}$.

4.5 Inverse Estimates

In this section we discuss the relations among various norms on a finite-element space.

Let K be a bounded domain in $\mathbb{R}^n$. If v is a function defined on K, then $\hat{v}$ is defined on $\widehat{K} = \{(1/\operatorname{diam} K)\, x : x \in K\}$ by

$$(4.5.1) \qquad \hat{v}(\hat{x}) = v\left((\operatorname{diam} K)\hat{x}\right) \quad \forall \hat{x} \in \widehat{K}.$$

It is clear that $v \in W_r^k(K)$ iff $\hat{v} \in W_r^k(\widehat{K})$ and

$$(4.5.2) \qquad |\hat{v}|_{W_r^k(\widehat{K})} = (\operatorname{diam} K)^{k-(n/r)}\, |v|_{W_r^k(K)}.$$

If $\mathcal{P}$ is a vector space of functions defined on K, then $\widehat{\mathcal{P}} := \{\hat{v} : v \in \mathcal{P}\}$.

(4.5.3) Lemma. *Let $\rho h \leq \operatorname{diam} K \leq h$, where $0 < h \leq 1$, and $\mathcal{P}$ be a finite-dimensional subspace of $W_p^l(K) \cap W_q^m(K)$, where $1 \leq p \leq \infty$, $1 \leq q \leq \infty$ and $0 \leq m \leq l$. Then there exists $C = C(\widehat{\mathcal{P}}, \widehat{K}, l, p, q, \rho)$ such that for all $v \in \mathcal{P}$, we have*

$$(4.5.4) \qquad \|v\|_{W_p^l(K)} \leq C\, h^{m-l+n/p-n/q} \|v\|_{W_q^m(K)}.$$

Proof. We will use C to represent a generic constant depending only on $\widehat{\mathcal{P}}, \widehat{K}, l, p, q$ and ρ.

We first establish (4.5.4) for the case $m = 0$. For any finite-dimensional space $\mathcal{P}$ satisfying the conditions of the lemma, we have by the equivalence of norms that

$$(4.5.5) \qquad \|\hat{v}\|_{W_p^l(\widehat{K})} \leq C\, \|\hat{v}\|_{L^q(\widehat{K})} \quad \forall v \in \mathcal{P}.$$

Therefore, (4.5.2) implies that

$$(4.5.6) \qquad |v|_{W_p^j(K)}(\operatorname{diam} K)^{j-n/p} \leq C\, \|v\|_{L^q(K)}(\operatorname{diam} K)^{-n/q}$$

for $0 \leq j \leq l$, from which we deduce that

(4.5.7) $|v|_{W_p^j(K)} \le C\, h^{-j+n/p-n/q} \|v\|_{L^q(K)}$ for $0 \le j \le l$.

Since $h \le 1$, we have

(4.5.8) $\|v\|_{W_p^j(K)} \le C\, h^{-j+n/p-n/q} \|v\|_{L^q(K)}$ for $0 \le j \le l$,

which is just (4.5.4) when $m = 0$ if we take $j = l$.

For the case of general $m \le l$, we argue as follows. For $l - m \le k \le l$ and $|\alpha| = k$, we may write $D^\alpha v = D^\beta D^\gamma v$ for $|\beta| = l - m$ and $|\gamma| = k+m-l$:

$$\|D^\alpha v\|_{L^p(K)} \le \|D^\gamma v\|_{W_p^{l-m}(K)}$$
$$\le C h^{-(l-m)+\frac{n}{p}-\frac{n}{q}} \|D^\gamma v\|_{L^q(K)} \qquad \text{(by 4.5.8 for } D^\gamma \mathcal{P}\text{)}$$
$$\le C h^{-(l-m)+\frac{n}{p}-\frac{n}{q}} |v|_{W_q^{k+m-l}(K)}.$$

Since $|\alpha| = k$ was arbitrary, we have

(4.5.9) $|v|_{W_p^k(K)} \le C\, h^{-(l-m)+n/p-n/q} |v|_{W_q^{k+m-l}(K)}$

for any k satisfying $l - m \le k \le l$. In particular, this implies that

(4.5.10) $|v|_{W_p^k(K)} \le C\, h^{-(l-m)+n/p-n/q} \|v\|_{W_q^m(K)}$

for k satisfying $l - m \le k \le l$, since the latter implies $k + m - l \le m$. The estimate (4.5.4) now follows from (4.5.8) with $j = l - m$ and (4.5.10).   □

The following theorem is a global version of Lemma 4.5.3.

(4.5.11) Theorem. *Let $\{\mathcal{T}^h\}$, $0 < h \le 1$, be a quasi-uniform family of subdivisions of a polyhedral domain $\Omega \subseteq \mathbb{R}^n$. Let $(K, \mathcal{P}, \mathcal{N})$ be a reference finite element such that $\mathcal{P} \subseteq W_p^l(K) \cap W_q^m(K)$ where $1 \le p \le \infty$, $1 \le q \le \infty$ and $0 \le m \le l$. For $T \in \mathcal{T}^h$, let $(T, \mathcal{P}_T, \mathcal{N}_T)$ be the affine-equivalent element, and $V^h = \{v : v \text{ is measurable and } v|_T \in \mathcal{P}_T \ \forall T \in \mathcal{T}^h\}$. Then there exists $C = C(l, p, q, \rho)$ such that*

$$
(4.5.12) \quad \left[\sum_{T \in \mathcal{T}^h} \|v\|_{W_p^l(T)}^p \right]^{1/p} \le C h^{m-l+\min(0,\frac{n}{p}-\frac{n}{q})} \left[\sum_{T \in \mathcal{T}^h} \|v\|_{W_q^m(T)}^q \right]^{1/q}
$$

for all $v \in V^h$. When $p = \infty$ (respectively, $q = \infty$), $\left[\sum_{T \in \mathcal{T}^h} \|v\|_{W_p^l(T)}^p \right]^{1/p}$ (respectively, $\left[\sum_{T \in \mathcal{T}^h} \|v\|_{W_q^l(T)}^q \right]^{1/q}$) is interpreted as $\max_{T \in \mathcal{T}^h} \|v\|_{W_\infty^l(T)}$ (respectively, $\max_{T \in \mathcal{T}^h} \|v\|_{W_\infty^m(T)}$).

Proof. We first observe that Lemma 4.5.3 and the quasi-uniformity of $\{\mathcal{T}^h\}$ imply that

(4.5.13) $\|v\|_{W_p^l(T)} \le C(\widehat{\mathcal{P}}_T, \widehat{T}, l, p, q, \rho) h^{m-l+n/p-n/q} \|v\|_{W_q^m(T)}$.

for all $T \in \mathcal{T}^h$ and all $v \in \mathcal{P}_T$.

Also, an argument similar to the one in the proof of Proposition 4.4.11 shows that

$$(4.5.14) \qquad C(\widehat{\mathcal{P}}_T, \widehat{T}, l, p, q, \rho) \leq \zeta(a_T) \, C(l, p, q, \rho),$$

where $Ax = a_T x + b_T$ is the affine transformation that maps $\widehat{K}$ to $\widehat{T}$, and ζ is a positive function which depends continuously on $a_T \in GL(\mathbb{R}^n)$. Since $\{\mathcal{T}^h\}$ is non-degenerate, the argument in the proof of Theorem 4.4.20 shows that $\{a_T : T \in \mathcal{T}^h, 0 < h \leq 1\}$ is a compact subset of $GL(\mathbb{R}^n)$. Therefore, from (4.5.13) and (4.5.14) we deduce that

$$(4.5.15) \qquad \|v\|_{W_p^l(T)} \leq C(l, p, q, \rho) \, h^{m-l+n/p-n/q} \, \|v\|_{W_q^m(T)}$$

for all $T \in \mathcal{T}^h$ and all $v \in \mathcal{P}_T$ (also see exercise 4.x.13). From now on we will use C to denote a generic constant depending only on l, p, q, and ρ.

For $p = \infty$, inequality (4.5.12) follows immediately from the estimate (4.5.15).

Assume that $p < \infty$. From (4.5.15) we obtain

$$(4.5.16) \qquad \left(\sum_{T \in \mathcal{T}^h} \|v\|_{W_p^l(T)}^p \right)^{1/p} \leq C \, h^{m-l+n/p-n/q} \left(\sum_{T \in \mathcal{T}^h} \|v\|_{W_q^m(T)}^p \right)^{1/p}$$

for all $v \in V^h$.

If $p \geq q$, then (cf. exercise 4.x.7)

$$(4.5.17) \qquad \left(\sum_{T \in \mathcal{T}^h} \|v\|_{W_q^m(T)}^p \right)^{1/p} \leq \left(\sum_{T \in \mathcal{T}^h} \|v\|_{W_q^m(T)}^q \right)^{1/q}.$$

Inequality (4.5.12) in this case follows immediately from (4.5.16) and (4.5.17).

If $p < q$, then Hölder's inequality implies (cf. exercise 4.x.8)

$$(4.5.18) \qquad \left(\sum_{T \in \mathcal{T}^h} \|v\|_{W_q^m(T)}^p \right)^{\frac{1}{p}} \leq \left(\sum_{T \in \mathcal{T}^h} 1 \right)^{\frac{1}{p} - \frac{1}{q}} \left(\sum_{T \in \mathcal{T}^h} \|v\|_{W_q^m(T)}^q \right)^{\frac{1}{q}}.$$

It follows from the quasi-uniformity of $\{\mathcal{T}^h\}$ that

$$(4.5.19) \qquad \sum_{T \in \mathcal{T}^h} 1 \leq C \, h^{-n}.$$

Inequality (4.5.12) in this case follows from estimates (4.5.16), (4.5.18) and (4.5.19). $\qquad \square$

(4.5.20) *Remark.* Theorem 4.5.11 is applicable to both conforming and non-conforming finite elements. In the case of conforming finite elements, one can replace the summations in (4.5.12) by globally-defined norms.

4.6 Tensor-product Polynomial Approximation

We showed that approximation of order h^m can be achieved as long as polynomials of degree $m - 1$ are used. These results apply to both the standard tensor-product polynomial spaces and the serendipity elements as well. However, one may wonder what the extra terms (of degree $\geq m$) in the tensor-product polynomial spaces provide in terms of approximation, if they do not affect the order, h^m, of approximation. There is a significant effect, but it is only visible in the norm appearing in the error term. Define $A = \{me^i \; : \; i = 1, \ldots, n\}$ where the multi-indices e^i are the standard basis vectors, e^i_j being the Kronecker delta. Let A^0 denote the set of multi-indices corresponding to tensor-product polynomials of degree less than m; note that this can be characterized as

$$(4.6.1) \qquad A^0 = \left\{\beta \; : \; D^\alpha x^\beta = 0 \quad \forall \alpha \in A\right\}.$$

Define

$$(4.6.2) \qquad Q^A u(x) = \int_B \sum_{\alpha \in A^0} \frac{1}{\alpha!} D^\alpha u(y)(x - y)^\alpha \phi(y)\, dy.$$

As before, we have

$$(4.6.3) \qquad \|Q^A u\|_{W^k_\infty(\Omega)} \leq C_{m,n,\rho} \|u\|_{L^1(B)}.$$

(4.6.4) Proposition. *The remainder* $R^A u := u - Q^A u$ *satisfies*

$$(4.6.5) \qquad R^A u(x) = \sum_{\alpha \in A} \int_{C_x} \tilde{k}_\alpha(x, z) D^\alpha u(z)\, dz,$$

where

$$(4.6.6) \qquad \left|\left(\frac{\partial}{\partial x}\right)^\beta \left(\frac{\partial}{\partial y}\right)^\gamma \tilde{k}_\alpha(x, y)\right| \leq C \left|x - y\right|^{|\alpha| - n - |\beta| - |\gamma|}.$$

Proof. We may write

$$u(x) = Q^{n \times m} u(x) + R^{n \times m} u(x)$$

$$= Q^A u(x) + \int_B \sum_{\substack{\alpha \notin A^0 \\ |\alpha| < n \times m}} \frac{1}{\alpha!} D^\alpha u(y)(x-y)^\alpha \phi(y) \, dy$$

(4.6.7)

$$+ n \times m \sum_{|\alpha| = n \times m} \int_{C_x} k_\alpha(x, z) D^\alpha u(z) \, dz.$$

Note that $|\alpha| = n \times m$ implies that $\alpha \notin A^0$. For any $\alpha \notin A^0$, we may write $\alpha = \beta + \gamma$ for some $\beta \in A$ and γ a multi-index, that is $\gamma_i \geq 0 \quad \forall i$. Integrating by parts, we find

(4.6.8) $$u(x) = Q^A u(x) + \sum_{\alpha \in A} \int_{C_x} \tilde{k}_\alpha(x, z) D^\alpha u(z) \, dz.$$

Estimate (4.6.6) follows from the fact that

$$\left| \left(\frac{\partial}{\partial x} \right)^\beta \left(\frac{\partial}{\partial y} \right)^\gamma k_\alpha(x, y) \right| \leq C \left| x - y \right|^{|\alpha| - n - |\beta| - |\gamma|},$$

which can be proved as in the proof of Proposition 4.2.8.   □

From Lemma 4.3.1 it follows that

(4.6.9) $$|u - Q^A u|_{W^k_\infty(\Omega)} \leq C_{m,n,\gamma} \, d^{m-k-n/p} \left(\sum_{i=1,\ldots,n} \left\| \frac{\partial^m u}{\partial x_i^m} \right\|^p_{L^p(\Omega)} \right)^{1/p}$$

where d is the diameter of Ω, provided that $m - k - n/p > 0$ when $p > 1$ or $m - k - n \geq 0$ when $p = 1$. From Lemma 4.3.6 follows, for $0 \leq k < m$,

(4.6.10) $$|u - Q^A u|_{W^k_p(\Omega)} \leq C_{m,n,\gamma} \, d^{m-k} \left(\sum_{i=1,\ldots,n} \left\| \frac{\partial^m u}{\partial x_i^m} \right\|^p_{L^p(\Omega)} \right)^{1/p}.$$

Thus, the proof of Theorem 4.4.4 yields the following.

(4.6.11) Theorem. *Let $(K, \mathcal{P}, \mathcal{N})$ be the tensor-product finite element of order $m - 1$, let $h = \mathrm{diam}\,(K)$ and let $\mathcal{I}$ denote the corresponding interpolant. Then for $u \in W^m_p(K)$, we have*

(4.6.12) $$|u - \mathcal{I}u|_{W^k_\infty(K)} \leq C_{m,n} h^{m-k-n/p} \left(\sum_{i=1,\ldots,n} \left\| \frac{\partial^m u}{\partial x_i^m} \right\|^p_{L^p(K)} \right)^{1/p},$$

provided that $m - k - n/p > 0$ when $p > 1$ or $m - k - n \geq 0$ when $p = 1$, and

$$(4.6.13) \qquad |u - \mathcal{I}u|_{W_p^k(K)} \leq C_{m,n} h^{m-k} \left(\sum_{i=1,\ldots,n} \left\| \frac{\partial^m u}{\partial x_i^m} \right\|_{L^p(K)}^p \right)^{1/p},$$

for $0 \leq k < m$.

Since the tensor-product finite elements generate C^0 elements, we have the following global estimate for the interpolant.

(4.6.14) Theorem. *Let $\mathcal{I}^h$ denote the interpolant for tensor-product finite elements of order $m - 1$ on a rectangular subdivision of Ω of maximum mesh-size h. Then for $u \in W_p^m(\Omega)$ we have*

$$(4.6.15) \qquad |u - \mathcal{I}^h u|_{L^\infty(\Omega)} \leq C_{m,n} h^{m-n/p} \left(\sum_{i=1,\ldots,n} \left\| \frac{\partial^m u}{\partial x_i^m} \right\|_{L^p(\Omega)}^p \right)^{1/p},$$

provided that $m - n/p > 0$ when $p > 1$ or $m - n \geq 0$ when $p = 1$, and

$$(4.6.16) \qquad |u - \mathcal{I}^h u|_{W_p^k(\Omega)} \leq C_{m,n} h^{m-k} \left(\sum_{i=1,\ldots,n} \left\| \frac{\partial^m u}{\partial x_i^m} \right\|_{L^p(\Omega)}^p \right)^{1/p},$$

for $0 \leq k \leq 1$.

Note that the above estimates definitely do not hold with ordinary polynomial approximation. For example, take $m = 2$ and $u(x, y) = xy$. The right-hand sides in the estimates above are all zero, so we would have $u = \mathcal{I}^h u$. But this is impossible for linear interpolation; we must include the additional terms available in bilinear approximation.

Serendipity elements are more complicated due, in part, to the freedom available in their definition. Consider quadratic serendipity elements in two dimensions. The appropriate space of shape functions, $\mathcal{P}$, is any set containing all quadratic polynomials which can be determined uniquely by the edge nodes. One way to define such a set is to let

$$(4.6.17) \qquad A = \{(3,0), (0,3), (2,2)\}$$

and let $\mathcal{P}$ be the set of polynomials with basis given by

$$A^0 = \{\beta : D^\alpha x^\beta = 0 \quad \forall \alpha \in A\}$$
$$= \{1, x, y, x^2, xy, y^2, x^2 y, xy^2\}$$

(no $x^2 y^2$ term).

First, let us see that $\mathcal{P}$ is a valid space of shape functions for a quadratic serendipity element. Suppose that the element domain, K, is the unit square, and suppose that all of the serendipity (edge) nodal values of $P \in \mathcal{P}$ vanish. Then $P = cx(1 - x)y(1 - y) = cx^2 y^2 + \cdots$. Thus, we must have $c = 0$. Therefore, the serendipity nodal variables, $\mathcal{N}$ (depicted in Fig.3.12), determine $\mathcal{P}$, and $(K, \mathcal{N}, \mathcal{P})$ is a well-defined finite element.

Using A^0 above, we can define Q^A as in (4.6.3) and obtain error estimates as in Theorems 4.6.11 and 4.6.14.

(4.6.18) Theorem. *Let $(K, \mathcal{P}, \mathcal{N})$ be the two-dimensional quadratic serendipity finite element defined above using A as given in (4.6.17), and let $\mathcal{I}$ denote the corresponding interpolant. Then for $u \in W_p^4(K)$ we have*

$$|u - \mathcal{I}u|_{W_\infty^k(K)} \le C \left(\sum_{\alpha \in A} (\text{diam } K)^{p(|\alpha|-k)-2} \|D^\alpha u\|_{L^p(K)}^p \right)^{1/p},$$

provided that $3 - k - 2/p > 0$ when $p > 1$ or $k \le 1$ when $p = 1$, and

$$|u - \mathcal{I}u|_{W_p^k(K)} \le C \left(\sum_{\alpha \in A} (\text{diam } K)^{p(|\alpha|-k)} \|D^\alpha u\|_{L^p(K)}^p \right)^{1/p},$$

for $0 \le k \le 2$.

(4.6.19) Theorem. *Let $\mathcal{I}^h$ denote the interpolant for the quadratic serendipity finite element defined above using A as given in (4.6.17), on a rectangular subdivision of Ω of maximum mesh-size h. Then for $u \in W_p^4(\Omega)$ we have*

$$|u - \mathcal{I}^h u|_{L^\infty(\Omega)} \le C \left(\sum_{\alpha \in A} h^{p|\alpha|-2} \|D^\alpha u\|_{L^p(\Omega)}^p \right)^{1/p}$$

and

$$|u - \mathcal{I}^h u|_{W_p^k(\Omega)} \le C \left(\sum_{\alpha \in A} h^{p(|\alpha|-k)} \|D^\alpha u\|_{L^p(\Omega)}^p \right)^{1/p},$$

for $0 \le k \le 1$.

Analogous results can be obtained for cubic serendipity elements by choosing

(4.6.20) $A = \{(4,0), (0,4), (3,3), (3,2), (2,3), (2,2)\}$

except that the range of k is increased by one in the analog of Theorem 4.6.18 and more smoothness is required of u. Let $\mathcal{P}$ be the space with basis given by $\{x^\beta : \beta \in A^0\}$ where

$$A^0 = \{\beta : D^\alpha x^\beta = 0 \quad \forall \alpha \in A\}.$$

To see that $\mathcal{P}$ is a valid space of shape functions for a cubic serendipity element, suppose that all of the serendipity (edge) nodal values (as depicted in Fig. 3.22) of $P \in \mathcal{P}$ vanish. Then $P = x(1 - x)y(1 - y)B(x, y)$ where B is bilinear. Since we must have $D^\alpha P = 0$ for all $\alpha \in A$, we conclude that $B \equiv 0$ (start with $\alpha = (3, 3)$ and then continue with decreasing $|\alpha|$).

(4.6.21) Theorem. *Let $(K, \mathcal{P}, \mathcal{N})$ be the two-dimensional cubic serendipity finite element defined above using A as given in (4.6.20), and let $\mathcal{I}$ denote the corresponding interpolant. Then for $u \in W_p^6(K)$ we have*

$$|u - \mathcal{I}u|_{W_\infty^k(K)} \leq C \left(\sum_{\alpha \in A} (\operatorname{diam} K)^{p(|\alpha|-k)-2} \|D^\alpha u\|_{L^p(K)}^p \right)^{1/p},$$

provided that $4 - k - 2/p > 0$ when $p > 1$ or $k \leq 2$ when $p = 1$, and

$$|u - \mathcal{I}u|_{W_p^k(K)} \leq C \left(\sum_{\alpha \in A} (\operatorname{diam} K)^{p(|\alpha|-k)} \|D^\alpha u\|_{L^p(K)}^p \right)^{1/p},$$

for $0 \leq k \leq 3$. Furthermore, if $\mathcal{I}^h$ denotes the corresponding interpolant on a rectangular subdivision of Ω of maximum mesh-size h, then

$$|u - \mathcal{I}^h u|_{L^\infty(\Omega)} \leq C \left(\sum_{\alpha \in A} h^{p|\alpha|-2} \|D^\alpha u\|_{L^p(\Omega)}^p \right)^{1/p}$$

and

$$|u - \mathcal{I}^h u|_{W_p^k(\Omega)} \leq C \left(\sum_{\alpha \in A} h^{p(|\alpha|-k)} \|D^\alpha u\|_{L^p(\Omega)}^p \right)^{1/p},$$

for $0 \leq k \leq 1$.

4.7 Isoparametric Polynomial Approximation

When using high-order elements for problems with curved boundaries, it is essential to include some way of approximating the boundary conditions accurately. One of the most effective ways in engineering practice is using isoparametric elements, which involve a more general piecewise-polynomial change of variables in the definition of the approximating spaces than we have considered so far. In this approach, the element basis functions ϕ_N^e, defined on a given element domain, K_e, are related to the reference basis functions, defined on the reference element domain, K, via a polynomial mapping, $\xi \to F(\xi)$, of K to K_e:

(4.7.1) $$\phi_N^e(x) = \phi_N \left(F^{-1}(x) \right).$$

This is analogous to having affine-equivalent elements, except now the mapping is allowed to be more general. Elements where the mapping, F, comes from the same finite element space are called "isoparametric."

More precisely, we have a base polyhedral domain, $\widetilde{\Omega} \subseteq \mathrm{I\!R}^n$, and a base finite element space, $\widetilde{V}_h$, defined on $\widetilde{\Omega}$. We construct (by some means) a one-to-one continuous mapping $\widetilde{F} : \widetilde{\Omega} \longrightarrow \mathrm{I\!R}^n$ where each component $\widetilde{F}_i \in \widetilde{V}_h$. The resulting space,

$$(4.7.2) \qquad V_h := \left\{ v \left(\widetilde{F}^{-1}(x) \right) \; : \; x \in \widetilde{F}(\widetilde{\Omega}), \; v \in \widetilde{V}_h \right\},$$

is called an isoparametric-equivalent finite element space.

The approximation theory for isoparametric-equivalent spaces is quite simple, invoking only the chain rule. Thus, all of the results derived in the chapter so far remain valid, the only modification being that the constants now depend on the Jacobian, $J_{\widetilde{F}}$, of the mapping, $\widetilde{F}$. Some care must be exercised to obtain mappings that have regular Jacobians (Ciarlet & Raviart 1972b). However, the details of how to do this in two and three dimensions have been presented in (Lenoir 1986).

For example, let Ω be a smooth domain in two or three dimensions, and let Ω_h be a polyhedral approximation to it, e.g., defined by a piecewise linear interpolation of the boundary having facets of size at most h. Let $\mathcal{T}^h$ denote corresponding triangulations consisting of simplices of size at most h. Then it is possible to construct piecewise polynomial mappings, F^h, of degree $k-1$ which
1. equal the identity map away from the boundary of Ω_h,
2. have the property that the distance from any point on $\partial\Omega$ to the closest point on $\partial F^h(\Omega_h)$ is at most Ch^k and
3. $\|J_{F^h}\|_{W_\infty^k(\Omega_h)} \le C$ and $\left\|J_{F^h}^{-1}\right\|_{W_\infty^k(\Omega_h)} \le C$, independent of h.

Note that Ω is only approximated by $F^h(\Omega_h)$, not equal. However, using the extension result in (1.45), we can easily view that all functions being approximated are defined on $F^h(\Omega_h)$ as needed. Using the techniques in Sect. 4.4, we easily prove the following.

(4.7.3) Theorem. Let $\{\mathcal{T}^h\}$, $0 < h \le 1$, be a non-degenerate family of subdivisions of a family of polyhedral approximations, Ω_h, of a Lipschitz domain Ω. Suppose that piecewise polynomial mappings, F^h, of degree $m-1$ exist, which satisfy properties 1–3 above. Let $(K, \mathcal{P}, \mathcal{N})$ be a C^0 reference element, satisfying the conditions of Theorem 4.4.4 for some m, p. Suppose that all $(T, \mathcal{P}_T, \mathcal{N}_T)$, for all $T \in \mathcal{T}^h$, $0 < h \le 1$, are affine-interpolation equivalent to $(K, \mathcal{P}, \mathcal{N})$. Then there exists a positive constant C depending on the reference element, n, m, p and the number ρ in (4.4.16) such that for $0 \le s \le 1$,

$$(4.7.4) \qquad \|v - \mathcal{I}^h v\|_{W_p^s(F^h(\Omega_h))} \le C\, h^{m-s} \, |v|_{W_p^m(F^h(\Omega_h))},$$

and provided $m > s + n/p$ ($m \ge s + n$ if $p = 1$)

(4.7.5) $$\|v - \mathcal{I}^h v\|_{W_{\infty}^s(F^h(\Omega_h))} \leq C\, h^{m-s-n/p}\, |v|_{W_p^m(F^h(\Omega_h))}\,.$$

Here $\mathcal{I}^h v$ denotes the isoparametric interpolant defined by $\mathcal{I}^h v\left(F^h(x)\right) = \widetilde{\mathcal{I}^h}\widetilde{v}(x)$ for all $x \in \Omega_h$ where $\widetilde{v}(x) = v(F^h(x))$ for all $x \in \Omega_h$ and $\widetilde{\mathcal{I}^h}$ is the global interpolant for the base finite element space, $\widetilde{V}_h$.

We note that isoparametric C^r, $r \geq 1$, elements are not of much practical value. The problem is not that the elements would not remain C^r under a C^r mapping (they will), but rather that, in mapping a polyhedral domain to a smooth domain, a C^1 mapping is inappropriate. Thus, the above result has been restricted to the case $r = 0$.

4.8 Interpolation of Non-smooth Functions

In order to be defined, the interpolants in the previous sections all require a certain amount of smoothness on the part of the function being approximated. In Chapter 14, we present one way to extend such approximation results using Banach space interpolation theory. Here we illustrate a constructive approach based on the ideas in (Clement 1975 and Scott & Zhang 1990 & 1992). We will not consider the most general form that such an approach can take, but we will give the simplest version and refer to (Scott & Zhang 1990 & 1992 and Girault & Scott 2002) for more results.

We begin with a re-interpretation of the global interpolant $\mathcal{I}^h$ as follows. For each node, z, there is a corresponding nodal variable, N_z. This could be a point evaluation, derivative evaluation, integral on an edge, and so forth. For a smooth function, we can write

(4.8.1) $$\mathcal{I}^h u = \sum_z N_z(u)\Phi_z$$

where Φ_z denotes the global basis function given by

$$\Phi_z|_K = \phi_{N_z}^K$$

and $\phi_{N_z}^K$ is a local nodal basis function on K; $\Phi_z|_K := 0$ if z is not a node associated with K.

We will say that "K contains z" if N_z is one of the nodal variables for the element with element domain K. By abuse of notation, we will think of z encoding both the place of the nodal variable as well as some descriptor. For example, Hermite nodes have multiple nodes z at the same "place" (e.g., a vertex).

We now pick, for each node z, a particular element domain K_z containing z, and we then pick a subset $\widetilde{K}_z \subset K_z$. But there is extreme flexibility in the choice of both K_z and $\widetilde{K}_z$. The domain $\widetilde{K}_z$ will be used in forming our new interpolant; the action of N_z will be represented as an integral over $\widetilde{K}_z$.

To simplify arguments, we assume that the elements in question are hierarchical in the sense that restrictions to $\widetilde{K}_z$ of functions in the element based on K_z are themselves appropriate functions for an element with a subset of the nodes for the element on K_z. More precisely, suppose that the basic element is $(K_z, \mathcal{P}_z, \mathcal{N}_z)$ and let $\widetilde{\mathcal{P}}_z$ denote the restrictions of the element functions in $\mathcal{P}_z$ to $\widetilde{K}_z$:

$$\widetilde{\mathcal{P}}_z = \left\{ f|_{\widetilde{K}_z} \ : \ f \in \mathcal{P}_z \right\}.$$

Then we assume that $(\widetilde{K}_z, \widetilde{\mathcal{P}}_z, \widetilde{\mathcal{N}}_z)$ is a well defined finite element for some $\widetilde{\mathcal{N}}_z \subset \mathcal{N}_z$. The latter condition implies that

(4.8.2) $\quad \phi_N^{K_z}|_{\widetilde{K}_z} = \phi_N^{\widetilde{K}_z} \quad \forall N \in \widetilde{\mathcal{N}}_z \quad$ and $\quad \phi_N^{K_z}|_{\widetilde{K}_z} \equiv 0 \quad \forall N \in \mathcal{N}_z \backslash \widetilde{\mathcal{N}}_z.$

Let us present the details in the case that the basic elements are simplices, as the terminology is simpler in this case. For each z, we let K_z be a simplex containing z, and we pick $\widetilde{K}_z$ to be a (closed) simplicial subcomplex of K_z, i.e., a vertex, edge, face, etc., of K_z. But we allow the dimension of $\widetilde{K}_z$ to be anything:

$$0 \leq \dim(\widetilde{K}_z) \leq d$$

where d is the dimension of the underlying domain Ω. We only require that z lie in $\widetilde{K}_z$. Thus we have

$$z \in \widetilde{K}_z \subset K_z \quad \text{and} \quad 0 \leq \dim(\widetilde{K}_z) \leq \dim(K_z).$$

Three examples which fit into this framework are noteworthy. The first is the standard Lagrange, Hermite, Argyris, etc., interpolant (4.8.1) in which $\dim(\widetilde{K}_z) = 0$ for every z: $\widetilde{K}_z = \{z\}$. A second example is $\widetilde{K}_z = K_z$, so that $\dim(\widetilde{K}_z) = d$, for every z, which allows for the maximum amount of smoothing (Clement 1975). This does not uniquely determine $\widetilde{K}_z$; one of the elements to which z belongs would still have to be picked. Finally, it is often useful to choose $\widetilde{K}_z \subset \partial\Omega$ whenever $z \in \partial\Omega$ (Scott & Zhang 1990).

For each node z, consider the corresponding local nodal basis for $\widetilde{\mathcal{P}}_z$:

$$\left\{ \phi_N^{\widetilde{K}_z} \ : \ N \in \widetilde{\mathcal{N}}_z \right\}.$$

Let

(4.8.3) $$\left\{ \psi_N^{\widetilde{K}_z} \ : \ N \in \widetilde{\mathcal{N}}_z \right\}$$

denote the $L^2(\widetilde{K}_z)$-dual basis for $\widetilde{\mathcal{P}}_z$:

(4.8.4) $$\int_{\widetilde{K}_z} \phi_N^{\widetilde{K}_z}(x) \psi_M^{\widetilde{K}_z}(x) \, dx = \delta_{MN} \quad \forall M, N \in \widetilde{\mathcal{N}}_z$$

(δ_{MN} is the Kronecker delta). Equivalently, $\{\psi_N \; : \; N \in \mathcal{N}\}$ is the representation in $L^2(\widetilde{K}_z)$ of $\{N \in \widetilde{\mathcal{N}}_z\}$ guaranteed by the Riesz Representation Theorem 2.4.2. We can extend the set of functions (4.8.3) to a set $\left\{\psi_N^{\widetilde{K}_z} \; : \; N \in \mathcal{N}_z\right\}$ by setting $\psi_N \equiv 0$ for $N \in \mathcal{N}_z \backslash \widetilde{\mathcal{N}}_z$. Then we have

$$(4.8.5) \qquad \int_{\widetilde{K}_z} \phi_N^{K_z}(x)\psi_M^{\widetilde{K}_z}(x)\,dx = \delta_{MN} \quad \forall M, N \in \mathcal{N}_z$$

by (4.8.2).

We are now in a position to define a global interpolant for more general functions, say $u \in L^1(\widetilde{K}_z)$ for all nodes z:

$$(4.8.6) \qquad \widetilde{\mathcal{I}}^h u = \sum_z \left(\int_{\widetilde{K}_z} \psi_{N_z}^{\widetilde{K}_z}(x)u(x)\,dx \right) \Phi_z.$$

Then we have the following result.

(4.8.7) Theorem. $\widetilde{\mathcal{I}}^h$ *is a projection which equals $\mathcal{I}^h$ on V_h, the space spanned by $\{\Phi_z\}$:*

$$\widetilde{\mathcal{I}}^h v = v \quad \forall v \in V_h.$$

Proof. This follows from (4.8.5) since it implies that

$$\int_K \psi_{N_z}^{\widetilde{K}_z}(x)v(x)\,dx = N_z(v) \quad \forall v \in V_h. \qquad \square$$

Another key property of $\widetilde{\mathcal{I}}^h$ is that it can be constructed to satisfy boundary conditions. We state the result in the case of Lagrange elements for the simple Dirichlet boundary condition. See (Girault & Scott 2002) for Hermite elements and Dirichlet boundary conditions involving derivatives.

(4.8.3.8) Theorem. *Suppose that $\widetilde{K}_z \subset \partial\Omega$ for all nodes $z \in \partial\Omega$, and suppose that all of the elements are Lagrange elements. Then $\widetilde{\mathcal{I}}^h v = 0$ on $\partial\Omega$ as long as $v = 0$ on $\partial\Omega$.*

We now outline estimates for the approximation error $v - \widetilde{\mathcal{I}}^h v$; for complete details, see (Scott & Zhang 1990) for Lagrange elements and (Girault & Scott 2002) for Hermite elements. Let us assume that all elements have shape functions $\mathcal{P}_K$ containing $\mathcal{P}_{m-1}$. Corresponding to Lemma 4.4.1 we have, for arbitrary $K \in \mathcal{T}^h$ and $1 \le p \le \infty$,

$$\left\|\widehat{\widetilde{\mathcal{I}}^h v}\right\|_{W_p^m(\widehat{K})} \le C \|\widehat{v}\|_{L^1(\widehat{S_K})}$$

where S_K is a domain made of the elements in $\mathcal{T}^h$ neighboring K

$$(4.8.9) \qquad S_K = \text{interior} \left(\cup \left\{ \overline{K_i} \,:\, \overline{K_i} \cap \overline{K} \neq \emptyset, \; K_i \in \mathcal{T}^h \right\} \right)$$

and the "hats" refer to the corresponding domains (and functions) scaled by the diameter of K, cf. (4.3.11). For any $0 \le s \le k \le m$ we have using the techniques of proof of Theorem 4.4.4 and (Scott & Zhang 1990)

$$(4.8.10) \qquad \begin{aligned} \left\| v - \widetilde{\mathcal{I}}^h v \right\|_{W_p^s(K)} &\le \left\| v - Q^{k-1} v \right\|_{W_p^s(K)} \\ &\quad + \left\| \widetilde{\mathcal{I}}^h (v - Q^{k-1} v) \right\|_{W_p^s(K)} \\ &\le C h_K^{k-s} |v|_{W_p^k(S_K)} \end{aligned}$$

where $h_K := \operatorname{diam} K$ (this holds for $s = k = 0$ by setting $Q^{-1} \equiv 0$). If $\mathcal{T}^h$ is non-degenerate then

$$(4.8.11) \quad \sup \left\{ \text{cardinality} \left\{ K \in \mathcal{T}^h \,:\, \text{interior } K \subset S_K \right\} \,:\, K \in \mathcal{T}^h \right\} \le C$$

where C is a constant depending only on ρ in (4.4.16). Therefore, the following holds.

(4.8.12) Theorem. *Suppose all elements' sets of shape functions contain all polynomials of degree less than m and $\mathcal{T}^h$ is non-degenerate. Let $v \in W_p^k(\Omega)$ for $0 \le k \le m$ and $1 \le p \le \infty$. Then*

$$(4.8.14) \qquad \left(\sum_{K \in \mathcal{T}^h} h_K^{p(s-k)} \left\| v - \widetilde{\mathcal{I}}^h v \right\|_{W_p^s(K)}^p \right)^{1/p} \le C \, |v|_{W_p^k(\Omega)},$$

for $0 \le s \le k$, where $\widetilde{\mathcal{I}}^h$ is defined in (4.8.6).

Letting $s = k$ and applying the triangle inequality, the following corollary is derived.

(4.8.15) Corollary. *Under the conditions of Theorem 4.8.12*

$$(4.8.16) \qquad \left(\sum_{K \in \mathcal{T}^h} \left\| \widetilde{\mathcal{I}}^h v \right\|_{W_p^k(K)}^p \right)^{1/p} \le C \, |v|_{W_p^k(\Omega)}.$$

Recalling that $h = \max_{K \in \mathcal{T}^h} \operatorname{diam}(K)$, the statement of Theorem 4.8.12 can be simplified as follows:

$$(4.8.17) \qquad \left(\sum_{K \in \mathcal{T}^h} \left\| v - \widetilde{\mathcal{I}}^h v \right\|_{W_p^s(K)}^p \right)^{1/p} \le C h^{k-s} \, |v|_{W_p^k(\Omega)},$$

for $0 \le s \le k \le m$.

4.9 A Discrete Sobolev Inequality

Let Ω be a polygonal domain in $\mathbb{R}^2$ and $\mathcal{T}^h$ be a quasi-uniform triangulation of Ω. Denote by $V_h \subset H^1(\Omega)$ the $\mathcal{P}_m$ Lagrange finite element space associated with $\mathcal{T}^h$. Note that on the one hand, Example 1.4.3 indicates that in general $H^1(\Omega) \not\subset L^\infty(\Omega)$, and on the other hand functions in V_h are continuous on $\overline{\Omega}$ by construction. This conflict is reconciled by the discrete Sobolev inequality, whose proof below is essentially the same as the one in (Bramble, Pasciak & Schatz 1986).

(4.9.1) Lemma. (Discrete Sobolev Inequality) *The following estimate holds:*

$$(4.9.2) \qquad \|v\|_{L^\infty(\Omega)} \le C(1 + |\ln h|)^{1/2}\|v\|_{H^1(\Omega)} \qquad \forall\, v \in V_h,$$

where the positive constant C is independent of h.

Proof. First we observe that Ω has the cone property, i.e., each point $x \in \Omega$ is the vertex of a cone K_x congruent to the cone (or sector) K defined in polar coordinates by

$$K = \{(r, \theta) : 0 < r < d < \infty, 0 < \theta < \omega < 2\pi\}.$$

Without loss of generality we may assume that $h < d/2$ (cf. exercise 4.x.17).

Let T be a triangle of $\mathcal{T}^h$ and c be the centroid of T. For simplicity we may take c to be the origin and K_c to be K. The quasi-uniformity of $\mathcal{T}^h$ implies that there exists a number η which is independent of T and h such that $0 < \eta < 1$ and the cone

$$K_\eta = \{(r, \theta) : 0 < r < \eta h, 0 < \theta < \omega\}$$

is a subset of T.

Let $v \in V_h$ be arbitrary and $\alpha = v(c)$. It follows from the fundamental theorem of calculus that

$$\alpha = v(r, \theta) - \int_0^r \frac{\partial v}{\partial r}(\rho, \theta)\, d\rho \qquad \text{for} \quad \frac{d}{2} < r < d,$$

and hence

$$(4.9.3) \qquad \alpha^2 \le 2v^2(r, \theta) + 2\left(\int_0^r \frac{\partial v}{\partial r}(\rho, \theta)\, d\rho\right)^2 \qquad \text{for} \quad \frac{d}{2} < r < d.$$

We can estimate the integral on the right-hand side of (4.9.3) by

$$(4.9.4) \qquad \int_0^r \frac{\partial v}{\partial r}(\rho, \theta)\, d\rho = \int_0^{\eta h} \frac{\partial v}{\partial r}(\rho, \theta)\, d\rho + \int_{\eta h}^r \frac{\partial v}{\partial r}(\rho, \theta)\, d\rho$$

$$\le \eta h |v|_{W^1_\infty(T)} + \left[\int_{\eta h}^r \left(\frac{\partial v}{\partial r}(\rho, \theta)\right)^2 \rho\, d\rho\right]^{1/2} \sqrt{\ln(d/\eta h)}.$$

Combining (4.9.3) and (4.9.4) we obtain

$$\alpha^2 \int_0^\omega \int_{d/2}^d r\,dr d\theta \leq 2 \int_0^\omega \int_{d/2}^d v^2(r,\theta) r\,dr d\theta + 4(\eta h)^2 |v|^2_{W^1_\infty(T)} \int_0^\omega \int_{d/2}^d r\,dr d\theta$$

$$+ 4\ln(d/\eta h) \int_0^\omega \int_{d/2}^d \left[\int_{\eta h}^r \left(\frac{\partial v}{\partial r}(\rho,\theta) \right)^2 \rho\,d\rho d\theta \right] r\,dr\,,$$

which implies, by the inverse estimate (4.5.4),

(4.9.5) $$|v(c)| \leq C_1 (1 + |\ln h|)^{1/2} \|v\|_{H^1(\Omega)}\,,$$

where the constant C_1 is independent of h and c.

Let x be an arbitrary point in T. The inverse estimate (4.5.4) implies that

(4.9.6) $$|v(x) - v(c)| \leq h|v|_{W^1_\infty(T)} \leq C_2 |v|_{H^1(T)}\,,$$

where the positive constant C_2 is independent of h and x.

The estimate (4.9.2) follows from (4.9.5), (4.9.6) and the arbitrariness of T and x.   □

(4.9.7.) *Remark.* A proof of the discrete Sobolev inequality in the context of finite difference grid functions can be found in (Bramble 1966).

(4.9.8) *Remark.* The estimate (4.9.2) is sharp (cf. (Brenner & Sung 2000)).

4.x Exercises

4.x.1 Show that for $|\alpha| \leq m-1$, $D_x^\alpha T_y^m u(x) = T_y^{m-|\alpha|} D_x^\alpha u(x)$ $\forall\, u \in C^{|\alpha|}(B)$ (cf. the proof of Proposition 4.1.17).

4.x.2 Prove the form of Taylor's Theorem given in (4.2.1).

4.x.3 Verify the claims in (4.3.12) and (4.3.13). (Hint: show that $T_{d\,\hat{y}}^m u(d\,\hat{x}) = \hat{T}_{\hat{y}}^m \hat{u}(\hat{x})$ where

$$\hat{T}_{\hat{y}}^m \hat{u}(\hat{x}) = \sum_{|\alpha| < m} \frac{1}{\alpha!} D_{\hat{y}}^\alpha \hat{u}(\hat{y})(\hat{x} - \hat{y})^\alpha.)$$

4.x.4 Complete the proof of the general case of Theorem 4.4.4 by using a homogeneity argument.

4.x.5 Prove that a family $\{\mathcal{T}^h\}$ of triangulations is quasi-uniform if and only if it is non-degenerate and there exist positive constants c and

C, independent of h, such that $c \operatorname{diam} K_1 \le \operatorname{diam} K_2 \le C \operatorname{diam} K_1$ for any $K_1, K_2 \in \mathcal{T}^h$.

4.x.6 Prove the inequality (4.4.22).

4.x.7 Let $\{a_m\}_{m=1}^{\infty}$ be a sequence of nonnegative numbers. Show that if $1 \le q \le p$, then $\left(\sum_{m=1}^{\infty} a_m^p\right)^{1/p} \le \left(\sum_{m=1}^{\infty} a_m^q\right)^{1/q}$.

4.x.8 Let $\{a_m\}_{m=1}^{M}$ be a finite sequence of nonnegative numbers. Show that if $p < q \le \infty$, then

$$\left(\sum_{m=1}^{M} a_m^p\right)^{1/p} \le M^{1/p - 1/q} \left(\sum_{m=1}^{M} a_m^q\right)^{1/q} \qquad \text{if } q < \infty$$

and

$$\left(\sum_{m=1}^{M} a_m^p\right)^{1/p} \le M^{1/p} \max_{1 \le m \le M} a_m \qquad \text{if } q = \infty.$$

4.x.9 (Minimum angle condition) Let $\{\mathcal{T}^h\}$ be a family of triangulations of $\Omega \subseteq \mathbb{R}^2$. Show that $\{\mathcal{T}^h\}$ is non-degenerate if and only if all the angles of the triangles in $\{\mathcal{T}^h\}$ are bounded below by a positive constant.

4.x.10 (Maximum angle condition) Let $\mathcal{T}$ be a triangulation of a polygonal domain $\Omega \subseteq \mathbb{R}^2$, and $V_h = \{v \in C^0(\overline{\Omega}) : v|_T \text{ is linear for all } T \in \mathcal{T}\}$. Let $u \in H^2(\Omega)$ and $\mathcal{I}u \in V_h$ be the linear interpolant of u. Show that

$$\|u - \mathcal{I}u\|_{L^2(\Omega)} + h\,|u - \mathcal{I}u|_{H^1(\Omega)} \le \zeta(\theta)\,h^2\,|u|_{H^2(\Omega)}$$

where $h = \max_{T \in \mathcal{T}} \operatorname{diam}(T)$, θ is the maximum angle in $\mathcal{T}$, and ζ is an increasing positive function defined on $[\pi/3, \pi)$.

4.x.11 Derive estimates analogous to those in Sect. 4.6 for serendipity elements in three dimensions.

4.x.12 Prove the following Sobolev inequality:

$$\|v\|_{L^q(\Omega)} \le \|v\|_{W_p^m(\Omega)} \qquad \text{for} \quad m + \frac{n}{q} \ge \frac{n}{p}$$

provided $q < \infty$. (Hint: show that a Riesz potential of order m maps $L^p(\Omega)$ to $L^q(\Omega)$ for $m + \frac{n}{q} \ge \frac{n}{p}$, cf. the proof of (4.3.4) and (Sobolev 1963 & 1991).)

4.x.13 Let $\mathcal{P}$ be a finite dimensional space of polynomials, and suppose K is any open set containing a ball of radius $\rho \operatorname{diam} K$ with $\rho > 0$. Prove (4.5.4) holds with C independent of $\widehat{K}$. (Hint: let $B_{\rho h} \subset K \subset B_h$ be balls and replace (4.5.5) by $\|v\|_{W_p^l(B_h)} \le C\,\|v\|_{L^q(B_{\rho h})} \quad \forall v \in \mathcal{P}$.)

4.x.14 Prove that a family $\{\mathcal{T}^h\}$ of non-degenerate triangulations is locally quasi-uniform, in two or higher dimensions (Hint: neighboring elements are all connected to each other via a sequence of elements with common faces.),

4.x.15 Consider the stiffness matrix K for piecewise linear functions on a quasi-uniform mesh in one dimension as in Sect. 0.5. Prove that the condition number (Isaacson & Keller 1966) of K is bounded by $\mathcal{O}(h^{-2})$. (Hint: use inverse estimates.)

4.x.16 Complete the density argument for the proof of Proposition 4.3.2. (Hint: R^m is a continuous operator from W_p^m to L^p by construction and by Corollary 4.1.15. For a Cauchy sequence of smooth functions in W_p^m, R^m applied to the sequence will be Cauchy in both L^∞ and L^p. Show that the limits are the same.)

4.x.17 Show that the estimate (4.9.2) is trivial if $h \geq d/2$. (Hint: Apply the inverse estimate (4.5.4).)

4.x.18 Show that the function u in Lemma 4.3.4 is continuous on $\overline{\Omega}$ and

$$\|u\|_{L^\infty(\overline{\Omega})} \leq C_{m,n,\gamma,d}\, \|u\|_{W_p^m(\Omega)},$$

if Ω satisfies the segment condition. (Hint: Use the density of $C^\infty(\overline{\Omega})$ stated in Remark 1.3.4 and the arguments in the proof of Lemma 4.3.4.)

4.x.19 Let $\hat{K}$ be the unit square and K be a convex quadrilateral. Show that there exists a diffeomorphism $F : \hat{K} \longrightarrow K$ such that the components of F belong to Q_1.

4.x.20 Let $\{\mathcal{T}^h\}$ be a non-degenerate family of triangulations of the polygonal domain Ω by convex quadrilaterals, V_h be the Q_1-isoparametric finite element space associated with $\mathcal{T}^h$ (cf. Section 4.7 and exercise 4.x.19), and $\mathcal{I}^h : H^2(\Omega) \longrightarrow V_h$ be the the nodal interpolation operator. Show that

$$\|u - \mathcal{I}^h u\|_{L^2(\Omega)} + h|u - \mathcal{I}^h u|_{H^1(\Omega)} \leq Ch^2 |u|_{H^2(\Omega)} \qquad \forall\, u \in H^2(\Omega).$$

(More on quadrilateral elements can be found in (Girault & Raviart 1979) and (Arnold, Boffi & Falk 2000).)

Chapter 5

n-Dimensional Variational Problems

We now give several examples of higher-dimensional variational problems that use the theory developed in previous chapters. The basic notation is provided by the Sobolev spaces developed in Chapter 1. We combine the existence theory of Chapter 2 together with the approximation theory of Chapters 3 and 4 to provide a complete theory for the discretization process. Several examples will be fully developed in the text, and several others are found in the exercises. Throughout this chapter, we assume that the domain Ω is bounded.

5.1 Variational Formulation of Poisson's Equation

To begin with, consider Poisson's equation

$$(5.1.1) \qquad\qquad -\Delta u \;=\; f \quad \text{in} \quad \Omega$$

where Δ denotes the Laplace operator

$$\Delta \;:=\; \sum_{i=1}^{n} \frac{\partial^2}{\partial x_i^2}.$$

Augmenting this equation, we again consider boundary conditions of two types:

$$u \;=\; 0 \quad \text{on} \quad \Gamma \subset \partial\Omega \qquad\qquad \text{(Dirichlet)}$$

(5.1.2)

$$\frac{\partial u}{\partial \nu} \;=\; 0 \quad \text{on} \quad \partial\Omega \backslash \Gamma \qquad\qquad \text{(Neumann)}$$

where $\frac{\partial u}{\partial \nu}$ denotes the derivative of u in the direction normal to the boundary, $\partial\Omega$. That is, we assume that $\partial\Omega$ is Lipschitz continuous, we let ν denote the outward unit normal vector to $\partial\Omega$, which is by assumption in $L^\infty(\partial\Omega)^n$, and we set

$$\frac{\partial u}{\partial \nu} = \nu \cdot \nabla u.$$

To begin with, we assume that Γ is closed and has nonzero measure. Later, we will return to the case when Γ is empty, the pure Neumann case.

To formulate the variational equivalent of ((5.1.1), (5.1.2)), we define a variational space that incorporates the essential, i.e., Dirichlet, part of boundary condition (5.1.2):

$$(5.1.3) \qquad V := \left\{ v \in H^1(\Omega) \; : \; v|_\Gamma = 0 \right\},$$

where we note that $v|_\Gamma = 0$ is to be interpreted, using the trace theorem, in $L^2(\partial\Omega)$. That is, we think of it as meaning $v \cdot \chi_\Gamma = 0$ where χ is the usual characteristic function. The appropriate bilinear form for the variational problem is determined, as in the one dimensional cases, by multiplying Poisson's equation by a suitably smooth function, integrating over Ω and then integrating by parts. To prepare for this last step, we develop some of the standard theorems of advanced calculus in the setting of Sobolev spaces. In the following, we assume that Ω is a Lipschitz domain and ν denotes the outward unit normal to $\partial\Omega$. Moreover, for any linear space B, we denote by B^n the linear space of n-tuples of members of B (if B is a normed space, then we give B^n a norm defined as an appropriate combination of the norms on each component separately).

(5.1.4) Proposition. *Let* $\mathbf{u} \in W_1^1(\Omega)^n$. *Then*

$$\int_\Omega \nabla \cdot \mathbf{u}\, dx = \int_{\partial\Omega} \mathbf{u} \cdot \nu\, ds.$$

Proof. Recall this result for smooth functions and use a density argument as in the proof of Proposition 1.6.3. Note that the trace theorem implies that, if

$$C^\infty(\overline{\Omega})^n \ni \mathbf{u}_k \to \mathbf{u} \quad \text{in} \quad W_1^1(\Omega)^n,$$

then $\mathbf{u}_k \cdot \nu$ converges in $L^1(\partial\Omega)$ to $\mathbf{u} \cdot \nu$ because $\nu_i \in L^\infty(\partial\Omega)$ for all $i = 1, \ldots, n$ (Hölder's inequality). $\qquad\square$

(5.1.5) Proposition. *Let* $v, w \in H^1(\Omega)$. *Then, for* $i = 1, \ldots, n$,

$$\int_\Omega \left(\frac{\partial v}{\partial x_i} \right) w\, dx = -\int_\Omega v \left(\frac{\partial w}{\partial x_i} \right) dx + \int_{\partial\Omega} vw\nu_i\, ds.$$

Proof. Apply the previous proposition to $\mathbf{u} := vw e_i$, which is in $W_1^1(\Omega)^n$ by Schwarz' inequality and exercise 5.x.2. $\qquad\square$

(5.1.6) Proposition. *Let* $u \in H^2(\Omega)$ *and* $v \in H^1(\Omega)$. *Then*

$$\int_\Omega (-\Delta u)v\, dx = \int_\Omega \nabla u \cdot \nabla v\, dx - \int_{\partial\Omega} \frac{\partial u}{\partial \nu} v\, ds.$$

Proof. Apply the previous proposition to $v := -\frac{\partial u}{\partial x_i}$ and $w := v$, and sum over i.   □

Using the latter proposition, if $u \in H^2(\Omega)$ satisfies Poisson's equation (5.1.1) with boundary conditions (5.1.2) and $v \in V$, then

$$(f,v) = \int_\Omega (-\Delta u) v\, dx = \int_\Omega \nabla u \cdot \nabla v\, dx - \int_{\partial\Omega} v \frac{\partial u}{\partial \nu}\, ds$$

$$= \int_\Omega \nabla u \cdot \nabla v\, dx := a(u,v).$$

The boundary term vanishes for $v \in V$ because either v or $\frac{\partial u}{\partial \nu}$ is zero on any part of the boundary. Thus, we have proved the following.

(5.1.7) Proposition. *Let $u \in H^2(\Omega)$ solve Poisson's equation (5.1.1) (this implies $f \in L^2(\Omega)$) with boundary conditions (5.1.2). Then u can be characterized via*

(5.1.8) $u \in V$ satisfies $a(u,v) = (f,v)$ $\forall v \in V.$

The companion result, namely that a solution to the variational problem (5.1.8) solves Poisson's equation, can be proved in a similar fashion to the proof given for Theorem 0.1.4.

(5.1.9) Proposition. *Let $f \in L^2(\Omega)$ and suppose that $u \in H^2(\Omega)$ solves the variational equation (5.1.8). Then u solves Poisson's equation (5.1.1) with boundary conditions (5.1.2).*

Proof. The Dirichlet boundary condition on u follows since $u \in V$. Using Proposition 5.1.6 with $v \in \mathcal{D}(\Omega) \subset V$ and (5.1.8), we find

$$\int_\Omega (f + \Delta u) v\, dx = (f,v) - \int_\Omega \nabla u \cdot \nabla v\, dx = (f,v) - a(u,v) = 0.$$

Since $\mathcal{D}(\Omega)$ is dense in $L^2(\Omega)$ (cf. exercise 2.x.14), the differential equation (5.1.1) is satisfied in $L^2(\Omega)$ (compare this with the proof of Theorem 0.1.4.) Moreover, Proposition 5.1.6 then implies that

$$0 = (f,v) - a(u,v) = \int_\Omega (-\Delta u) v\, dx - \int_\Omega \nabla u \cdot \nabla v\, dx = \int_{\partial\Omega} v \frac{\partial u}{\partial \nu}\, ds$$

for all $v \in V$. The Neumann boundary condition on u follows if we can show that $v|_{\partial\Omega\setminus\Gamma}$ can be chosen arbitrarily with $v \in V$. This is a bit technical, since we are only assuming that $\partial\Omega$ is Lipschitz and that Γ is a closed subset of it. But this means at least that, for any point P in $\partial\Omega\setminus\Gamma$ there is a neighborhood, N, of P in $\partial\Omega\setminus\Gamma$ that can be written as a graph of a Lipschitz function ϕ:

$$N = \{(\hat{x}, \phi(\hat{x})) \; : \; \hat{x} \in \omega\}$$

where ω is an open subset of $\mathbb{R}^{n-1}$ and $\hat{x} := (x_1, \dots, x_{n-1})$. Moreover, we can assume that

$$\{(x_1, \dots, x_n) \; : \; \hat{x} \in \omega, \; -\epsilon < x_n - \phi(\hat{x}) < 0\} \subset \Omega$$

where $\epsilon > 0$ depends only on Ω. The boundary integral over N can be written as

$$\int_N v \frac{\partial u}{\partial \nu} \, ds = \int_\omega \left(v \frac{\partial u}{\partial \nu} \right) (\hat{x}, \phi(\hat{x})) \sqrt{1 + |\nabla \phi(\hat{x})|^2} \, d\hat{x}.$$

Let $w \in \mathcal{D}(\omega)$ and set

$$v(\hat{x}, t + \phi(\hat{x})) := w(\hat{x})(1 + t/\epsilon) \quad \forall \hat{x} \in \omega, \; -\epsilon < t < 0$$

with v defined to be zero elsewhere. Then $v \in V$ and

$$0 = \int_{\partial \Omega} v \frac{\partial u}{\partial \nu} \, ds = \int_N v \frac{\partial u}{\partial \nu} \, ds = \int_\omega w(\hat{x}) \frac{\partial u}{\partial \nu} (\hat{x}, \phi(\hat{x})) \sqrt{1 + |\nabla \phi(\hat{x})|^2} \, d\hat{x}.$$

Since the L^∞ function $\sqrt{1 + |\nabla \phi(\hat{x})|^2}$ is bounded below by 1 and w was arbitrary, we conclude that $\frac{\partial u}{\partial \nu}\big|_N = 0$ (cf. exercise 1.x.39). Since $\partial \Omega \backslash \Gamma$ is covered by such neighborhoods, N, we conclude that the Neumann condition holds on all of $\partial \Omega \backslash \Gamma$. $\qquad \square$

5.2 Variational Formulation of the Pure Neumann Problem

In the previous section, we introduced the variational formulation for Poisson's equation with a combination of boundary conditions, and they all contained some essential (i.e., Dirichlet) component. The situation for the case of pure Neumann (or natural) boundary conditions

(5.2.1) $$\frac{\partial u}{\partial \nu} = 0 \quad \text{on} \quad \partial \Omega$$

(i.e., when $\Gamma = \emptyset$) is a bit different, just as in the one-dimensional case (cf. exercise 0.x.3). In particular, solutions are unique only up to an additive constant, and they can exist only if the right-hand side f in (5.1.1) satisfies

(5.2.2) $$\int_\Omega f(x) \, dx = \int_\Omega -\Delta u(x) \, dx$$
$$= \int_\Omega \nabla u(x) \cdot \nabla 1 \, dx - \int_{\partial \Omega} \frac{\partial u}{\partial \nu} \, ds = 0$$

using Proposition 5.1.6. We will see later that such behavior is typical for *elliptic* operators, such as $-\Delta + \lambda I$, in the case when λ is a simple eigenvalue (in this case, $\lambda = 0$). A variational space appropriate for the present case is

$$(5.2.3) \qquad V = \left\{ v \in H^1(\Omega) : \int_\Omega v(x)\, dx = 0 \right\}.$$

For any integrable function g, we define the *mean*, $\overline{g}$, of g as follows:

$$(5.2.4) \qquad \overline{g} := \frac{1}{\text{meas}\,(\Omega)} \int_\Omega g(x)\, dx.$$

For any $v \in H^1(\Omega)$, note that $v - \overline{v} \in V$. Using the same techniques as in the previous section, we can prove the following proposition.

(5.2.5) Proposition. *Let $u \in H^2(\Omega)$ solve Poisson's equation (5.1.1) with pure Neumann boundary conditions ((5.1.2) with $\Gamma = \emptyset$ or (5.2.1); this implies $f \in L^2(\Omega)$ satisfies (5.2.2)). Then $u - \overline{u}$ satisfies the variational formulation (5.1.8) with V defined as in (5.2.3).*

The companion to this result is more complicated than its counterpart in the previous section.

(5.2.6) Proposition. *Let $f \in L^2(\Omega)$ and suppose that $u \in H^2(\Omega)$ solves the variational equation (5.1.8) with V defined as in (5.2.3). Then u solves Poisson's equation (5.1.1) with a right-hand-side given by*

$$\tilde{f}(x) := f(x) - \overline{f} \quad \forall x \in \Omega$$

with boundary conditions (5.2.1).

(5.2.7) Remark. The statements of the equivalence of the original and the variational problems are similar to the previous section, except that the definition of V has changed and constraints appear on f and u. Note that the Riesz Representation Theorem guarantees a solution for *any* f as soon as we know $a(\cdot, \cdot)$ to be coercive. Since $v \in V$ implies $\int_\Omega v(x)\, dx = 0$, we have

$$\int_\Omega v(x)\tilde{f}(x)\, dx = \int_\Omega v(x)f(x)\, dx \quad \forall v \in V,$$

so that the variational problems for f and $\tilde{f}$ are identical.

Proof. The previous remark shows that the variational problem is the same whether we use f or $\tilde{f}$. Thus, it suffices to suppose that f has mean zero and to verify (5.1.1) and (5.2.1) in this case. Using the argument of the previous section, we see that

$$\int_\Omega (-\Delta u - f)v\, dx = 0 \quad \forall v \in \tilde{\mathcal{D}}(\Omega) := \left\{ \phi \in \mathcal{D}(\Omega) : \overline{\phi} = 0 \right\}.$$

It is easy to see that the closure of $\tilde{\mathcal{D}}(\Omega)$ in $L^2(\Omega)$ is the subspace

$$\tilde{L}^2(\Omega) := \left\{ \phi \in L^2(\Omega) : \bar{\phi} = 0 \right\}$$

(cf. exercise 5.x.1). Thus, we conclude that the equation $-\Delta u = f + \delta$ holds (in $L^2(\Omega)$) where $\delta = -\overline{\Delta u}$ since $\int_\Omega \delta v(x)\,dx = 0$ for $v \in V$. Using Proposition 5.1.6 again, we find

$$0 = (f, v) - a(u, v) = \int_\Omega (-\Delta u - \delta) v\, dx - a(u, v)$$

$$= \int_\Omega (-\Delta u) v\, dx - a(u, v) = -\int_{\partial\Omega} v \frac{\partial u}{\partial \nu}\, ds \quad \forall v \in V.$$

But clearly $v|_{\partial\Omega}$ can be chosen arbitrarily while keeping $v \in V$: let $w \in H^1(\Omega)$ be arbitrary and set $v := w - \overline{w}\phi$ where $\phi \in \mathcal{D}(\Omega)$ satisfies $\bar{\phi} = 1$. Thus, we conclude that (5.2.1) holds. Finally, applying (5.1.6) one more time with $v \equiv 1$ shows then that $\delta = 0$.   □

5.3 Coercivity of the Variational Problem

To apply the theory developed in Chapter 2, we must see that the variational form $a(\cdot, \cdot)$ introduced in the previous two sections is coercive on the corresponding spaces V (that it is bounded on all of $H^1(\Omega)$ is obvious). Coercivity in the case of pure Neumann boundary conditions follows from the approximation theory in Chapter 4. If Ω is a bounded domain that can be written as a finite union of domains that are star-shaped with respect to a ball as described there, then

$$(5.3.1) \qquad \inf_{r \in \mathbb{R}} \|v - r\|_{L^2(\Omega)} \leq C_\Omega \, |v|_{H^1(\Omega)} \quad \forall v \in H^1(\Omega).$$

But we know that

$$\inf_{r \in \mathbb{R}} \|v - r\|_{L^2(\Omega)} = \|v - \bar{v}\|_{L^2(\Omega)}$$

where $\bar{v}$ denotes the mean of v on Ω, defined in (5.2.4). With V as defined in (5.2.3), $v \in V$ implies that $\bar{v} = 0$, so $\|v\|_{L^2(\Omega)} \leq C_\Omega \, |v|_{H^1(\Omega)} = C_\Omega \sqrt{a(v, v)}$. Squaring this and adding $a(v, v)$ to both sides yields the desired coercivity inequality, which we state as the following proposition.

(5.3.2) Proposition. *Let Ω be such that (5.3.1) holds, e.g., a finite union of domains that are star-shaped with respect to a ball. Let V be defined by (5.2.3) and let C_Ω be the approximation constant in (5.3.1). Then*

$$\|v\|^2_{H^1(\Omega)} \leq \left(1 + C_\Omega^2\right) a(v, v) \quad \forall v \in V.$$

Now let us consider the verification of coercivity in the case of Dirichlet boundary conditions, with V as defined in (5.1.3). We do so using a *compactness argument*, as follows. Suppose that $a(\cdot, \cdot)$ were *not* coercive on V. Then it would be possible to find a sequence $\{v_j \in V : j = 1, 2, \ldots\}$ such that $a(v_j, v_j)/\|v_j\|^2_{H^1(\Omega)} \to 0$ as $j \to \infty$. Defining $w_j = v_j/\|v_j\|_{H^1(\Omega)}$, we find a sequence in V such that

$$\|w_j\|_{H^1(\Omega)} = 1 \quad \text{and} \quad a(w_j, w_j) \to 0 \quad \text{as} \quad j \to \infty.$$

Applying the coercivity result from the pure Neumann case, we find

$$\|w_j - \overline{w}_j\|^2_{H^1(\Omega)} \leq Ca(w_j - \overline{w}_j, w_j - \overline{w}_j) = Ca(w_j, w_j) \to 0 \text{ as } j \to \infty.$$

Moreover, all of the mean values $\overline{w}_j$ satisfy

$$|\overline{w}_j| = \frac{1}{|\Omega|}\left|\int_\Omega w_j(x)\, dx\right| \leq |\Omega|^{-1/2}\|w_j\|_{L^2(\Omega)} \leq |\Omega|^{-1/2},$$

where $|\Omega| := \text{meas}\,(\Omega)$, by Hölder's inequality. Since the numbers $\overline{w}_j$ range over a closed, bounded (hence compact) set, there must be a subsequence $\overline{w}_{j_k}$ that converges to some number $r_0 \in [-|\Omega|^{-1/2}, |\Omega|^{-1/2}]$. We conclude that

$$\|w_{j_k} - r_0\|_{H^1(\Omega)} \leq \|w_{j_k} - \overline{w}_{j_k}\|_{H^1(\Omega)} + \|\overline{w}_{j_k} - r_0\|_{H^1(\Omega)} \to 0 \text{ as } k \to \infty.$$

That is, $w_{j_k} \to r_0$ in $H^1(\Omega)$, where r_0 is a constant. The trace theorem implies then that $w_{j_k}|_\Gamma \to r_0$ in $L^2(\Gamma)$. But $w_{j_k}|_\Gamma = 0$, and $\text{meas}\,(\Gamma) > 0$, so $r_0 = 0$. Now we can not have $w_{j_k} \to 0$ in $H^1(\Omega)$ and $\|w_{j_k}\|_{H^1(\Omega)} = 1$ at the same time. Thus, our original assumption must be false, and indeed $a(\cdot, \cdot)$ must be coercive. We formalize this in the following.

(5.3.3) Proposition. *Let Ω be as in Proposition 5.3.2, and suppose Γ is closed and $\text{meas}\,(\Gamma) > 0$. Let V be defined by (5.1.3). Then there is a constant C depending only on Ω and Γ such that*

$$\|v\|^2_{H^1(\Omega)} \leq Ca(v, v) \quad \forall v \in V.$$

(5.3.4) Remark. Note that the constant C in Proposition 5.3.3 was not determined constructively, so we have no estimate of it. This is one shortcoming of a compactness argument; it demonstrates existence, but provides no size estimates.

This estimate is also true in an L^p context as well. The special case $\Gamma = \partial\Omega$ is frequently used and is known by a special name. Recall the notation $\mathring{W}^1_p(\Omega)$ (cf. (1.6.7)), for functions in $W^1_p(\Omega)$ satisfying a Dirichlet boundary condition on $\partial\Omega$.

(5.3.5) Proposition. (Poincaré's Inequality) *Let Ω be as in Proposition 5.3.2. Then there is a constant $C < \infty$ such that*

$$\|v\|_{W_p^1(\Omega)} \leq C\,|v|_{W_p^1(\Omega)} \quad \forall v \in \mathring{W}_p^1(\Omega).$$

5.4 Variational Approximation of Poisson's Equation

Let $\mathcal{T}^h$ denote a subdivision of Ω and let $\mathcal{I}^h$ denote a global interpolator for a family of finite elements based on the components of $\mathcal{T}^h$. Let us suppose that $\mathcal{I}^h u$ is continuous, i.e., that the family of elements involved are C^0. Further, suppose that the corresponding shape functions have an approximation order, m, that is

(5.4.1) $$\left\| u - \mathcal{I}^h u \right\|_{H^1(\Omega)} \leq C h^{m-1}\, |u|_{H^m(\Omega)}\,.$$

In order to approximate the variational problem (5.1.8) with variational space (5.1.3), we need to insure two properties of the corresponding space, V_h. In order to apply Céa's Theorem 2.8.1, we must have

(5.4.2) $$V_h \subset V.$$

In order to use the approximation theory derived in Chapter 4, we need to have

(5.4.3) $$\mathcal{I}^h\left(V \cap C^k(\Omega)\right) \subset V_h$$

(k is the highest order of differentiation in the definition of $\mathcal{I}^h$). If both of these conditions hold, then the following is an immediate consequence of (2.8.1):

(5.4.4) Theorem. *Suppose that conditions (5.4.1), (5.4.2) and (5.4.3) hold. Then the unique solution, $u_h \in V_h$, to the variational problem*

$$a(u_h, v) = (f, v) \quad \forall v \in V_h$$

satisfies

$$\|u - u_h\|_{H^1(\Omega)} \leq C h^{m-1}\, |u|_{H^m(\Omega)}\,.$$

The requirements (5.4.2) and (5.4.3) place a constraint on the subdivision in the case that Γ is neither empty nor all of the boundary. In such a case, it is necessary to choose the mesh so that it aligns properly with the points where the boundary conditions change from Dirichlet to Neumann. For example, in two dimensions, if one uses Lagrange elements and insures that the points where the boundary conditions change are vertices in the triangulation, then defining

$$V_h := \mathcal{I}^h \left(V \cap C^0(\Omega) \right)$$

is equivalent to defining V_h to be the space of piecewise polynomials that vanish on edges contained in Γ. Since we have chosen the mesh so that the edges contained in Γ form a subdivision of the latter, it follows that (5.4.2) holds. On the other hand, if the set of edges where functions in V_h vanish is too small, we fail to obtain (5.4.2). If the set of edges where functions in V_h vanish is too big, (5.4.3) fails to hold. In the case of pure Dirichlet data, i.e., $\Gamma = \partial\Omega$, then V_h is just the set of piecewise polynomials that vanish on the entire boundary. In the case of pure Neumann data, i.e., $\Gamma = \emptyset$, V_h is the entire set of piecewise polynomials with no constraints at the boundary.

We now consider error estimates for $u - u_h$ in the L^2 norm. To estimate $\|u - u_h\|_{L^2(\Omega)}$, we use a "duality" argument that is similar to the one given in Chapter 0. Let w be the solution of

$$-\Delta w = e \quad \text{in} \quad \Omega$$

(5.4.5)

$$w = 0 \quad \text{on} \quad \Gamma \subset \partial\Omega \quad \& \quad \frac{\partial w}{\partial \nu} = 0 \quad \text{on} \quad \partial\Omega \backslash \Gamma.$$

where $e := u - u_h$. The variational formulation of this problem is: find $w \in V$ such that

$$(5.4.6) \qquad\qquad a(w, v) = (e, v) \quad \forall v \in V.$$

Since $u - u_h \in V'$, the solution exists uniquely. Therefore,

$$\begin{aligned}
\|u - u_h\|_{L^2(\Omega)}^2 &= (u - u_h, u - u_h) \\
&= a(w, u - u_h) \\
&= a(u - u_h, w - \mathcal{I}^h w) \\
&\leq C \|u - u_h\|_{H^1(\Omega)} \|w - \mathcal{I}^h w\|_{H^1(\Omega)} \\
&\leq Ch \|u - u_h\|_{H^1(\Omega)} |w|_{H^2(\Omega)}.
\end{aligned}$$

We now suppose that the equivalent problems (5.4.5) and (5.4.6) have the property that

$$(5.4.7) \qquad\qquad |w|_{H^2(\Omega)} \leq C \|e\|_{L^2(\Omega)}.$$

Such a condition will be discussed in the following section. With such a condition holding, we thus have proved the following.

(5.4.8) Theorem. *Suppose that conditions (5.4.1), (5.4.2), (5.4.3) and (5.4.7) hold. Then*

$$\begin{aligned}
\|u - u_h\|_{L^2(\Omega)} &\leq Ch \|u - u_h\|_{H^1(\Omega)} \\
&\leq Ch^m |u|_{H^m(\Omega)}.
\end{aligned}$$

Inhomogeneous boundary conditions are easily treated. For example, suppose that we wish to solve (5.1.1) with boundary conditions

$$(5.4.9) \qquad u = g_D \quad \text{on} \quad \Gamma \subset \partial\Omega \quad \& \quad \frac{\partial u}{\partial \nu} = g_N \quad \text{on} \quad \partial\Omega\backslash\Gamma$$

where g_D and g_N are given. For simplicity, let us assume that g_D is defined on all of Ω, with $g_D \in H^1(\Omega)$ and that $g_N \in L^2(\partial\Omega\backslash\Gamma)$. Define V to be the space (5.1.3). Then the variational formulation of ((5.1.1), (5.4.9)) is as follows: find u such that $u - g_D \in V$ and such that

$$(5.4.10) \qquad a(u,v) = (f,v) + \int_{\partial\Omega\backslash\Gamma} g_N v\, ds \quad \forall v \in V.$$

This is well-posed since the linear form

$$F(v) := (f,v) + \int_{\partial\Omega\backslash\Gamma} g_N v\, ds$$

is well defined (and continuous) for all $v \in V$.

The equivalence of these formulations follows from (5.1.6):

$$\int_\Omega (-\Delta u)v\, dx = \int_\Omega \nabla u \cdot \nabla v\, dx\, ds - \int_{\partial\Omega} v\frac{\partial u}{\partial \nu}\, ds$$

$$= a(u,v) - \int_{\partial\Omega\backslash\Gamma} v\frac{\partial u}{\partial \nu}\, ds$$

for any $v \in V$. Thus, if u solves ((5.1.1), (5.4.9)) then (5.4.10) follows as a consequence. Conversely, if u solves (5.4.10) then choosing v to vanish near $\partial\Omega$ shows that (5.1.1) holds, and thus

$$\int_{\partial\Omega\backslash\Gamma} g_N v\, ds - \int_{\partial\Omega\backslash\Gamma} v\frac{\partial u}{\partial \nu}\, ds = 0 \quad \forall v \in V.$$

Choosing v as in the proof of (5.1.9), (5.4.9) follows.

The finite element approximation of (5.4.10) involves, typically, the use of an interpolant, $\mathcal{I}^h g_D$, of the Dirichlet data. We pick a subspace V_h of V just as before, and we seek u_h such that $u_h - \mathcal{I}^h g_D \in V_h$ and such that

$$(5.4.11) \qquad a(u_h,v) = (f,v) + \int_{\partial\Omega\backslash\Gamma} g_N v\, ds \quad \forall v \in V_h.$$

We leave it to the reader (in exercise 5.x.10) to prove the analogs of (5.4.4) and (5.4.8) for the solution to (5.4.11). We note that further approximation may be necessary in order to compute efficiently (and accurately) the right-hand side in (5.4.11). This can also be done using interpolants of f and g_N (cf. exercise 5.x.11).

5.5 Elliptic Regularity Estimates

Consider the validity of the estimate (5.4.7) in this section. To begin with, we motivate it by studying the problem on all of $\mathbb{R}^n$. We will freely use properties of the Fourier transform,

$$\hat{v}(\xi) := \int_{\mathbb{R}^n} v(x)e^{-ix\cdot\xi}\,dx,$$

(cf. Stein & Weiss 1971). This operator *diagonalizes* differential operators, that is,

$$\widehat{D^\alpha v}(\xi) = (i\xi)^\alpha \hat{v}(\xi).$$

The condition (5.4.7) means that we can bound all second derivatives in terms of a particular combination of them. This follows because

$$\widehat{D^\alpha v}(\xi) = (i\xi)^\alpha \hat{v}(\xi) = -\frac{(i\xi)^\alpha}{|\xi|^2}\widehat{\Delta v}(\xi).$$

The function $\frac{\xi^\alpha}{|\xi|^2}$ is bounded by 1 if $|\alpha| = 2$, so

$$\left\|\widehat{D^\alpha v}\right\|_{L^2(\mathbb{R}^n)} \leq \left\|\widehat{\Delta v}\right\|_{L^2(\mathbb{R}^n)}.$$

Since the Fourier transform is an isometry on L^2, it follows that

$$\|D^\alpha v\|_{L^2(\mathbb{R}^n)} \leq \|\Delta v\|_{L^2(\mathbb{R}^n)} \quad \forall |\alpha| = 2,$$

proving (5.4.7) in this special case.

If Ω has a smooth boundary, and Γ is either empty or all of $\partial\Omega$, then (5.4.7) is known to hold (Friedman 1976, Rauch 1991). In case $n = 2$ and Ω is convex, with Γ again either empty or all of $\partial\Omega$, then (5.4.7) is known to hold (Grisvard 1985). More generally, the following estimate holds (assuming Ω is bounded):

$$(5.5.1) \qquad \|v\|_{W_p^2(\Omega)} \leq \|\Delta v\|_{L^p(\Omega)}, \quad 1 < p < \mu,$$

where μ depends on $\partial\Omega$ (Dauge 1988). When Ω is neither smooth nor convex, then (5.4.7) does not hold, as we now show by example.

(5.5.2) Example. Let Ω be a sector of the unit disk with angle π/β:

$$(5.5.3) \qquad \Omega := \{(r,\theta) \ : \ 0 < r < 1, \quad 0 < \theta < \pi/\beta\}.$$

If $\beta < 1$, then Ω is not convex. Let $v(r,\theta) = r^\beta \sin\beta\theta = \Im(z^\beta)$. Being the imaginary part of a complex analytic function, v is harmonic in Ω. Define $u(r,\theta) = (1 - r^2)v(r,\theta)$. Note that u is zero on $\partial\Omega$. Computing, we find

$$\Delta u = (1 - r^2)\Delta v + 2\nabla(1 - r^2) \cdot \nabla v + v\Delta(1 - r^2)$$
$$= -4r\frac{\partial v}{\partial r} - 4v = -(4\beta + 4)v.$$

Since v is bounded, it is in $L^p(\Omega)$ for all $1 \le p \le \infty$. However,

$$\frac{\partial^2 u}{\partial r^2} \approx r^{\beta - 2} \quad \text{near} \quad r = 0,$$

so for $\beta < 1$, u cannot be in $H^2(\Omega)$. Note that since $\beta \ge 1/2$ in any case, we always have $u \in W_p^2(\Omega)$ for some $p > 1$ depending only on β.

(5.5.4) Example. In the case that the boundary is regular, but boundary conditions change from Dirichlet to Neumann type, there is also a lack of regularity. Consider the domain $\Omega = \{(r, \theta) : 0 < r < 1, \quad 0 < \theta < \pi\}$, and suppose that

$$\Gamma := \{(x, 0) : 0 \le x \le 1\} \cup \{(1, \theta) : 0 \le \theta \le \pi\}.$$

We want to study the regularity of the solution to

$$\Delta u = f \quad \text{in} \quad \Omega$$
$$u = 0 \quad \text{on} \quad \Gamma \quad \& \quad \frac{\partial u}{\partial \nu} = 0 \quad \text{on} \quad \partial\Omega \backslash \Gamma.$$

For motivation, consider the domain $\tilde{\Omega}$ defined by reflecting Ω with respect to x-axis. Functions satisfying the above boundary conditions can be extended by $\tilde{v}(x, y) = v(x, -y)$. Across the segment $\{(x, 0) : -1 \le x \le 0\}$, the extension will be C^1 (assuming that $v \in C^1(\overline{\Omega})$ to start with). On the other hand, $\tilde{v}$ will vanish on

$$\{(x, 0) : 0 \le x \le 1\} \cup \{(1, \theta) : 0 \le \theta < 2\pi\}.$$

These are precisely the boundary conditions in Example 5.5.2 with $\beta = 1/2$, a slit domain. If we pick the solution to that problem, namely $u(r, \theta) = (1 - r^2)r^{1/2}\sin\theta/2$, then this solves the above system of equations with $f = -6r^{1/2}\sin\theta/2 \in L^2(\Omega)$. Since the singularity of u is $r^{1/2}$ the solution will not, in general, be in $H^2(\Omega)$ for $f \in L^2(\Omega)$.

(5.5.5) Example. Finally, for completeness, we remark that the restriction $p \ne 1, \infty$ in (5.5.1) is sharp. We give an example to show it is false for $p = \infty$. Let $u(x, y) = xy\log r$, where $r^2 = x^2 + y^2$. Then

$$\Delta u = xy\Delta\log r + 2\nabla xy \cdot \nabla \log r + \log r \, \Delta xy$$
$$= 2\nabla xy \cdot \nabla \log r$$
$$= 2(y, x) \cdot (x, y)/r^2$$
$$= 4xy/r^2.$$

Thus, $\Delta u \in L^\infty(\Omega)$, where Ω is, say, the unit disk, and $u = 0$ on $\partial\Omega$. However, it is easy to verify that

$$\frac{\partial^2 u}{\partial x \partial y} = \log r + v$$

where $v \in L^\infty(\Omega)$. Thus, u does not lie in $W^2_\infty(\Omega)$.

5.6 General Second-Order Elliptic Operators

Let us now consider more general elliptic operators and their associated variational forms. Let $a_{ij} \in L^\infty(\Omega)$, $i, j = 1, \ldots, n$ and $a_{ij} = a_{ji}$. The matrix of coefficients, (a_{ij}), is said to be *uniformly elliptic* provided there is a positive constant, α, such that

(5.6.1) $$\sum_{i,j=1}^n a_{ij}(x)\xi_i\xi_j \geq \alpha \sum_{i=1}^n \xi_i^2 \quad \forall \xi \in \mathbb{R}^n \quad a.e. \quad \text{in} \quad \Omega,$$

that is, if the matrix (a_{ij}) is uniformly (a.e.) positive definite. The elliptic operator associated with an elliptic coefficient matrix is defined via

(5.6.2) $$Au(x) := -\sum_{i,j=1}^n \frac{\partial}{\partial x_j}\left(a_{ij}(x)\frac{\partial u}{\partial x_i}(x)\right).$$

If, for example, $(a_{ij}(x))$ is the identity matrix for all $x \in \Omega$, then we obtain the Laplace operator studied earlier, and the constant α may be taken to be one. The natural bilinear form associated with (5.6.2) is

(5.6.3) $$a(u, v) := \int_\Omega \sum_{i,j=1}^n a_{ij}(x)\frac{\partial u}{\partial x_i}(x)\frac{\partial v}{\partial x_j}(x)\,dx.$$

Note that we do not assume any smoothness of the coefficients, so that even discontinuous coefficients are allowed. This causes no difficulty for the variational formulation, but the differentiation of discontinuous functions in (5.6.2) would require further justification (e.g., via the variational formulation itself). The following estimate follows simply by replacing ξ_i by $\frac{\partial v}{\partial x_i}$ in (5.6.1) and integrating.

(5.6.4) **Lemma.** *Suppose the coefficient matrix satisfies* (5.6.1). *Then the associated bilinear form* (5.6.3) *satisfies*

$$a(v, v) \geq \alpha \,|v|^2_{H^1(\Omega)} \quad \forall v \in H^1(\Omega)$$

where α is the constant in (5.6.1).

The natural boundary conditions associated with the general form (5.6.3) are more complex than for example (5.2.1). Applying (5.1.5) with $v := -a_{ji}\frac{\partial u}{\partial x_j}$, and summing over i, yields (after switching i and j)

$$\int_{\Omega} Au\, w\, dx = a(u,w) - \int_{\partial\Omega} \sum_{i,j=1}^{n} a_{ij}\frac{\partial u}{\partial x_i}\nu_j w.$$

Thus, the natural boundary condition associated with (5.6.3), expressed in operator form, is

(5.6.5)
$$\sum_{i,j=1}^{n} a_{ij}\frac{\partial u}{\partial x_i}\nu_j = 0.$$

The variational form (5.6.3) is symmetric. In general, nonsymmetric problems arise in practice. A problem in the differential form

(5.6.6)
$$Au(x) := \\ -\sum_{i,j=1}^{n} \frac{\partial}{\partial x_j}\left(a_{ij}(x)\frac{\partial u}{\partial x_i}(x)\right) + \sum_{k=1}^{n} b_k(x)\frac{\partial u}{\partial x_k}(x) + b_0(x)u(x)$$

is naturally expressed using the variational form

(5.6.7)
$$a(u,v) := \int_{\Omega} \sum_{i,j=1}^{n} a_{ij}\frac{\partial u}{\partial x_i}\frac{\partial v}{\partial x_j} + \sum_{k=1}^{n} b_k\frac{\partial u}{\partial x_k}v + b_0 uv\, dx.$$

The question of coercivity of such a variational form is answered in part by the following.

(5.6.8) Theorem. (Gårding's Inequality)   *Suppose (5.6.1) holds and that the coefficients $b_k \in L^{\infty}(\Omega)$, $k = 0,\ldots,n$. Then there is a constant, $K < \infty$, such that*

$$a(v,v) + K\,\|v\|_{L^2(\Omega)}^2 \geq \frac{\alpha}{2}\|v\|_{H^1(\Omega)}^2 \quad \forall v \in H^1(\Omega),$$

where α is the constant in (5.6.1).

This means that, although $a(\cdot,\cdot)$ itself may not be coercive, adding a sufficiently large constant times the L^2-inner-product makes it coercive over all of $H^1(\Omega)$.

Proof. By (5.6.4), we have

$$a(v,v) + K\,\|v\|_{L^2(\Omega)}^2$$

$$\geq \alpha|v|_{H^1(\Omega)}^2 + \int_{\Omega} \sum_{k=1}^{n} b_k(x)\frac{\partial v}{\partial x_k}(x)v(x) + (b_0(x) + K)v(x)^2\, dx.$$

By Hölder's inequality,

$$\left| \int_\Omega \sum_{k=1}^n b_k(x) \frac{\partial v}{\partial x_k}(x) v(x)\, dx \right| \leq \int_\Omega \sum_{k=1}^n |b_k(x)| \left| \frac{\partial v}{\partial x_k}(x) \right| |v(x)|\, dx$$

$$\leq \sum_{k=1}^n \|b_k\|_{L^\infty(\Omega)} \int_\Omega \left| \frac{\partial v}{\partial x_k}(x) \right| |v(x)|\, dx$$

$$\leq \sum_{k=1}^n \|b_k\|_{L^\infty(\Omega)} \left\| \frac{\partial v}{\partial x_k} \right\|_{L^2(\Omega)} \|v\|_{L^2(\Omega)}$$

$$\leq B\, |v|_{H^1(\Omega)}\, \|v\|_{L^2(\Omega)}\, ,$$

where

(5.6.9)
$$B^2 := \sum_{k=1}^n \|b_k\|_{L^\infty(\Omega)}^2\, .$$

Therefore,

$$a(v,v) + K \|v\|_{L^2(\Omega)}^2$$

$$\geq \alpha |v|_{H^1(\Omega)}^2 - B\, |v|_{H^1(\Omega)}\, \|v\|_{L^2(\Omega)} + \int_\Omega (b_0(x) + K) v(x)^2\, dx$$

$$\geq \alpha |v|_{H^1(\Omega)}^2 - B\, |v|_{H^1(\Omega)}\, \|v\|_{L^2(\Omega)} + (\beta + K) \|v\|_{L^2(\Omega)}^2\, ,$$

where

(5.6.10)
$$\beta := \operatorname{ess\,inf} \{ b_0(x)\ :\ x \in \Omega \}\, .$$

From the arithmetic-geometric mean inequality (0.9.5), we conclude that

$$a(v,v) + K \|v\|_{L^2(\Omega)}^2 \geq \frac{\alpha}{2} \left(|v|_{H^1(\Omega)}^2 + \|v\|_{L^2(\Omega)}^2 \right),$$

provided

(5.6.11)
$$K \geq \frac{\alpha}{2} + \frac{B^2}{2\alpha} - \beta.$$

Note that K need not be positive, if $\beta > 0$. □

We may interpret Gårding's inequality as follows. Given an elliptic operator, A, there is a constant, K_0, such that for all $K > K_0$, the variational problem for $A + K$ is coercive on all of H^1, and thus any boundary condition of the form (5.1.2) leads to a well-posed problem. In general, for $K \leq K_0$, there may be further constraints, as occurs with the pure Neumann problem studied in Sect. 5.2. In particular, one can show (Agmon 1965) that there is a set of values $K_1 > K_2 > \cdots$ (the *eigenvalues* of A) with no finite accumulation point such that there is a unique solution for the variational problem associated with $A+K$, provided only that $K \neq K_i$ for any i. Moreover, the problems $A + K_i$ each can be solved subject to a finite number of linear constraints, similar to the ones in Sect. 5.2. Moreover, provided all the coefficients and the boundary are sufficiently smooth, one has a smooth solution in the case that the boundary conditions are either pure Neumann (5.2.1) or pure Dirichlet ($\Gamma = \partial\Omega$ in (5.1.2)) and

$$(5.6.12) \qquad \|u\|_{W_p^k(\Omega)} \leq C_{K,k,\Omega,A,p} \|(A + K)u\|_{W_p^{k-2}(\Omega)}, \quad \forall\, 1 < p < \infty.$$

In general, the index k may depend on the degree of smoothness of the various ingredients in the problem, but if all are C^∞ then k may be taken arbitrarily large. In the case of Ω being a convex polygon in $\mathbb{R}^2$, (5.6.12) is known to hold for $k \leq 2$ (Grisvard 1985).

5.7 Variational Approximation of General Elliptic Problems

In the previous section we observed that there can be well-posed elliptic problems for which the corresponding variational problem is not coercive, although a suitably large additive constant can always make it coercive. In this setting, it is possible to give estimates for finite element approximations of the variation problems, although slightly more complicated arguments must be given. To begin with, let us assume that we have a variational form, $a(\cdot, \cdot)$, satisfying the following properties. We assume that $a(\cdot, \cdot)$ is continuous on $H^1(\Omega)$,

$$(5.7.1) \qquad |a(u, v)| \leq C_1 \|u\|_{H^1(\Omega)} \|v\|_{H^1(\Omega)} \quad \forall u, v \in H^1(\Omega),$$

and that a suitable additive constant, $K \in \mathbb{R}$, makes it coercive,

$$(5.7.2) \qquad a(v, v) + K(v, v) \geq \alpha \|v\|_{H^1(\Omega)}^2 \quad \forall v \in H^1(\Omega).$$

We assume that there is some $V \subset H^1(\Omega)$ such that there is a unique solution, u, to the variational problem

$$a(u, v) = (f, v) \quad \forall v \in V$$

as well as to the adjoint variational problem

$$a(v, u) = (f, v) \quad \forall v \in V$$

and that, in both cases, the regularity estimate

$$(5.7.3) \qquad |u|_{H^2(\Omega)} \leq C_R \|f\|_{L^2(\Omega)}$$

holds for all $f \in L^2(\Omega)$.

Let V_h be a finite element subspace of V, as described in Sect. 5.4, and define $u_h \in V_h$ via

(5.7.4) $$a(u_h, v) = (f, v) \quad \forall v \in V_h.$$

On the face of it, (5.7.4) need not have a unique solution, since $a(\cdot, \cdot)$ may not be coercive. Indeed, we will not be able to guarantee that it does for all subspaces, V_h. However, we assume that the following holds:

(5.7.5) $$\inf_{v \in V_h} \|u - v\|_{H^1(\Omega)} \le C_A h \, |u|_{H^2(\Omega)} \quad \forall u \in H^2(\Omega).$$

The following should be compared with Céa's Theorem 2.8.1.

(5.7.6) Theorem. *Under conditions* (5.7.1), (5.7.2), (5.7.3) *and* (5.7.5), *there are constants* h_0 *and* C, *such that for all* $h \le h_0$, *there is a unique solution to* (5.7.4) *satisfying*

$$\|u - u_h\|_{H^1(\Omega)} \le C \inf_{v \in V_h} \|u - v\|_{H^1(\Omega)},$$

where we may take $C = 2C_1/\alpha$, *and*

$$\|u - u_h\|_{L^2(\Omega)} \le C_1 C_A C_R h \, \|u - u_h\|_{H^1(\Omega)}.$$

In particular, we may take

$$h_0 = (\alpha/2K)^{1/2}/C_1 C_A C_R.$$

Proof. We begin by deriving an estimate for any solution to (5.7.4) that may exist. Note that in any case, we always have

(5.7.7) $$a(u - u_h, v) = 0 \quad \forall v \in V_h.$$

From (5.7.2), (5.7.7), and (5.7.1), it thus follows that, for any $v \in V_h$,

$$
\begin{aligned}
\alpha \|u - u_h\|_{H^1(\Omega)}^2 &\le a(u - u_h, u - u_h) + K(u - u_h, u - u_h) \\
&= a(u - u_h, u - v) + K \|u - u_h\|_{L^2(\Omega)}^2 \\
&\le C_1 \|u - u_h\|_{H^1(\Omega)} \|u - v\|_{H^1(\Omega)} + K \|u - u_h\|_{L^2(\Omega)}^2.
\end{aligned}
$$

We apply standard duality techniques to bound $\|u - u_h\|_{L^2(\Omega)}$. Let w be the solution (guaranteed by condition (5.7.3)) to the adjoint problem

$$a(v, w) = (u - u_h, v) \quad \forall v \in V.$$

Then, for any $w_h \in V_h$,

$$
\begin{aligned}
(u - u_h, u - u_h) &= a(u - u_h, w) \\
&= a(u - u_h, w - w_h) && \text{(by 5.7.7)} \\
&\le C_1 \|u - u_h\|_{H^1(\Omega)} \|w - w_h\|_{H^1(\Omega)}. && \text{(by 5.7.1)}
\end{aligned}
$$

Therefore, for appropriate choice of w_h,

$$(u - u_h, u - u_h)$$
$$\leq C_1 C_A h \, \|u - u_h\|_{H^1(\Omega)} \, |w|_{H^2(\Omega)} \qquad \text{(by 5.7.5)}$$
$$\leq C_1 C_A C_R h \, \|u - u_h\|_{H^1(\Omega)} \, \|u - u_h\|_{L^2(\Omega)} \qquad \text{(by 5.7.3)}.$$

Therefore,

(5.7.8) $$\|u - u_h\|_{L^2(\Omega)} \leq Ch \, \|u - u_h\|_{H^1(\Omega)}$$

where $C = C_1 C_A C_R$. Applying this above, we find

$$\alpha \, \|u - u_h\|_{H^1(\Omega)}^2$$
$$\leq C_1 \, \|u - u_h\|_{H^1(\Omega)} \, \|u - v\|_{H^1(\Omega)} + KC^2 h^2 \, \|u - u_h\|_{H^1(\Omega)}^2 .$$

Thus, for $h \leq h_0$, where $h_0 = (\alpha/2K)^{1/2}/C_1 C_A C_R$, we find

(5.7.9) $$\alpha \, \|u - u_h\|_{H^1(\Omega)} \leq 2C_1 \, \|u - v\|_{H^1(\Omega)} \quad \forall v \in V_h.$$

So far, we have been operating under the assumption of the existence of a solution, u_h. Now we consider the question of its existence and uniqueness. Since (5.7.4) is a finite dimensional system having the same number of unknowns as equations, existence and uniqueness are equivalent. Nonuniqueness would imply the existence of a nontrivial solution, u_h, for $f \equiv 0$. In such a case, we have $u \equiv 0$ by (5.7.3). But (5.7.9) then implies that $u_h \equiv 0$ as well, provided h is sufficiently small. In particular, this says that (5.7.4) has unique solutions for h sufficiently small, since $f = 0$ implies $u_h = 0$.

Finally, for $h \leq h_0$, we conclude that the unique solution to (5.7.4) satisfies (5.7.9) and (5.7.8), thus completing the proof.   □

5.8 Negative-Norm Estimates

The previous section presented, as a byproduct, L^2 error estimates which are improved over H^1 estimates by a factor of h. One can, in some circumstances, continue to get improved error estimates in lower (negative) norms. Having a higher power of h in an estimate in a negative norm may be interpreted as saying that the error is oscillatory. Recall that the H^{-s} norm is defined by

$$\|u\|_{H^{-s}(\Omega)} = \sup_{0 \neq v \in H^s(\Omega)} \frac{(u, v)}{\|v\|_{H^s(\Omega)}}.$$

We assume that the regularity estimate

(5.8.1) $$\|u\|_{H^{s+2}(\Omega)} \leq C_R \|f\|_{H^s(\Omega)}$$

and the approximation estimate

$$(5.8.2) \qquad \inf_{v \in V_h} \|u - v\|_{H^1(\Omega)} \le Ch^{s+1} \|u\|_{H^{s+2}(\Omega)}$$

hold for some $s \ge 0$.

(5.8.3) Theorem. *Under conditions* (5.7.1), (5.7.2), (5.8.1) *and* (5.8.2), *there are constants h_0 and C such that for all $h \le h_0$, the solution to* (5.7.4) *satisfies*

$$\|u - u_h\|_{H^{-s}(\Omega)} \le Ch^{s+1} \|u - u_h\|_{H^1(\Omega)} .$$

Proof. We again apply duality techniques. For arbitrary $\phi \in H^s(\Omega)$, we need to estimate $(u - u_h, \phi)$. Let w be the solution (guaranteed by condition (5.8.1)) to the adjoint problem

$$a(v, w) = (v, \phi) \quad \forall v \in V.$$

Then, for any $w_h \in V_h$,

$$
\begin{aligned}
(u - u_h, \phi) &= a(u - u_h, w) \\
&= a(u - u_h, w - w_h) && \text{(by 5.7.7)} \\
&\le C \|u - u_h\|_{H^1(\Omega)} \|w - w_h\|_{H^1(\Omega)} . && \text{(by 5.7.1)}
\end{aligned}
$$

Therefore, for appropriate choice of w_h, we have

$$
\begin{aligned}
(u - u_h, \phi) &\le Ch^{s+1} \|u - u_h\|_{H^1(\Omega)} \|w\|_{H^{s+2}(\Omega)} && \text{(by 5.8.2)} \\
&\le Ch^{s+1} \|u - u_h\|_{H^1(\Omega)} \|\phi\|_{H^s(\Omega)} . && \text{(by 5.8.1)}
\end{aligned}
$$

Taking the supremum over ϕ completes the estimate.   □

The theorem implies that the approximation oscillates around the correct values, since "weighted averages" of the form $(u - u_h, \phi)/\|\phi\|_{H^1(\Omega)}$ are smaller by a factor, h, than $u - u_h$ itself.

The crucial requirement for the improved rate of convergence in a negative norm is the regularity estimate (5.8.1). When this does not hold (see Example 5.5.2), improvement will not occur. We give a simple explanation of this here. Using (5.7.4), we find

$$
\begin{aligned}
(u - u_h, f) &= a(u, u - u_h) \\
&= a(u - u_h, u - u_h) \\
&\ge \alpha \|u - u_h\|_{H^1(\Omega)}^2 .
\end{aligned}
$$

If $f \in H^s(\Omega)$, then we have

$$\|u - u_h\|_{H^{-s}(\Omega)} = \sup_{0 \neq \phi \in H^s(\Omega)} \frac{(u - u_h, \phi)}{\|\phi\|_{H^s(\Omega)}}$$

(5.8.4)
$$\geq \frac{(u - u_h, f)}{\|f\|_{H^s(\Omega)}}$$

$$\geq \alpha \|u - u_h\|_{H^1(\Omega)}^2 / \|f\|_{H^s(\Omega)},$$

so that the negative norms of the error can never be smaller than the square of the $H^1(\Omega)$-norm of the error. This is often called the "pollution effect" of the singular point on the boundary (see the article by Wahlbin in Ciarlet & Lions 1991).

We will now show that $\|u - u_h\|_{H^1(\Omega)} \geq ch^\beta$ for the problem in Example 5.5.2. In general, if V_h consists of piecewise polynomials of degree $\leq k + 1$,

$$|u - u_h|_{H^1(\Omega)}^2 \geq \inf_{v \in V_h} \sum_{T \in \mathcal{T}^h} \int_T |\nabla u - \nabla v|^2 \, dx$$

$$\geq \sum_{T \in \mathcal{T}^h} \inf_{\underset{\sim}{v} \in \mathcal{P}_k^2} \int_T |\nabla u - \underset{\sim}{v}|^2 \, dx$$

$$\geq \inf_{\underset{\sim}{v} \in \mathcal{P}_k^2} \int_T |\nabla u - \underset{\sim}{v}|^2 \, dx$$

for any $T \in \mathcal{T}^h$, where $\mathcal{P}_k$ denotes polynomials of degree $\leq k$. Choosing T to be an element with a vertex at the origin, we find

$$\inf_{\underset{\sim}{v} \in \mathcal{P}_k^2} \int_T |\nabla (r^\beta \sin \beta \theta) - \underset{\sim}{v}|^2 \, dx \geq c_0 \operatorname{diam}(T)^{2\beta}$$

by a homogeneity argument (exercises 5.x.20 and 5.x.21). Since $u - r^\beta \sin \beta \theta = r^{2+\beta} \sin \beta \theta$, we have

$$\inf_{\underset{\sim}{v} \in \mathcal{P}_k^2} \int_T |\nabla u - \underset{\sim}{v}|^2 \, dx \geq c_0 \operatorname{diam}(T)^{2\beta} - c_1 \operatorname{diam}(T)^{2+2\beta}$$

by the triangle inequality. If $\mathcal{T}^h$ is quasi-uniform, we conclude that

(5.8.5) $$\|u - u_h\|_{H^1(\Omega)} \geq c_2 h^\beta$$

where $c_2 > 0$, provided h is sufficiently small.

5.9 The Plate-Bending Biharmonic Problem

So far, we have considered only second-order elliptic equations. In later chapters, we will consider systems of second-order equations, but here we will briefly describe how the theory developed so far can be applied to higher-order equations. We will restrict our attention to one problem of physical interest, the model of plate bending given by the biharmonic equation. Let $a(\cdot, \cdot)$ be the bilinear form defined on $H^2(\Omega)$ given by

$$a(u, v) := \int_\Omega \Delta u\, \Delta v - (1 - \nu)\, (2u_{xx}v_{yy} + 2u_{yy}v_{xx} - 4u_{xy}v_{xy})\, dxdy$$

where ν is a physical constant known as Poisson's ratio. In the model for the bending of plates, ν is restricted to the range $[0, \frac{1}{2}]$. However, $a(\cdot, \cdot)$ is known (Agmon 1965) to satisfy a Gårding-type inequality,

(5.9.1) $$a(v, v) + K\, \|v\|^2_{L^2(\Omega)} \geq \alpha\, \|v\|^2_{H^2(\Omega)} \quad \forall v \in H^2(\Omega),$$

where $\alpha > 0$ and $K < \infty$, for all $-3 < \nu < 1$. Note that for $\nu = 1$, such an inequality cannot hold as $a(v, v)$ vanishes in that case for all harmonic functions, v.

A simple coercivity estimate can be derived for $0 < \nu < 1$, as follows. Write

$$a(v, v)$$
$$= \int_\Omega \nu\, (v_{xx} + v_{yy})^2 + (1 - \nu) \left((v_{xx} - v_{yy})^2 + 4v_{xy}^2 \right) dxdy$$

(5.9.2) $$\geq \min\{\nu, 1 - \nu\} \int_\Omega (v_{xx} + v_{yy})^2 + (v_{xx} - v_{yy})^2 + 4v_{xy}^2\, dxdy$$

$$= 2\min\{\nu, 1 - \nu\} \int_\Omega v_{xx}^2 + v_{yy}^2 + 2v_{xy}^2\, dxdy$$

$$= 2\min\{\nu, 1 - \nu\}\, |v|^2_{H^2(\Omega)}\, .$$

From (5.9.2), it follows that $a(\cdot, \cdot)$ is coercive over any closed subspace, $V \subset H^2(\Omega)$, such that $V \cap \mathcal{P}_1 = \emptyset$. For if not, there would be a sequence of $v_j \in H^2(\Omega)$ such that $a(v_j, v_j) < 1/j$ and $\|v_j\|_{H^2(\Omega)} = 1$. The set of linear polynomials, $Q^2 v_j$, is bounded:

$$\|Q^2 v_j\|_{H^2(\Omega)} \leq C\, \|v_j\|_{H^2(\Omega)} = C$$

and hence (see exercise 5.x.17) a subsequence, $\{Q^2 v_{j_k}\}$, converges to some fixed linear polynomial, L. Since

$$\|v_j - Q^2 v_j\|_{H^2(\Omega)} \leq C\, |v_j|_{H^2(\Omega)} \to 0,$$

we conclude, as before, that v_{j_k} converges to L. But since $V \cap \mathcal{P}_1 = \emptyset$, we reach a contradiction. Thus, there must be a constant $\alpha > 0$ such that

(5.9.3) $$a(v, v) \geq \alpha\, \|v\|^2_{H^2(\Omega)} \quad \forall v \in V.$$

For $F \in H^2(\Omega)'$ and $V \subset H^2(\Omega)$, we consider the problem: find $u \in V$ such that

(5.9.4) $a(u, v) = F(v) \quad \forall v \in V.$

As a consequence of the Riesz Representation Theorem (cf. (2.5.6)), we have the following.

(5.9.5) Theorem. *If $V \subset H^2(\Omega)$ is a closed subspace such that $V \cap \mathcal{P}_1 = \emptyset$ and (5.9.2) holds, then (5.9.4) has a unique solution.*

If u is sufficiently smooth, then integration by parts can be carried out, say with $v \in C_0^\infty(\Omega)$, to determine the differential equation satisfied. For example, if $F(v) := \int_\Omega f(x, y) v(x, y) \, dx dy$, where $f \in L^2(\Omega)$, then under suitable conditions on Ω (Blum and Rannacher 1980) we have $u \in H^4(\Omega)$. Note that when integrating by parts, all of the terms multiplied by $1 - \nu$ cancel, as they all yield various versions of the cross derivative u_{xxyy} (cf. exercise 5.x.4). Thus, we find that $\Delta^2 u = f$ holds in the L^2 sense, independent of the choice of ν.

Various boundary conditions are of physical interest. Let V^{ss} denote the subset of $H^2(\Omega)$ consisting of functions which vanish (to first-order only) on $\partial\Omega$, i.e.,

$$V^{\mathrm{ss}} = \left\{ v \in H^2(\Omega) \; : \; v = 0 \text{ on } \partial\Omega \right\}.$$

With this choice for V in (5.9.4), the resulting model is called the "simply-supported" plate model, since the displacement, u, is held fixed (at a height of zero), yet the plate is free to rotate at the boundary. The "clamped" plate model consists of choosing $V^{\mathrm{c}} = \mathring{H}^2(\Omega)$, the subset of $H^2(\Omega)$ consisting of functions which vanish to second order on $\partial\Omega$:

$$V^{\mathrm{c}} = \left\{ v \in H^2(\Omega) \; : \; v = \frac{\partial v}{\partial \nu} = 0 \text{ on } \partial\Omega \right\}.$$

Here, the rotation of the plate is also prescribed at the boundary.

In the simply-supported case ($V = V^{\mathrm{ss}}$), there is another, *natural* boundary condition that holds. In this sense, this problem has a mixture of Dirichlet and Neumann boundary conditions, but they hold on all of $\partial\Omega$. The natural boundary condition is found using integration by parts, but with v having an arbitrary, nonzero normal derivative on $\partial\Omega$. One finds (Bergman & Schiffer 1953) that the "bending moment" $\Delta u + (1 - \nu) u_{tt}$ must vanish on $\partial\Omega$, where u_{tt} denotes the second directional derivative in the tangential direction. These results are summarized in the following.

(5.9.6) Theorem. *Suppose that V is any closed subspace satisfying $\mathring{H}^2(\Omega) \subset V \subset H^2(\Omega)$. If $f \in L^2(\Omega)$, and if $u \in H^4(\Omega)$ satisfies (5.9.4) with $F(v) = (f, v)$, then u satisfies*

$$\Delta^2 u = f$$

in the $L^2(\Omega)$ sense. For $V = V^c$, u satisfies

$$u = \frac{\partial u}{\partial \nu} = 0 \text{ on } \partial\Omega$$

and for $V = V^{ss}$, u satisfies

$$u = \Delta u + (1 - \nu)u_{tt} = 0 \text{ on } \partial\Omega.$$

To approximate (5.9.4), we need a subspace V_h of $H^2(\Omega)$. For example, we could take a space based on the Argyris elements (cf. (3.2.10) and Proposition 3.3.17). With either choice of V as above, if we choose V_h to satisfy the corresponding boundary conditions, we obtain the following.

(5.9.7) Theorem. *If $V_h \subset V$ is based on Argyris elements of order $k \geq 5$ then there is a unique $u_h \in V_h$ such that*

$$a(u_h, v) = F(v) \quad \forall v \in V_h.$$

Moreover,

$$\|u - u_h\|_{H^2(\Omega)} \leq C \inf_{v \in V_h} \|u - v\|_{H^2(\Omega)}$$
$$\leq Ch^{k-1}\|u\|_{H^{k+1}(\Omega)}.$$

Since the Argyris interpolant is defined for $u \in H^4(\Omega)$, the above result also holds for k replaced by any integer s in the range $3 \leq s \leq k$. Provided sufficient regularity holds (Blum & Rannacher 1980) for the solution of (5.9.4), namely

(5.9.8) $$\|u\|_{H^{s+1}(\Omega)} \leq C\|f\|_{H^{s-3}(\Omega)},$$

then we also have the following negative-norm estimate (cf. exercise 5.x.16).

(5.9.9) Theorem. *Assuming (5.9.8) holds for $s = m+3$, with $0 \leq m \leq k-3$, we have*

$$\|u - u_h\|_{H^{-m}(\Omega)} \leq Ch^{m+2} \|u - u_h\|_{H^2(\Omega)}.$$

It is interesting to note that we do not immediately get an estimate in the H^1 norm for the error $u - u_h$. In Chapter 14, techniques will be presented to obtain such estimates. In addition, it will also be shown how to relax the above condition $3 \leq s \leq k$. That is, an estimate of the form

$$\|u - u_h\|_{H^2(\Omega)} \leq Ch^{s-1}\|u\|_{H^{s+1}(\Omega)}$$

will be proved for any $1 \leq s \leq k$. For more details regarding the biharmonic equation model for plate bending, see the survey (Scott 1976).

5.x Exercises

5.x.1 Prove that any function in $\tilde{L}^2(\Omega) := \{\phi \in L^2(\Omega) : \int_\Omega \phi \, dx = 0\}$ can be written as a limit of functions in $\tilde{\mathcal{D}}(\Omega) := \{\phi \in \mathcal{D}(\Omega) : \int_\Omega \phi \, dx = 0\}$. (Hint: pick a sequence in $\mathcal{D}(\Omega)$ that converges in L^2 and modify it to get a mean-zero sequence.)

5.x.2 Let $f, g \in H^1(\Omega)$. Show that $D_w^\alpha(fg)$ exists for all $|\alpha| = 1$ and equals $D_w^\alpha f g + f D_w^\alpha g$. (Hint: Prove it first for $g \in C^\infty(\Omega) \cap H^1(\Omega)$ and then use the density result (1.3.4).)

5.x.3 State and prove appropriate L^p ($p \neq 2$) versions of Propositions 5.1.5 and 5.1.6.

5.x.4 Show that if $u \in W_p^k(\Omega)$ and $|\alpha| \leq k$ then $D^\alpha u \in W_p^{k-|\alpha|}(\Omega)$. Prove that if $|\alpha| + |\beta| \leq k$ then $D^\beta D^\alpha u = D^\alpha D^\beta u$.

5.x.5 For solving the potential flow equation ($\Delta u = 0$), the following method has been proposed by E. Wu, *A cubic triangular element with local continuity — an application in potential flow,* **Int. J. Num. Meth. Eng. 17** (1981), 1147–1159. For element, we take K to be a triangle, $\mathcal{P}$ to be cubic polynomials, v, such that $\int_{\partial K} \frac{\partial v}{\partial \nu} \, ds = 0$ and $\mathcal{N}$ to be evaluation of v and its gradient at each vertex. Note that for $v \in \mathcal{P}$, the vector field ∇v has zero net flux through K (whatever goes in must also come out). Prove that this element is well defined. Using this element, derive results analogous to those in Sect. 5.4, namely (5.4.4) with $m = 4$, for solving $\Delta u = 0$ with Neumann boundary conditions. (Hint: First prove the element $(K, \mathcal{P}_3, \tilde{\mathcal{N}})$ where

$$\tilde{\mathcal{N}} = \mathcal{N} \cup \{v \longrightarrow \int_{\partial K} \frac{\partial v}{\partial \nu} \, ds\}$$

to be well defined, then prove estimates for $u - \tilde{\mathcal{I}}^h u$ and observe that $\Delta u = 0$ implies that

$$\int_{\partial K} \frac{\partial u}{\partial \nu} \, ds = 0 \quad \forall K.)$$

5.x.6 Prove that the reflection used in Example 5.5.4 maps the space

$$\left\{ v \in H^2(\Omega) : v|_\Gamma = 0, \; \frac{\partial v}{\partial \nu}\Big|_{\partial \Omega \setminus \Gamma} = 0 \right\}$$

to $H^2(\tilde{\Omega})$.

5.x.7 Prove that the variational problem, $a(u, v) = (f, v) \quad \forall v \in V$, with form (5.6.7) and using $V = H^1(\Omega)$ is equivalent to $Au = f$, where A is defined by (5.6.6), with boundary conditions (5.6.5).

5.x.8 Formulate the duality arguments for the case of a general second-order elliptic Dirichlet problem, with form (5.6.7). (Hint: consider the Dirichlet problem for the *adjoint* operator

$$A^t v := - \sum_{i,j=1}^{n} \frac{\partial}{\partial x_i} \left(a_{ij} \frac{\partial v}{\partial x_j} \right) - \sum_{k=1}^{n} \frac{\partial}{\partial x_k} (b_k v) + b_0 v.$$

State the needed regularity of all elliptic problems involved.)

5.x.9 Determine the appropriate natural boundary conditions associated with the adjoint problem in exercise 5.x.8. Formulate the duality arguments for the case of a general second-order elliptic Neumann problem, with form (5.6.7) and space $V = H^1(\Omega)$.

5.x.10 Let u_h be the solution of (5.4.11). Show that

$$|u - u_h|_{H^1(\Omega)} \le \inf_{v \in V_h} |u - g_D - v|_{H^1(\Omega)} + 2|g_D - \mathcal{I}^h g_D|_{H^1(\Omega)}.$$

Assuming the elliptic regularity (5.4.7) is valid, show also that

$$\|u - u_h\|_{L^2(\Omega)} \le C \left[h \inf_{v \in V_h} |u - g_D - v|_{H^1(\Omega)} + \|g_D - \mathcal{I}^h g_D\|_{L^2(\Omega)} \right].$$

5.x.11 Prove that analogs of (5.4.4) and (5.4.8) hold for the solution to a perturbed version of (5.4.11) in which f and g_N are replaced by their respective interpolants.

5.x.12 Consider the equation (5.1.1) with $u = 0$ on Γ and the *Robin* boundary condition

$$\alpha u + \frac{\partial u}{\partial \nu} = 0 \quad \text{on} \quad \partial \Omega \backslash \Gamma,$$

where α is a constant.
1) Derive a variational formulation for this problem and give conditions on α and f (if any are required) that guarantee it has a unique solution in H^1. (Hint: multiply by v and integrate by parts, converting the $\frac{\partial u}{\partial \nu} v$ term to uv using the boundary condition.)
2) Prove that if $u \in H^2$ solves your variational problem, with $f \in L^2$, then it solves the original differential problem in 5.x.12.
3) Prove error estimates for the variational approximation using piecewise linear functions on a general mesh in both H^1 and L^2. (Assume H^2 regularity for the problem).
4) Compute the "difference stencil" corresponding to using piecewise linear functions on a regular mesh on $\Omega = [0,1] \times [0,1]$ in the variational approximation (cf. Sect. 0.5). (Hint: away from the boundary it is the standard 5-point difference stencil, if the mesh consists of 45° right triangles.)

5.x.13 Prove that $\|u\|_{L^p(\Omega)} \le C\,|u|_{W_p^1(\Omega)}$ for $u \in \mathring{W}_p^1(\Omega)$. (Hint: see the proof of Proposition 5.3.3.)

5.x.14 Consider the equation (5.1.1) with periodic boundary conditions on $\Omega = [0,1] \times [0,1]$. Show that the variational formulation is similar to the Neumann problem, except that the variational spaces have periodicity imposed.

5.x.15 Prove Proposition 5.3.3 for the bilinear form (5.6.3).

5.x.16 Prove (5.9.9). (Hint: follow the techniques in Sect. 5.8.)

5.x.17 Prove that a set of polynomials of degree $< N$, that is closed and bounded in a Sobolev norm, is a compact set. (Hint: show that the set corresponds to a closed, bounded subset of Euclidean space by writing the norm as a metric on the coefficients of the polynomials.)

5.x.18 Formulate inhomogeneous boundary value problems for the plate bending problem.

5.x.19 Use (5.9.1) to prove (5.9.2) for $-3 < \nu < 1$. (Hint: use Rellich's Lemma (Agmon 1965) which says that bounded subsets of $H^m(\Omega)$, $m \ge 1$, are precompact in $L^2(\Omega)$.)

5.x.20 Let $T_h = \{(r,\theta) : 0 < r < h,\ \theta_0 < \theta < \theta_1\}$. Prove that

$$\inf_{v \in \mathcal{P}_k^2} \int_{T_h} |\nabla(r^\beta \sin \beta\theta) - \underset{\sim}{v}|^2 \, dx = c_{k,\beta}\, h^{2\beta},$$

where $\mathcal{P}_k$ denotes polynomials of degree at most k and $c_{k,\beta} > 0$ depends only on k and β. (Hint: by a dilation of coordinates, show that this holds with

$$c_{k,\beta} = \inf_{v \in \mathcal{P}_k^2} \int_{T_1} |\nabla(r^\beta \sin \beta\theta) - \underset{\sim}{v}|^2 \, dx.$$

Prove that $c_{k,\beta} = 0$ leads to the contradiction $\nabla(r^\beta \sin \beta\theta) \in \mathcal{P}_k^2$ by a compactness argument.)

5.x.21 Prove that, for any $f \in L^p(\Omega)$ and any $\tilde{\Omega} \subset \Omega$,

$$\inf_{v \in W} \int_\Omega |f - v|^p \, dx \ge \inf_{v \in W} \int_{\tilde{\Omega}} |f - v|^p \, dx$$

for any subset $W \subset L^p(\Omega)$.

Chapter 6

Finite Element Multigrid Methods

The multigrid method provides an optimal order algorithm for solving elliptic boundary value problems. The error bounds of the approximate solution
obtained from the full multigrid algorithm are comparable to the theoretical
bounds of the error in the finite element method, while the amount of computational work involved is proportional only to the number of unknowns
in the discretized equations.

The multigrid method has two main features: smoothing on the current
grid and error correction on a coarser grid. The smoothing step has the effect
of damping out the oscillatory part of the error. The smooth part of the
error can then be accurately corrected on the coarser grid.

The discussion in this chapter is based on the papers (Bank & Dupont
1981) and (Bramble & Pasciak 1987). It is restricted to a model problem
for simplicity. We recommend the books (Hackbusch 1985), (McCormick
1987), and (Bramble 1993), the survey article (Bramble & Zhang 2000),
and the references therein for the general theory of multigrid methods.

In practice, higher-order finite element equations would be preconditioned by a low-order solver, so we will only consider the piecewise linear
case. Similarly, we restrict to the two-dimensional case to minimize techni-
calities. For theory regarding the general case, see (Scott & Zhang 1992).

6.1 A Model Problem

Let $\Omega \subseteq \mathbb{R}^2$ be a convex polygon and

$$(6.1.1) \qquad a(u,v) = \int_\Omega (\alpha \nabla u \cdot \nabla v + \beta \, u \, v) \, dx,$$

where α and β are smooth functions such that for α_0, α_1, $\beta_1 \in \mathbb{R}^+$ we have
$\alpha_0 \le \alpha(x) \le \alpha_1$ and $0 \le \beta(x) \le \beta_1$ for all x in Ω. We consider the Dirichlet
problem:
Find $u \in V := \mathring{H}^1(\Omega)$ such that

$$(6.1.2) \qquad a(u,v) = (f,v) \quad \forall v \in V$$

where $f \in L^2(\Omega)$.

Elliptic regularity (cf. (Grisvard 1985)) implies that $u \in H^2(\Omega) \cap \mathring{H}^1(\Omega)$. To approximate u, we consider a sequence of triangulations $\mathcal{T}_k$ of Ω determined as follows. Suppose $\mathcal{T}_1$ is given and let $\mathcal{T}_k$, $k \geq 2$, be obtained from $\mathcal{T}_{k-1}$ via a "regular" subdivision: edge midpoints in $\mathcal{T}_{k-1}$ are connected by new edges to form $\mathcal{T}_k$. Let V_k denote C^0 piecewise linear functions with respect to $\mathcal{T}_k$ that vanish on $\partial\Omega$. Note that

$$\text{(6.1.3)} \qquad \mathcal{T}_k \supset \mathcal{T}_{k-1} \Longrightarrow V_{k-1} \subset V_k$$

for all $k \geq 1$.

The discretized problem is:
Find $u_k \in V_k$ such that

$$\text{(6.1.4)} \qquad a(u_k, v) = (f, v) \qquad \forall\, v \in V_k.$$

Let h_k be the mesh size of $\mathcal{T}_k$, i.e., $h_k := \max_{T \in \mathcal{T}_k} \operatorname{diam} T$. Note that for any $T \in \mathcal{T}_{k-1}$, the four subtriangles in $\mathcal{T}_k$ of T are all similar to T and have half the size of T. Thus,

$$\text{(6.1.5)} \qquad h_k = \frac{1}{2} h_{k-1}.$$

Similarly, $\{\mathcal{T}_k\}$ is quasi-uniform, since each $T \in \mathcal{T}_k$ is an exact replica of some $\hat{T} \in \mathcal{T}_1$ in miniature, with the size ratio exactly h_k/h_1. Thus, we have from Theorem 5.7.6 and Theorem 4.4.20,

$$\text{(6.1.6)} \qquad \|u - u_k\|_{H^s(\Omega)} \leq C\, h_k^{2-s}\, \|u\|_{H^2(\Omega)} \quad s = 0, 1,\ k = 1, 2, \ldots.$$

Throughout this chapter, C, with or without subscripts, denotes a generic constant independent of k.

Let $n_k = \dim V_k$. The goal of the multigrid method is to calculate $\hat{u}_k \in V_k$ in $\mathcal{O}(n_k)$ operations such that

$$\text{(6.1.7)} \qquad \|u_k - \hat{u}_k\|_{H^s(\Omega)} \leq C\, h_k^{2-s}\, \|u\|_{H^2(\Omega)}, \quad s = 0, 1,\ k = 1, 2, \ldots.$$

The $\mathcal{O}(n_k)$ operation count means that the multigrid method is asymptotically optimal.

(6.1.8) Remark. Full elliptic regularity is not essential for the convergence analysis (cf. (Bank & Dupont 1981), (Bramble, Pasciak, Wang & Xu 1991), (Zhang 1992), (Bramble & Pasciak 1993)) and (Brenner 2002). It is assumed here for simplicity.

6.2 Mesh-Dependent Norms

(6.2.1) Definition. *The **mesh-dependent inner product** $(\cdot, \cdot)_k$ on V_k is defined by*

$$(6.2.2) \qquad (v, w)_k := h_k^2 \sum_{i=1}^{n_k} v(p_i)\, w(p_i),$$

where $\{p_i\}_{i=1}^{n_k}$ is the set of internal vertices of $\mathcal{T}_k$.

The operator $A_k : V_k \longrightarrow V_k$ is defined by

$$(6.2.3) \qquad (A_k v, w)_k = a(v, w) \quad \forall\, v, w \in V_k.$$

In terms of the operator A_k, the discretized equation (6.1.4) can be written as

$$(6.2.4) \qquad A_k u_k = f_k,$$

where $f_k \in V_k$ satisfies

$$(6.2.5) \qquad (f_k, v)_k = (f, v) \quad \forall\, v \in V_k.$$

Since A_k is symmetric positive definite with respect to $(\cdot, \cdot)_k$, we can define a scale of **mesh-dependent norms** $\|\cdot\|_{s,k}$ in the following way:

$$(6.2.6) \qquad \|v\|_{s,k} := \sqrt{(A_k^s v, v)_k}$$

where A_k^s denotes the s power of the symmetric, positive definite operator A_k (Halmos 1957).

The mesh-dependent norms $\|\cdot\|_{s,k}$ will play an important role in the convergence analysis. Observe that the energy norm $\|\cdot\|_E = \sqrt{a(\cdot, \cdot)}$ coincides with the $\|\cdot\|_{1,k}$ norm on V_k. Similarly, $\|\cdot\|_{0,k}$ is the norm associated with the mesh-dependent inner product (6.2.2). The following lemma shows that $\|\cdot\|_{0,k}$ is equivalent to the L^2-norm.

(6.2.7) Lemma. *There exist positive constants C_1 and C_2 such that*

$$C_1 \|v\|_{L^2(\Omega)} \le \|v\|_{0,k} \le C_2 \|v\|_{L^2(\Omega)} \quad \forall\, v \in V_k.$$

Proof. Using a quadrature formula for quadratic polynomials on triangles, we have

$$\|v\|_{L^2(\Omega)}^2 = \sum_{T \in \mathcal{T}_k} \int_T v^2 \, dx$$

$$= \sum_{T \in \mathcal{T}_k} \frac{|T|}{3} \left(\sum_{i=1}^3 v^2(m_i) \right) \quad \text{(see exercise 6.x.11)}$$

where $|T|$ denotes the area of T. By definition,

$$\|v\|_{0,k}^2 = (v,v)_k$$
$$= h_k^2 \sum_{i=1}^{n_k} v^2(p_i).$$

The fact that $\{\mathcal{T}_k\}$ is quasi-uniform implies that $|T|$ is equivalent to h_k^2. Since v is linear, we have

$$\begin{pmatrix} v(m_1) \\ v(m_2) \\ v(m_3) \end{pmatrix} = \begin{pmatrix} 0 & \frac{1}{2} & \frac{1}{2} \\ \frac{1}{2} & 0 & \frac{1}{2} \\ \frac{1}{2} & \frac{1}{2} & 0 \end{pmatrix} \begin{pmatrix} v(p_1) \\ v(p_2) \\ v(p_3) \end{pmatrix},$$

where p_1, p_2, and p_3 are the vertices of T, and m_i is the midpoint of the edge opposite p_i, $i = 1, 2, 3$. The lemma now follows.   □

The spectral radius, $\Lambda(A_k)$, of A_k is estimated in the next lemma.

(6.2.8) Lemma. $\Lambda(A_k) \leq C\, h_k^{-2}$.

Proof. Let λ be an eigenvalue of A_k with eigenvector ϕ. We have

$$a(\phi,\phi) = (A_k\phi,\phi)_k$$
$$= \lambda\,(\phi,\phi)_k$$
$$= \lambda\,\|\phi\|_{0,k}^2.$$

Therefore,

$$\lambda = \frac{a(\phi,\phi)}{\|\phi\|_{0,k}^2}.$$

Since $\|\phi\|_{0,k}^2$ is equivalent to $\|\phi\|_{L^2(\Omega)}^2$, the lemma follows from (4.5.3):

$$(6.2.9) \qquad \|v\|_E \leq C\, h_k^{-1}\|v\|_{L^2(\Omega)} \quad \forall\, v \in V_k.$$

□

We end this section with a useful lemma whose proof is left as an exercise (cf. exercise 6.x.1).

(6.2.10) Lemma. (Generalized Cauchy-Schwarz Inequality)

$$|a(v,w)| \leq \|v\|_{1+t,k}\, \|w\|_{1-t,k} \quad \forall\, v, w \in V_k$$

for any $t \in \mathbb{R}$.

6.3 The Multigrid Algorithm

The multigrid algorithm is an iterative solver. For each k there is an iteration scheme for solving the equation $A_k z = g$ on V_k. This scheme has two main features: smoothing on V_k and error correction on V_{k-1}, using the iteration scheme developed on V_{k-1}. The smoothing step has the effect of damping out the oscillatory part of the error, as shown in Fig. 6.1. A randomly chosen initial error is depicted together with three steps of the smoothing process (to be studied shortly) applied to this initial error. After a few such steps, the smooth part of the error is dominant and can be accurately captured on the coarser grid. This will be seen analytically in the proof of Theorem 6.5.7. Since the error correction is done on a coarser grid, less computational work is involved.

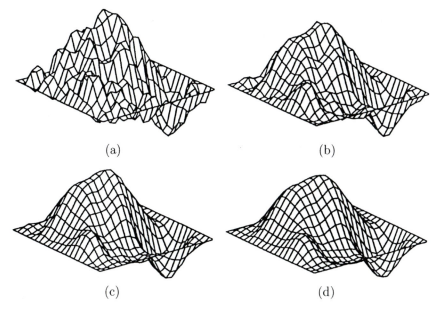

Fig. 6.1. Effect of smoothing step: (a) initial error, (b-d) one, two and three smoothing steps, resp., for the model problem (6.1.2) with $\alpha \equiv 1$ and $\beta \equiv 0$ in (6.1.1).

In order to describe the multigrid algorithm, we need to introduce the intergrid transfer operators.

(6.3.1) Definition. (Intergrid Transfer Operators) *The coarse-to-fine operator*

$$I_{k-1}^k : V_{k-1} \longrightarrow V_k$$

is taken to be the natural injection. In other words,

$$I_{k-1}^k v = v \quad \forall v \in V_{k-1}.$$

The fine-to-coarse intergrid transfer operator

$$I_k^{k-1} : V_k \longrightarrow V_{k-1}$$

is defined to be the transpose of I_{k-1}^k with respect to the $(\cdot, \cdot)_{k-1}$ and $(\cdot, \cdot)_k$ inner products. In other words,

$$(I_k^{k-1} w, v)_{k-1} = (w, I_{k-1}^k v)_k$$
$$= (w, v)_k \quad \forall \, v \in V_{k-1}, w \in V_k.$$

The kth Level Iteration. $MG(k, z_0, g)$ is the approximate solution of the equation

$$A_k z = g$$

obtained from the kth level iteration with initial guess z_0.
For $k = 1$, $MG(1, z_0, g)$ is the solution obtained from a direct method. In other words,

$$MG(1, z_0, g) = A_1^{-1} g.$$

For $k > 1$, $MG(k, z_0, g)$ is obtained recursively in three steps.

Presmoothing Step. For $1 \leq l \leq m_1$, let

$$z_l = z_{l-1} + \frac{1}{\Lambda_k} \Big(g - A_k z_{l-1} \Big),$$

where, from now on, we will assume that Λ_k denotes some upper-bound (Lemma 6.2.8) for the spectral radius of A_k satisfying

(6.3.2) $$\Lambda_k \leq C h_k^{-2}.$$

Error Correction Step. Let $\bar{g} := I_k^{k-1}(g - A_k z_{m_1})$ and $q_0 = 0$. For $1 \leq i \leq p$, let

$$q_i = MG(k - 1, q_{i-1}, \bar{g}).$$

Then

$$z_{m_1+1} := z_{m_1} + I_{k-1}^k q_p.$$

Postsmoothing Step. For $m_1 + 2 \leq l \leq m_1 + m_2 + 1$, let

$$z_l = z_{l-1} + \frac{1}{\Lambda_k}(g - A_k z_{l-1}).$$

Then the output of the kth level iteration is

$$MG(k, z_0, g) := z_{m_1+m_2+1}.$$

Here m_1 and m_2 are positive integers, and $p = 1$ or 2. When $p = 1$, this is called a $\mathcal{V}$-cycle method; $p = 2$ is called a $\mathcal{W}$-cycle method.

In the application of the kth level iteration to (6.2.4), we follow the following strategy. We take the initial guess to be $I_{k-1}^k \hat{u}_{k-1}$, where $\hat{u}_{k-1}$ is the approximate solution already obtained for the equation $A_{k-1} u_{k-1} = f_{k-1}$. Then we apply the kth level iteration r times. The full multigrid algorithm therefore consists of nested iterations.

The Full Multigrid Algorithm

For $k = 1$, $\hat{u}_1 = A_1^{-1} f_1$.

For $k \geq 2$, the approximate solutions $\hat{u}_k$ are obtained recursively from

$$u_0^k = I_{k-1}^k \hat{u}_{k-1}$$
$$u_l^k = MG(k, u_{l-1}^k, f_k), \quad 1 \leq l \leq r,$$
$$\hat{u}_k = u_r^k.$$

6.4 Approximation Property

We will now discuss one of the key ingredients for the convergence analysis of both the $\mathcal{W}$-cycle and $\mathcal{V}$-cycle algorithms.

(6.4.1) Definition. *Let* $P_k : V \longrightarrow V_k$ *be the orthogonal projection with respect to* $a(\cdot, \cdot)$. *In other words,* $P_k v \in V_k$ *and*

$$a(v, w) = a(P_k v, w) \qquad \forall\, w \in V_k.$$

The operator P_{k-1} relates the error after presmoothing in the kth level iteration scheme to the exact solution of the residual equation on the coarser grid. Recall that in the error correction step of the kth level iteration we had $\bar{g} = I_k^{k-1}(g - A_k z_{m_1})$. Let $q \in V_{k-1}$ satisfy the coarse grid residual equation $A_{k-1} q = \bar{g}$.

(6.4.2) Lemma. $q = P_{k-1}(z - z_{m_1})$.

Proof. For $w \in V_{k-1}$,

$$
\begin{aligned}
a(q, w) &= (A_{k-1} q, w)_{k-1} & \text{(by 6.2.3)} \\
&= (\bar{g}, w)_{k-1} & \text{(definition of } q) \\
&= (I_k^{k-1}(g - A_k z_{m_1}), w)_{k-1} & \text{(definition of } g) \\
&= (g - A_k z_{m_1}, w)_k & \text{(by 6.3.1)} \\
&= (A_k(z - z_{m_1}), w)_k & \text{(definition of } z) \\
&= a(z - z_{m_1}, w). & \text{(by 6.2.3)}
\end{aligned}
$$

$\square$

Therefore, if the residual equation is solved exactly on the coarser grid, the error after the correction step is $z - (z_{m_1} + q) = (I - P_{k-1})(z - z_{m_1})$. Hence, it is important to understand the operator $I - P_{k-1}$.

(6.4.3) Lemma. *There exists a positive constant C such that*

$$\|(I - P_{k-1})v\|_{0,k} \leq C\, h_k\, \|(I - P_{k-1})v\|_{1,k} \qquad \forall\, v \in V_k.$$

Proof. From Theorem 5.7.6 we know that

$$\|(I - P_{k-1})v\|_{L^2(\Omega)} \leq C\, h_k\, \|(I - P_{k-1})v\|_E = C\, h_k\, \|(I - P_{k-1})v\|_{1,k}.$$

Using Lemma 6.2.7 completes the proof.   □

We note the resemblance of the following result to, say, Theorem 5.4.4 with $m = 2$. However, the function, v, being approximated here does not lie in $H^2(\Omega)$.

(6.4.4) Corollary. (Approximation Property) *There exists a positive constant C such that*

$$\|(I - P_{k-1})v\|_{1,k} \leq C\, h_k\, \|v\|_{2,k} \qquad \forall\, v \in V_k.$$

Proof.

$$
\begin{aligned}
\|(I - P_{k-1})v\|_{1,k}^2 &= \|v - P_{k-1}v\|_E^2 \\
&= a(v - P_{k-1}v, v - P_{k-1}v) \\
&= a(v - P_{k-1}v, v) \\
&\leq \|v - P_{k-1}v\|_{0,k}\, \|v\|_{2,k} && \text{(by 6.2.10)} \\
&\leq C\, h_k\, \|v - P_{k-1}v\|_{1,k}\, \|v\|_{2,k}. && \text{(Lemma 6.4.3)}
\end{aligned}
$$

□

6.5 $\mathcal{W}$-cycle Convergence for the kth Level Iteration

In this section we will show that the kth level iteration scheme for a $\mathcal{W}$-cycle method has a contraction number bounded away from 1 if the number of smoothing steps is large enough. In the next section we will prove the same result for a $\mathcal{V}$-cycle method with one smoothing step, which implies that the number of smoothing steps in the $\mathcal{W}$-cycle method can also be taken to be one (cf. exercises 6.x.8 and 6.x.9). But, the weaker result in this section can be obtained with less effort, and the perturbation argument involved is also of independent interest.

For simplicity, we will consider the one-sided $\mathcal{W}$-cycle method, i.e., $p = 2$, $m_1 = m$ and $m_2 = 0$ in the algorithm in Sect. 6.3. Let $e_i = z - z_i$ for $0 \leq i \leq m + 1$ be the errors in the kth level iteration. Then from the smoothing step we have for $1 \leq l \leq m$

$$
\begin{aligned}
e_l &= z - z_l \\
&= z - z_{l-1} - \frac{1}{\Lambda_k} A_k(z - z_{l-1}) \\
&= e_{l-1} - \frac{1}{\Lambda_k} A_k e_{l-1} \\
&= (I - \frac{1}{\Lambda_k} A_k) e_{l-1}.
\end{aligned}
$$

(6.5.1)

(6.5.2) Definition. $R_k := I - \frac{1}{\Lambda_k} A_k : V_k \longrightarrow V_k$ *is the* **relaxation operator**.

Note that (cf. exercise 6.x.3)

(6.5.3)
$$
\begin{aligned}
\|R_k v\|_{s,k} &\leq \|v\|_{s,k} \quad \forall v \in V_k, s \in \mathbb{R} \quad \text{and} \\
(R_k v, v)_k &\leq (v, v)_k \quad \forall v \in V_k.
\end{aligned}
$$

It follows immediately that the effect of the smoothing step can be described by

(6.5.4)
$$
e_m = R_k^m e_0.
$$

We will first prove the convergence of a two-grid method. Let $q \in V_{k-1}$ satisfy $A_{k-1} q = \bar{g}$, and define the output of the two-grid method to be

(6.5.5)
$$
\hat{z}_{m+1} = z_m + q.
$$

Note that the q_2 in the multigrid algorithm in Sect. 6.3 is just an approximation of q by using the $k - 1$ level iteration twice. The final error of the two-grid algorithm is related to the initial error by

(6.5.6)
$$
\begin{aligned}
\hat{e}_{m+1} &= z - \hat{z}_{m+1} && \text{(by 6.5.5)} \\
&= z - z_m - q \\
&= e_m - q \\
&= e_m - P_{k-1} e_m && \text{(Lemma 6.4.2)} \\
&= (I - P_{k-1}) R_k^m e_0. && \text{(by 6.5.4)}
\end{aligned}
$$

We already have estimates for the operator $I - P_{k-1}$ (the approximation property). It remains to measure the effect of R_k^m.

(6.5.7) Theorem. (Smoothing Property for the $\mathcal{W}$-cycle Algorithm)

$$\|R_k^m v\|_{2,k} \leq C\, h_k^{-1}\, m^{-1/2}\, \|v\|_{1,k} \qquad \forall\, v \in V_k.$$

Proof. Let $0 < \lambda_1 \leq \lambda_2 \leq \ldots \leq \lambda_{n_k}$ be the eigenvalues of the operator A_k and ψ_i, $1 \leq i \leq n_k$, be the corresponding eigenvectors satisfying the orthonormal relation $(\psi_i, \psi_j)_k = \delta_{ij}$. We can write $v = \sum_{i=1}^{n_k} \nu_i\, \psi_i$. Hence,

$$R_k^m v = \left(I - \frac{1}{\varLambda_k} A_k\right)^m v$$

$$= \sum_{i=1}^{n_k} \left(1 - \frac{\lambda_i}{\varLambda_k}\right)^m \nu_i\, \psi_i.$$

Therefore,

$$\|R_k^m v\|_{2,k}^2 = \sum_{i=1}^{n_k} \left(1 - \frac{\lambda_i}{\varLambda_k}\right)^{2m} \nu_i^2\, \lambda_i^2$$

$$= \varLambda_k \left\{ \sum_{i=1}^{n_k} \left(1 - \frac{\lambda_i}{\varLambda_k}\right)^{2m} \left(\frac{\lambda_i}{\varLambda_k}\right) \lambda_i\, \nu_i^2 \right\}$$

$$\leq \varLambda_k \left\{ \sup_{0 \leq x \leq 1} (1 - x)^{2m}\, x \right\} \sum_{i=1}^{n_k} \lambda_i\, \nu_i^2 \qquad (|\lambda_i| \leq \varLambda_k)$$

$$\leq C\, h_k^{-2} m^{-1} \|v\|_{1,k}^2. \qquad\qquad \text{(by 6.3.2)}$$

Hence,

$$\|R_k^m v\|_{2,k} \leq C\, h_k^{-1}\, m^{-1/2} \|v\|_{1,k}.$$

$\square$

(6.5.8) Theorem. (Convergence of the Two-Grid Algorithm) *There exists a positive constant C independent of k such that*

$$\|\hat{e}_{m+1}\|_E \leq Cm^{-1/2} \|e_0\|_E.$$

Therefore, if m is large enough, the two-grid method is a contraction with contraction number independent of k.

Proof.

$$\|\hat{e}_{m+1}\|_E = \|(I - P_{k-1})R_k^m e_0\|_E \qquad \text{(by 6.5.6)}$$

$$\leq Ch_k \|R_k^m e_0\|_{2,k} \qquad\qquad \text{(by 6.4.4)}$$

$$\leq C\, h_k\, C\, h_k^{-1} m^{-1/2} \|e_0\|_{1,k} \qquad \text{(by 6.5.7)}$$

$$\leq Cm^{-1/2} \|e_0\|_E.$$

$\square$

We now use a perturbation argument to prove the convergence of the kth level iteration.

(6.5.9) Theorem. (Convergence of the kth Level Iteration) *For any $0 < \gamma < 1$, m can be chosen large enough such that*

$$(6.5.10) \qquad \|z - MG(k, z_0, g)\|_E \leq \gamma \|z - z_0\|_E, \quad \text{for } k = 1, 2, \dots.$$

Proof. Let C^* be the constant in Theorem 6.5.8 and let m be an integer greater than $(C^*/(\gamma - \gamma^2))^2$. Then (6.5.10) holds for this choice of m. The proof is by induction. For $k = 1$, (6.5.10) holds trivially since the left-hand side is 0. For $k > 1$, we have

$$
\begin{aligned}
z - MG(k, z_0, g) &= z - z_{m+1} \\
&= z - z_m - q_2 \\
&= \hat{e}_{m+1} + q - q_2.
\end{aligned}
$$

Therefore,

$$
\begin{aligned}
\|z - MG(k, z_0, g)\|_E \\
\leq \|\hat{e}_{m+1}\|_E + \|q - q_2\|_E \\
\leq C^* m^{-1/2} \|e_0\|_E + \gamma^2 \|q\|_E & \qquad \text{(ind. hyp. \& Thm. 6.5.8)} \\
\leq \left(C^* m^{-1/2} + \gamma^2 \right) \|e_0\|_E & \qquad \text{(by 6.4.2 \& 6.5.3)} \\
\leq \gamma \|e_0\|_E .
\end{aligned}
$$

$\square$

6.6 $\mathcal{V}$-cycle Convergence for the kth Level Iteration

In this section we will consider the symmetric $\mathcal{V}$-cycle algorithm. In other words, we consider the algorithm in Sect. 6.3 with $p = 1$ and $m_1 = m_2 = m$.

Since our goal is to prove convergence of this algorithm for any m, we can no longer just perform a two-grid analysis followed by a perturbation argument. We need to analyze directly the relation between the initial error $z - z_0$ and the final error $z - MG(k, z_0, g)$.

(6.6.1) Definition. *The error operator $E_k : V_k \longrightarrow V_k$ is defined recursively by*

$$
\begin{aligned}
E_1 &= 0 \\
E_k &= R_k^m \left[I - (I - E_{k-1}) P_{k-1} \right] R_k^m \quad k > 1.
\end{aligned}
$$

The following lemma shows that E_k relates the final error of the kth level iteration to the initial error.

(6.6.2) Lemma. *If $z, g \in V_k$ satisfy $A_k z = g$, then*

$$(6.6.3) \qquad E_k(z - z_0) = z - MG(k, z_0, g), \quad k \geq 1.$$

Proof. The proof is by induction. The case $k = 1$ is trivial since $MG(1, z_0, g) = A_1^{-1}g = z$. Assume (6.6.3) holds for $k-1$. Let $q \in V_{k-1}$ be the exact solution on the coarser grid of the residual equation, i.e., $A_{k-1}q = \overline{g}$. Since $q_1 = MG(k-1, 0, \overline{g})$, the induction hypothesis implies that $q - q_1 = E_{k-1}(q-0)$. Hence, by Lemma 6.4.2

$$(6.6.4) \qquad \begin{aligned} q_1 &= (I - E_{k-1})q \\ &= (I - E_{k-1})P_{k-1}(z - z_m). \end{aligned}$$

Therefore,

$$\begin{aligned} z - MG(k, z_0, g) &= z - z_{2m+1} \\ &= R_k^m (z - z_{m+1}) \\ &= R_k^m (z - (z_m + q_1)) \\ &= R_k^m (z - z_m - (I - E_{k-1})P_{k-1}(z - z_m)) \\ &= R_k^m (I - (I - E_{k-1})P_{k-1}) (z - z_m) \\ &= R_k^m (I - (I - E_{k-1})P_{k-1}) R_k^m (z - z_0) \\ &= E_k(z - z_0). \end{aligned}$$

$\square$

The operators E_k are clearly linear. It is less obvious that the E_k's are positive semi-definite with respect to $a(\cdot, \cdot)$. In order to prove this fact, which simplifies the convergence proof, we first need a lemma, whose proof is left as an exercise (cf. exercise 6.x.5).

(6.6.5) Lemma. *For all $v, w \in V_k$,*

$$a(R_k v, w) = a(v, R_k w).$$

(6.6.6) Proposition. *E_k is symmetric positive semi-definite with respect to $a(\cdot, \cdot)$ for $k \geq 1$.*

Proof. The proof is by induction. It is obvious for $k = 1$. Assume the proposition is true for $k - 1$. We first prove that E_k is symmetric with respect to $a(\cdot, \cdot)$:

$$a(E_k v, w) = a(R_k^m(I - P_{k-1} + E_{k-1}P_{k-1})R_k^m v, w) \quad \text{(by 6.6.1)}$$
$$= a((I - P_{k-1} + E_{k-1}P_{k-1})R_k^m v, R_k^m w)$$
$$\text{(Lemma 6.6.5 } m \text{ times)}$$
$$= a((I - P_{k-1})R_k^m v, R_k^m w)$$
$$\quad + a(E_{k-1}P_{k-1}R_k^m v, R_k^m w)$$
$$= a(R_k^m v, (I - P_{k-1})R_k^m w)$$
$$\quad + a(E_{k-1}P_{k-1}R_k^m v, R_k^m w)$$
$$= a(R_k^m v, (I - P_{k-1})R_k^m w)$$
$$\quad + a(E_{k-1}P_{k-1}R_k^m v, P_{k-1}R_k^m w) \quad \text{(by 6.4.1)}$$
$$= a(R_k^m v, (I - P_{k-1})R_k^m w)$$
$$\quad + a(P_{k-1}R_k^m v, E_{k-1}P_{k-1}R_k^m w)$$
$$\text{(induction hypothesis)}$$
$$= a(R_k^m v, (I - P_{k-1})R_k^m w)$$
$$\quad + a(R_k^m v, E_{k-1}P_{k-1}R_k^m w) \quad \text{(by 6.4.1)}$$
$$= a(R_k^m v, (I - P_{k-1} + E_{k-1}P_{k-1})R_k^m w)$$
$$= a(v, R_k^m(I - P_{k-1} + E_{k-1}P_{k-1})R_k^m w)$$
$$\text{(Lemma 6.6.5 again)}$$
$$= a(v, E_k w).$$

We next show that E_k is positive, semi-definite with respect to $a(\cdot, \cdot)$:

$$a(E_k v, v) = a(R_k^m(I - (I - E_{k-1})P_{k-1})R_k^m v, v)$$
$$= a((I - (I - E_{k-1})P_{k-1})R_k^m v, R_k^m v)$$
$$= a(R_k^m v, R_k^m v) - a((I - E_{k-1})P_{k-1}R_k^m v, R_k^m v)$$
$$= a(R_k^m v, R_k^m v) - a((I - E_{k-1})P_{k-1}R_k^m v, P_{k-1}R_k^m v) \quad \text{(by 6.4.1)}$$
$$= a(R_k^m v, R_k^m v) - a(P_{k-1}R_k^m v, P_{k-1}R_k^m v)$$
$$\quad + a(E_{k-1}P_{k-1}R_k^m v, P_{k-1}R_k^m v)$$
$$= a((I - P_{k-1})R_k^m v, (I - P_{k-1})R_k^m v)$$
$$\quad + a(E_{k-1}P_{k-1}R_k^m v, P_{k-1}R_k^m v)$$
$$\geq 0. \qquad\qquad\qquad\qquad \text{(induction hypothesis)}$$

$\square$

(6.6.7) Lemma.   **(Smoothing Property for the $\mathcal{V}$-Cycle)**

$$a((I - R_k)R_k^{2m} v, v) \leq \frac{1}{2m} a((I - R_k^{2m})v, v).$$

Proof. We first observe that for $0 \leq j \leq l$,

$$
\begin{aligned}
a\big((I - R_k)R_k^l v, v\big) &= \big(A_k(I - R_k)R_k^l v, v\big)_k \\
&= \big(A_k \Lambda_k^{-1} A_k R_k^l v, v\big)_k \\
&= \frac{1}{\Lambda_k}\big(R_k^l A_k v, A_k v\big)_k \\
&\leq \frac{1}{\Lambda_k}(R_k^j A_k v, A_k v)_k \qquad \text{(by 6.5.3)} \\
&= a\big((I - R_k)R_k^j v, v\big).
\end{aligned}
$$

(6.6.8)

Hence,

$$
\begin{aligned}
(2m)\, a\big((I - R_k)&R_k^{2m} v, v\big) \\
&= a\big((I - R_k)R_k^{2m} v, v\big) + a\big((I - R_k)R_k^{2m} v, v\big) + \dots \\
&\quad + a\big((I - R_k)R_k^{2m} v, v\big) \qquad\qquad \text{(2m terms)} \\
&\leq a\big((I - R_k)v, v\big) + a\big((I - R_k)R_k v, v\big) + \dots \\
&\quad + a\big((I - R_k)R_k^{2m-1} v, v\big) \qquad\qquad \text{(by 6.6.8)} \\
&\leq a\big((I - R_k^{2m})v, v\big). \qquad\qquad \text{(telescopic cancellation)}
\end{aligned}
$$

$\square$

(6.6.9) Proposition. *Let m be the number of smoothing steps. Then*

(6.6.10)
$$
a(E_k v, v) \leq \frac{C^*}{m + C^*} a(v, v) \quad \forall\, v \in V_k,
$$

where C^* is a positive constant independent of k.

Proof. We first estimate $a\big((I - P_{k-1})R_k^m v, (I - P_{k-1})R_k^m v\big)$.

(6.6.11)
$$
\begin{aligned}
a\big((I - P_{k-1})&R_k^m v, (I - P_{k-1})R_k^m v\big) \\
&= \|(I - P_{k-1})R_k^m v\|_E^2 \\
&\leq C\, h_k^2 \|R_k^m v\|_{2,k}^2 \qquad\qquad \text{(by 6.4.4)} \\
&= C\, h_k^2 (A_k^2 R_k^m v, R_k^m v)_k \qquad\qquad \text{(by 6.2.6)} \\
&= C\, h_k^2 a(A_k R_k^m v, R_k^m v) \qquad\qquad \text{(by 6.2.3)} \\
&= C\, h_k^2 \Lambda_k\, a\big((I - R_k)R_k^m v, R_k^m v\big) \qquad \text{(by 6.5.2)} \\
&\leq C\, a\big((I - R_k)R_k^m v, R_k^m v\big) \qquad\qquad \text{(by 6.2.8)} \\
&= C\, a\big((I - R_k)R_k^{2m} v, v\big) \qquad\qquad \text{(by 6.6.5)} \\
&\leq \frac{C}{m} a\big((I - R_k^{2m})v, v\big). \qquad\qquad \text{(by 6.6.7)}
\end{aligned}
$$

Let C^* be the constant C in the last inequality. Then (6.6.10) holds for $k \geq 1$. The proof is by induction. For $k = 1$, (6.6.10) is trivially true since

$E_1 = 0$. Assume that (6.6.10) holds for $k-1$. Then if we let $\gamma = C^*/(m+C^*)$ (therefore, $\gamma = (1-\gamma)C^*/m$),

$$
\begin{aligned}
a(E_k v, v) &= a(R_k^m v, R_k^m v) - a(P_{k-1} R_k^m v, P_{k-1} R_k^m v) \\
&\quad + a(E_{k-1} P_{k-1} R_k^m v, P_{k-1} R_k^m v) && \text{(by 6.6.1)} \\
&= a((I - P_{k-1}) R_k^m v, (I - P_{k-1}) R_k^m v) \\
&\quad + a(E_{k-1} P_{k-1} R_k^m v, P_{k-1} R_k^m v) && \text{(by 6.4.1)} \\
&\leq a((I - P_{k-1}) R_k^m v, (I - P_{k-1}) R_k^m v) \\
&\quad + \gamma\, a(P_{k-1} R_k^m v, P_{k-1} R_k^m v) && \text{(ind. hyp.)} \\
&= (1-\gamma) a\big((I - P_{k-1}) R_k^m v, (I - P_{k-1}) R_k^m v\big) \\
&\quad + \gamma\, a\big((I - P_{k-1}) R_k^m v, (I - P_{k-1}) R_k^m v\big) \\
&\quad + \gamma\, a(P_{k-1} R_k^m v, P_{k-1} R_k^m v) \\
&= (1-\gamma)\, a\big((I - P_{k-1}) R_k^m v, (I - P_{k-1}) R_k^m v\big) \\
&\quad + \gamma\, a(R_k^m v, R_k^m v) && \text{(by 6.4.1)} \\
&\leq (1-\gamma) C^* m^{-1} a\big((I - R_k^{2m}) v, v\big) \\
&\quad + \gamma a(R_k^m v, R_k^m v) && \text{(by 6.6.11)} \\
&= \gamma a\big((I - R_k^{2m}) v, v\big) + \gamma a(R_k^m v, R_k^m v) \\
&= \gamma a(v, v).
\end{aligned}
$$

$\square$

Proposition 6.6.9 implies immediately the following classical result by Braess and Hackbusch (cf. (Braess & Hackbusch 1983)).

(6.6.12) Theorem.   **(Convergence of the kth Level Iteration for the $\mathcal{V}$-cycle)** *Let m be the number of smoothing steps. Then*

$$
\|z - MG(k, z_0, g)\|_E \leq \frac{C^*}{m + C^*} \|z - z_0\|_E.
$$

Hence, the kth level iteration for any m is a contraction, with the contraction number independent of k.

Proof. From Lemma 6.6.2, it suffices to show

$$
\|E_k v\|_E \leq \frac{C^*}{m + C^*} \|v\|_E \quad \forall v \in V_k.
$$

In view of Proposition 6.6.6 we can apply the spectral theorem to E_k. Let $0 \leq \alpha_1 \leq \ldots \leq \alpha_{n_k}$ be the eigenvalues of E_k and $\eta_1, \eta_2, \ldots, \eta_{n_k}$ be the corresponding eigenvectors such that $a(\eta_i, \eta_j) = \delta_{ij}$.

Write $v = \sum \nu_i \eta_i$. Then $a(E_k v, v) = \sum \alpha_i \nu_i^2$. Therefore, (6.6.10) implies that $0 \leq \alpha_1 \leq \ldots \leq \alpha_{n_k} \leq C^*/(m + C^*)$. Hence,

$$\|E_k v\|_E^2 = a(E_k v, E_k v)$$

$$= \sum_{i=1}^{n_k} \alpha_i^2 \nu_i^2$$

$$\leq \left(\frac{C^*}{m + C^*}\right)^2 \sum \nu_i^2$$

$$= \left(\frac{C^*}{m + C^*}\right)^2 \|v\|_E^2.$$

$\square$

6.7 Full Multigrid Convergence Analysis and Work Estimates

The following theorem shows that the convergence of the full multigrid method is a simple consequence of the convergence of the kth level iteration.

(6.7.1) Theorem.   (Full Multigrid Convergence)   *If the kth level iteration is a contraction with a contraction number γ independent of k and if r is large enough, then there exists a constant $C > 0$ such that*

$$\|u_k - \hat{u}_k\|_E \leq C\, h_k\, |u|_{H^2(\Omega)}.$$

Proof. Define $\hat{e}_k := u_k - \hat{u}_k$. In particular, $\hat{e}_1 = 0$. We have

$$\|\hat{e}_k\|_E \leq \gamma^r \|u_k - \hat{u}_{k-1}\|_E$$
$$\leq \gamma^r \left\{\|u_k - u\|_E + \|u - u_{k-1}\|_E + \|u_{k-1} - \hat{u}_{k-1}\|_E\right\}$$
$$\leq \gamma^r \left[\tilde{C} h_k |u|_{H^2(\Omega)} + \|\hat{e}_{k-1}\|_E\right]. \qquad \text{(by 6.1.6 \& 6.1.5)}$$

By iterating the above inequality we have

$$\|\hat{e}_k\|_E \leq \tilde{C}\, h_k\, \gamma^r\, |u|_{H^2(\Omega)} + \tilde{C}\, h_{k-1}\, \gamma^{2r}\, |u|_{H^2(\Omega)} + \ldots + \tilde{C}\, h_1\, \gamma^{kr}\, |u|_{H^2(\Omega)}$$

$$\leq h_k\, |u|_{H^2(\Omega)}\, \frac{\tilde{C}\, \gamma^r}{1 - 2\, \gamma^r}$$

if $2\gamma^r < 1$. For such choice of r,

$$\|\hat{e}_k\|_E \leq C\, h_k\, |u|_{H^2(\Omega)}.$$

$\square$

We now turn our attention to the work estimates. First we obtain an asymptotic estimate for n_k ($=$ dim V_k). Let e_k^I denote the number of internal edges of $\mathcal{T}_k$, ν_k^I denote the number of internal vertices of $\mathcal{T}_k$, and t_k denote the number of triangles in $\mathcal{T}_k$. Euler's formula implies that

$$\nu_k^I - e_k^I + t_k = 1.$$

Therefore,
$$n_k = \nu_k^I = 1 + e_k^I - t_k.$$

The difference equations for e_k^I and t_k are

(6.7.2)
$$e_{k+1}^I = 2e_k^I + 3t_k$$
$$t_{k+1} = 4t_k.$$

By solving (6.7.2) (cf. exercise 6.x.10), we obtain

(6.7.3)
$$n_k \sim \frac{t_1}{8} 4^k.$$

(6.7.4) Proposition. *The work involved in the full multigrid algorithm is* $\mathcal{O}(n_k)$.

Proof. Let W_k denote the work in the kth level scheme. Together, the smoothing and correction steps yield

(6.7.5)
$$W_k \leq C\,\tilde{m}\,n_k + p\,W_{k-1},$$

where $\tilde{m} = m_1 + m_2$. Iterating (6.7.5) and using the fact that $p < 4$ we obtain

$$
\begin{aligned}
W_k &\leq C\tilde{m}n_k + p(C\tilde{m}n_{k-1}) + p^2(C\tilde{m}n_{k-2}) + \ldots + p^{k-1}(C\tilde{m}n_1) \\
&\leq C\tilde{m}4^k + pC\tilde{m}4^{k-1} + p^2C\tilde{m}4^{k-2} + \ldots + p^{k-1}C\tilde{m}4 \\
&\leq \frac{C\tilde{m}4^k}{1 - p/4} \leq C\,4^k \\
&\leq C\,n_k. \qquad \text{(by 6.7.3)}
\end{aligned}
$$

Hence,
$$W_k \leq C\,n_k.$$

Let $\hat{W}_k$ denote the work involved in obtaining $\hat{u}_k$ in the full multigrid algorithm. Then

(6.7.6)
$$\hat{W}_k \leq \hat{W}_{k-1} + r\,W_k.$$

Iterating (6.7.6), we have

$$
\begin{aligned}
\hat{W}_k &\leq rCn_k + rCn_{k-1} + \ldots + rCn_1 \\
&\leq rC(4^k + 4^{k-1} + \ldots + 4) \qquad \text{(by 6.7.3)} \\
&\leq \frac{rC4^k}{1 - 1/4} \\
&\leq C4^k \\
&\leq Cn_k. \qquad \text{(by 6.7.3)}
\end{aligned}
$$

$\square$

6.x Exercises

6.x.1 Prove Lemma 6.2.10. (Hint: see the proof of Theorem 6.5.7.)

6.x.2 Sketch the schedule for grids in the order they are visited for the $\mathcal{W}$-cycle and $\mathcal{V}$-cycle methods. Explain the reasons for the nomenclature.

6.x.3 Establish the inequalities in (6.5.3). (Hint: use the fact that A_k is self-adjoint in the inner-product (6.2.2) and that $(A_k v, v)_k \leq \Lambda_k(v, v)_k$ for all $v \in V_k$.)

6.x.4 Prove the convergence of the kth level iteration in the L^2-norm for the $\mathcal{W}$-cycle method when m is sufficiently large. (Hint: Use the smoothing property, approximation property and a duality argument.)

6.x.5 Prove Lemma 6.6.5. (Hint: show that A_k is self-adjoint in the inner-product $a(\cdot, \cdot)$.)

6.x.6 Show that the error operator E_k ($k > 1$), defined by 6.6.1, can also be written as

$$
\begin{aligned}
E_k = {} & \left[(I - P_k) + R_k^m P_k \right] \left[(I - P_{k-1}) + R_{k-1}^m P_{k-1} \right] \cdots \\
& \left[(I - P_2) + R_2^m P_2 \right] \left[I - P_1 \right] \left[(I - P_2) + R_2^m P_2 \right] \cdots \\
& \left[(I - P_{k-1}) + R_{k-1}^m P_{k-1} \right] \left[(I - P_k) + R_k^m P_k \right].
\end{aligned}
$$

6.x.7 Show that the error operator $\tilde{E}_k$ ($k > 1$) for the one-sided $\mathcal{V}$-cycle method with m presmoothing steps can be written as

$$
\begin{aligned}
\tilde{E}_k = {} & (I - P_1) \left[(I - P_2) + R_2^m P_2 \right] \cdots \\
& \left[(I - P_{k-1}) + R_{k-1}^m P_{k-1} \right] \left[(I - P_k) + R_k^m P_k \right],
\end{aligned}
$$

while the error operator for the one-sided $\mathcal{V}$-cycle method with m postsmoothing steps is $\tilde{E}_k^*$, the adjoint of $\tilde{E}_k$ in $a(\cdot, \cdot)$. Deduce that both methods converge for any m. (Hint: Use exercise 6.x.6 and Theorem 6.6.12).

6.x.8 Show that the error operator $\tilde{E}_k^w$ for the one-sided $\mathcal{W}$-cycle method with m presmoothing steps can be written as

$$
\tilde{E}_k^w = F_k \tilde{E}_k,
$$

where $\tilde{E}_k$ is defined as in 6.x.7 and $\|F_k\|_E \leq 1$. Deduce the convergence of the one-sided $\mathcal{W}$-cycle method for any m. (Hint: Use exercise 6.x.7.)

6.x.9 Show that the error operator E_k^w for the symmetric $\mathcal{W}$-cycle method with m presmoothing steps and m postsmoothing steps can be written as
$$E_k^W = \tilde{E}_k^* D_k \tilde{E}_k,$$
where $\tilde{E}_k^*$ and $\tilde{E}_k$ are defined in 6.x.7 and $\|D_k\|_E \leq 1$. Deduce the convergence of the symmetric $\mathcal{W}$-cycle method for any m. (Hint: Use 6.x.7.)

6.x.10 Solve (6.7.2) and establish (6.7.3).

6.x.11 Prove that the edge-midpoint quadrature rule used in the proof of Lemma 6.2.7 is exact for quadratic polynomials. Does an analogous result hold in three dimensions, with the quadrature points at the barycenters of the four faces?

6.x.12 Prove Lemma 6.2.7 in d dimensions. (Hint: use homogeneity and the fact that the Gram matrix for the Lagrange basis functions is positive definite.)

6.x.13 Establish the following additive expression for the error operator E_k ($k \geq 2$) of the kth level symmetric V-cycle algorithm.

$$E_k = \sum_{j=2}^{k} R_k^m I_{k-1}^k \cdots R_{j+1}^m I_j^{j+1} [R_j^m (Id_j - I_{j-1}^j P_j^{j-1}) R_j^m]$$
$$\times P_{j+1}^j R_{j+1}^m \cdots P_k^{k-1} R_k^m,$$

where $P_j^{j-1} : V_j \longrightarrow V_{j-1}$ is the restriction of the Ritz projection P_{j-1} to V_j and Id_j is the identity operator on V_j. Use this expression to give another proof of Proposition 6.6.6.

6.x.14 Show that the kth level multigrid iteration is a *linear* and *consistent* scheme, i.e., $MG(k, z_0, g) = B_k g + C_k z_0$ where $B_k, C_k : V_k \longrightarrow V_k$ are linear operators and $MG(k, z_0, Az_0) = z_0$.

6.x.15 Find a recursive definition of the operator B_k in exercise 6.x.14. (Hint: Use the recursive definition of $MG(k, z_0, g)$ in Section 6.3 and the fact that $B_k g = MG(k, 0, g)$.)

6.x.16 Show that the operators B_k and C_k in exercise 6.x.14 satisfy the relation $B_k A_k + C_k = Id_k$ (the identity operator on V_k) and deduce that
$$z - MG(k, z_0, g) = (Id_k - B_k A_k)(z - z_0),$$
where $A_k z = g$. (Hint: Use the consistency of $MG(k, z_0, g)$.)

Chapter 7

Additive Schwarz Preconditioners

The symmetric positive definite system arising from a finite element discretization of an elliptic boundary value problem can be solved efficiently using the preconditioned conjugate gradient method (cf. (Saad 1996)). In this chapter we discuss the class of additive Schwarz preconditioners, which has built-in parallelism and is particularly suitable for implementation on parallel computers. Many well-known preconditioners are included in this class, for example the hierarchical basis and BPX multilevel preconditioners, the two-level additive Schwarz overlapping domain decomposition preconditioner, and the BPS and Neumann-Neumann nonoverlapping domain decomposition preconditioners.

7.1 Abstract Additive Schwarz Framework

Let V be a finite dimensional vector space and V' be the dual space of V, i.e., the vector space of linear functionals on V. Let $\langle \cdot, \cdot \rangle$ be the canonical bilinear form on $V' \times V$ defined by

$$(7.1.1) \qquad \langle \alpha, v \rangle = \alpha(v) \qquad \forall \, \alpha \in V', v \in V .$$

(7.1.2) Definition. *A linear operator (transformation) $A : V \longrightarrow V'$ is symmetric positive definite* (SPD) *if*

$$(7.1.3) \qquad \langle Av_1, v_2 \rangle = \langle Av_2, v_1 \rangle \qquad \forall \, v_1, v_2 \in V,$$

$$(7.1.4) \qquad \langle Av, v \rangle > 0 \qquad \forall \, v \in V, v \neq 0 .$$

Similarly, a linear operator $B : V' \longrightarrow V$ is SPD if

$$(7.1.5) \qquad \langle \alpha_1, B\alpha_2 \rangle = \langle \alpha_2, B\alpha_1 \rangle \qquad \forall \, \alpha_1, \alpha_2 \in V',$$

$$(7.1.6) \qquad \langle \alpha, B\alpha \rangle > 0 \qquad \forall \, \alpha \in V', v \neq 0 .$$

Note that if $A : V \longrightarrow V'$ is SPD, then A is invertible and A^{-1} is a SPD operator from V' to V. The converse is also true.

Let $A : V \longrightarrow V'$ be SPD. In terms of certain auxiliary vector spaces V_j ($0 \leq j \leq J$), the SPD operators $B_j : V_j \longrightarrow V_j'$, and the linear operators $I_j : V_j \longrightarrow V$ that connect the auxiliary spaces to V, we can construct an abstract additive Schwarz preconditioner $B : V' \longrightarrow V$ for A:

$$(7.1.7) \qquad B = \sum_{j=0}^{J} I_j B_j^{-1} I_j^t,$$

where the transpose operator $I_j^t : V' \longrightarrow V_j'$ is defined by

$$(7.1.8) \qquad \langle I_j^t \alpha, v \rangle = \langle \alpha, I_j v \rangle \qquad \forall \alpha \in V', v \in V_j.$$

We assume that

$$(7.1.9) \qquad V = \sum_{j=0}^{J} I_j V_j.$$

(**7.1.10**) *Remark.* We can represent the operators B, B_j, I_j and I_j^t by matrices with respect to chosen bases for V and V_j, and the corresponding dual bases for V' and V_j'. Note that the matrix for B_j^{-1} is the inverse of the matrix for B_j, and the matrix for I_j^t is the transpose of the matrix for I_j. Therefore the matrix form of the additive Schwarz preconditioner is identical with (7.1.7), i.e., $\underset{\approx}{B} = \sum_{j=0}^{J} \underset{\approx}{I}_j \underset{\approx}{B}_j^{-1} \underset{\approx}{I}_j^t$.

(**7.1.11**) **Lemma.** $B : V' \longrightarrow V$ *is SPD.*

Proof. It follows easily from (7.1.7), (7.1.8) and the symmetric positive definiteness of B_j^{-1} that B is symmetric (cf. exercise 7.x.4) and

$$(7.1.12) \qquad \langle \alpha, B\alpha \rangle = \sum_{j=0}^{J} \langle I_j^t \alpha, B_j^{-1} I_j^t \alpha \rangle \geq 0 \qquad \forall \alpha \in V'.$$

Let $\alpha \in V'$ satisfy $\langle \alpha, B\alpha \rangle = 0$. It follows from (7.1.12) and the positive definiteness of the B_j's that

$$(7.1.13) \qquad I_j^t \alpha = 0 \qquad 0 \leq j \leq J.$$

Let $v \in V$ be arbitrary. In view of (7.1.9) we can write $v = \sum_{j=0}^{J} I_j v_j$, where $v_j \in V_j$ for $0 \leq j \leq J$, and then (7.1.13) implies

$$\langle \alpha, v \rangle = \langle \alpha, \sum_{j=0}^{J} I_j v_j \rangle = \sum_{j=0}^{J} \langle I_j^t \alpha, v_j \rangle = 0.$$

We conclude that $\alpha = 0$ and hence B is positive definite. $\qquad \square$

It follows that $B^{-1} : V \longrightarrow V'$ is SPD. The next lemma gives a characterization of the inner product $\langle B^{-1} \cdot, \cdot \rangle$.

(7.1.14) Lemma. *The following relation holds for $v \in V$:*

$$(7.1.15) \qquad \langle B^{-1}v, v \rangle = \min_{\substack{v = \sum_{j=0}^{J} I_j v_j \\ v_j \in V_j}} \sum_{j=0}^{J} \langle B_j v_j, v_j \rangle.$$

Proof. Since $B_j^{-1} : V_j' \longrightarrow V_j$ is SPD, the bilinear form $\langle \cdot, B_j^{-1} \cdot \rangle$ is an inner product on V_j', and the following Cauchy-Schwarz inequality holds:

$$(7.1.16) \qquad \langle \alpha_1, B_j^{-1} \alpha_2 \rangle \le \sqrt{\langle \alpha_1, B_j^{-1} \alpha_1 \rangle} \sqrt{\langle \alpha_2, B_j^{-1} \alpha_2 \rangle} .$$

Suppose $v = \sum_{j=0}^{J} I_j v_j$, where $v_j \in V_j$ for $0 \le j \le J$. We have

$$\langle B^{-1}v, v \rangle = \langle B^{-1}v, \sum_{j=0}^{J} I_j v_j \rangle$$

$$= \sum_{j=0}^{J} \langle I_j^t B^{-1}v, B_j^{-1} B_j v_j \rangle \qquad \text{(by 7.1.8)}$$

$$\le \sum_{j=0}^{J} \sqrt{\langle I_j^t B^{-1}v, B_j^{-1} I_j^t B^{-1}v \rangle} \sqrt{\langle B_j v_j, B_j^{-1} B_j v_j \rangle} \qquad \text{(by 7.1.16)}$$

$$\le \left(\sum_{j=0}^{J} \langle I_j^t B^{-1}v, B_j^{-1} I_j^t B^{-1}v \rangle \right)^{1/2} \left(\sum_{j=0}^{J} \langle B_j v_j, v_j \rangle \right)^{1/2}$$

$$= \left\langle B^{-1}v, \left(\sum_{j=0}^{J} I_j B_j^{-1} I_j^t \right) B^{-1}v \right\rangle^{1/2} \left(\sum_{j=0}^{J} \langle B_j v_j, v_j \rangle \right)^{1/2} \qquad \text{(by 7.1.8)}$$

$$= \langle B^{-1}v, v \rangle^{1/2} \left(\sum_{j=0}^{J} \langle B_j v_j, v_j \rangle \right)^{1/2} , \qquad \text{(by 7.1.7)}$$

which implies that

$$(7.1.17) \qquad \langle B^{-1}v, v \rangle \le \sum_{j=0}^{J} \langle B_j v_j, v_j \rangle .$$

On the other hand, for the special choice of

$$(7.1.18) \qquad v_j = B_j^{-1} I_j^t B^{-1}v ,$$

we have $v_j \in V_j$, $v = \sum_{j=0}^{J} I_j v_j$ (cf. exercise 7.x.5), and

$$\sum_{j=0}^{J} \langle B_j v_j, v_j \rangle = \sum_{j=0}^{J} \langle B_j B_j^{-1} I_j^t B^{-1} v, B_j^{-1} I_j^t B^{-1} v \rangle$$

(7.1.19)
$$= \left\langle B^{-1} v, \left(\sum_{j=0}^{J} I_j B_j^{-1} I_j^t \right) B^{-1} v \right\rangle \qquad \text{(by 7.1.8)}$$

$$= \langle B^{-1} v, v \rangle . \qquad \text{(by 7.1.7)}$$

The relation (7.1.15) follows from (7.1.17) and (7.1.19).   □

(7.1.20) Theorem. *The eigenvalues of BA are positive, and we have the following characterizations of the maximum and minimum eigenvalues :*

(7.1.21)
$$\lambda_{\max}(BA) = \max_{\substack{v \in V \\ v \neq 0}} \frac{\langle Av, v \rangle}{\min_{\substack{v = \sum_{j=0}^{J} I_j v_j \\ v_j \in V_j}} \sum_{j=0}^{J} \langle B_j v_j, v_j \rangle},$$

(7.1.22)
$$\lambda_{\min}(BA) = \min_{\substack{v \in V \\ v \neq 0}} \frac{\langle Av, v \rangle}{\min_{\substack{v = \sum_{j=0}^{J} I_j v_j \\ v_j \in V_j}} \sum_{j=0}^{J} \langle B_j v_j, v_j \rangle}.$$

Proof. First we observe that $BA : V \longrightarrow V$ is SPD with respect to the inner product $((\cdot, \cdot)) = \langle B^{-1} \cdot, \cdot \rangle$ (cf. exercise 7.x.6). Therefore the eigenvalues of BA are positive, and the Rayleigh quotient formula (cf. (Golub & Van Loan 1989)) implies that

$$\lambda_{\max}(BA) = \max_{\substack{v \in V \\ v \neq 0}} \frac{((BAv, v))}{((v, v))}$$

$$= \max_{\substack{v \in V \\ v \neq 0}} \frac{\langle Av, v \rangle}{\langle B^{-1} v, v \rangle}$$

$$= \max_{\substack{v \in V \\ v \neq 0}} \frac{\langle Av, v \rangle}{\min_{\substack{v = \sum_{j=0}^{J} I_j v_j \\ v_j \in V_j}} \sum_{j=0}^{J} \langle B_j v_j, v_j \rangle}.$$

The proof of (7.1.22) is similar.   □

(7.1.23) Remark. The abstract theory of additive Schwarz preconditioners has several equivalent formulations which are due to many authors (cf. (Dryja & Widlund 1987, 1989, 1990, 1992, 1995), (Nepomnyaschikh 1989), (Bjørstad & Mandel 1991), (Zhang 1991), (Xu 1992), (Dryja, Smith & Widlund 1994), (Griebel & Oswald 1995)).

Let Ω be a bounded polygonal domain in $\mathbb{R}^2$. In subsequent sections we will consider the model problem of finding $u \in \mathring{H}^1(\Omega)$ such that

(7.1.24) $$\int_{\Omega} \nabla u \cdot \nabla v \, dx = F(v) \qquad \forall\, v \in \mathring{H}^1(\Omega),$$

where $F \in [\mathring{H}^1(\Omega)]'$. The abstract theory of additive Schwarz preconditioners will be applied to various preconditioners for the system resulting from
finite element discretizations of (7.1.24).

Two multilevel preconditioners are discussed in Sects. 7.2 and 7.3, and
several domain decomposition preconditioners are discussed in Sects. 7.4
to 7.7. Applications to other domain decomposition preconditioners can
be found in (LeTallec 1994), (Chan & Matthew 1994), (Dryja, Smith &
Widlund 1994), (Smith, Bjørstad & Gropp 1996), (Xu & Zou 1998) and
(Quarteroni & Valli 1999).

In order to avoid the proliferation of constants, we shall use the nota-
tion $A \lesssim B$ ($B \gtrsim A$) to represent the statement $A \le \text{constant} \times B$, where
the constant is always independent of the mesh sizes of the triangulations
and the number of auxiliary subspaces. The statement $A \approx B$ is equivalent
to $A \lesssim B$ and $B \lesssim A$.

7.2 The Hierarchical Basis Preconditioner

Let $\mathcal{T}_1$ be a triangulation of Ω, and the triangulations $\mathcal{T}_2, \dots, \mathcal{T}_J$ be obtained
from $\mathcal{T}_1$ by regular subdivision. Therefore $h_j = \max_{T \in \mathcal{T}_j} \text{diam}\, T$ and $h_J = \max_{T \in \mathcal{T}_J} \text{diam}\, T$ are related by

(7.2.1) $$h_j = 2^{J-j} h_J \qquad \text{for} \quad 1 \le j \le J.$$

Let $V_j \subseteq \mathring{H}^1(\Omega)$ be the $\mathcal{P}_1$ finite element space associated with $\mathcal{T}_j$. The
discrete problem for (7.1.24) is to find $u_J \in V_J$ such that

(7.2.2) $$\int_{\Omega} \nabla u_J \cdot \nabla v \, dx = F(v) \qquad \forall\, v \in V_J.$$

Let the linear operator $A_J : V_J \longrightarrow V_J'$ be defined by

(7.2.3) $$\langle A_J w, v \rangle = \int_{\Omega} \nabla w \cdot \nabla v \, dx \quad \forall\, v, w \in V_J,$$

and $f_J \in V_J'$ be defined by $\langle f_J, v \rangle = F(v)$ for all $v \in V_J$. Then (7.2.2) can
be written as $A_J u_J = f_J$.

Our goal in this section is to study the *hierarchical basis preconditioner*
for A_J (cf. (Yserentant 1986)).

Let the subspace W_j of V_j ($2 \le j \le J$) be defined by

$$W_j = \{v \in V_j : v(p) = 0 \text{ for all the vertices } p \text{ of } \mathcal{T}_{j-1}\}.$$

It is easy to see that (cf. exercise 7.x.9)

(7.2.4) $$V_j = V_{j-1} \oplus W_j \,,$$

and hence

(7.2.5) $$V_J = W_1 \oplus \cdots \oplus W_J \,,$$

provided we take W_1 to be V_1. In fact, given $v \in V_J$, the decomposition (7.2.5) can be written explicitly as

(7.2.6) $$\begin{aligned} v = \Pi_1 v + (\Pi_2 v - \Pi_1 v) + (\Pi_3 v - \Pi_2 v) + \dots \\ + (\Pi_{J-1} v - \Pi_{J-2} v) + (v - \Pi_{J-1} v) \,, \end{aligned}$$

where $\Pi_j : C^0(\overline{\Omega}) \longrightarrow V_j$ is the nodal interpolation operator.

The auxiliary spaces $W_1, W_2, \ldots, W_J$ are connected to V_J by the natural injections $I_j : W_j \longrightarrow V_J$, and we can define SPD operators $B_j : W_j \longrightarrow W_j'$ by

(7.2.7) $$\langle B_j w_1, w_2 \rangle = \sum_{p \in \mathcal{V}_j \setminus \mathcal{V}_{j-1}} w_1(p) w_2(p) \qquad \forall\, w_1, w_2 \in W_j \,,$$

where $\mathcal{V}_j$ is the set of internal vertices of $\mathcal{T}_j$ (with $\mathcal{V}_0 = \emptyset$).

The hierarchical basis preconditioner for A_J is then given by

(7.2.8) $$B_{HB} = \sum_{j=1}^{J} I_j B_j^{-1} I_j^t \,.$$

Since (7.1.9) is implied by (7.2.5), B_{HB} can be analyzed using Theorem 7.1.20.

Let $v = \sum_{j=1}^{J} w_j$, where $w_j = \Pi_j v - \Pi_{j-1} v$, be the unique decomposition of $v \in V_J$ (cf. (7.2.6)). Here $\Pi_0 v$ is taken to be 0. Observe that, by a direct calculation (cf. exercise 7.x.10) we have

(7.2.9) $$\|w - \Pi_{j-1} w\|_{L^2(\Omega)} \lesssim h_j |w|_{H^1(\Omega)} \qquad \forall\, w \in V_j, 1 \le j \le J \,.$$

It follows from (7.2.9) and an inverse estimate (cf. (4.5.12)) that

(7.2.10) $$\begin{aligned} h_j^{-1} \|w_j\|_{L^2(\Omega)} = h_j^{-1} \|w_j - \Pi_{j-1} w_j\|_{L^2(\Omega)} \\ \lesssim |w_j|_{H^1(\Omega)} \lesssim h_j^{-1} \|w_j\|_{L^2(\Omega)} \,. \end{aligned}$$

Using (7.2.7), Lemma 6.2.7 and (7.2.10) we find the following relations for the denominator that appears in (7.1.21) and (7.1.22).

(7.2.11) $$\begin{aligned} \sum_{j=1}^{J} \langle B_j w_j, w_j \rangle = \sum_{j=1}^{J} h_j^{-2} \Big(h_j^2 \sum_{p \in \mathcal{V}_j \setminus \mathcal{V}_{j-1}} [w_j(p)]^2 \Big) \\ \approx \sum_{j=1}^{J} h_j^{-2} \|w_j\|_{L^2(\Omega)}^2 \approx \sum_{j=1}^{J} |w_j|_{H^1(\Omega)}^2 \,. \end{aligned}$$

The following lemma enables us to get a lower bound for $\lambda_{\min}(B_{HB} A_J)$.

(7.2.12) Lemma. *The following estimate holds for $1 \leq j \leq J$.*

$$(7.2.13) \qquad \|\Pi_j v - \Pi_{j-1} v\|_{L^2(\Omega)} \lesssim h_j \big(1 + \sqrt{J-j}\big)|v|_{H^1(\Omega)} \quad \forall v \in V_J.$$

Proof. Let $v_T = |T|^{-1} \int_T v \, dx$ be the average of v over a triangle $T \in \mathcal{T}_j$. We have

$$
\begin{aligned}
\|\Pi_j v - \Pi_{j-1} v\|_{L^2(\Omega)}^2 &= \|\Pi_j v - \Pi_{j-1} \Pi_j v\|_{L^2(\Omega)}^2 \\
&\lesssim h_j^2 \sum_{T \in \mathcal{T}_j} |\Pi_j v - v_T|_{H^1(T)}^2 && \text{(by 7.2.9)} \\
&\lesssim h_j^2 \sum_{T \in \mathcal{T}_j} \|\Pi_j v - v_T\|_{L^\infty(T)}^2 && \text{(by 4.5.3)} \\
&\lesssim h_j^2 \sum_{T \in \mathcal{T}_j} \|v - v_T\|_{L^\infty(T)}^2 \\
&\lesssim h_j^2 \sum_{T \in \mathcal{T}_j} \big(1 + |\ln(h_j/h_J)|\big)|v|_{H^1(T)}^2. && \text{(by 7.x.11)}
\end{aligned}
$$

The lemma then follows from (7.2.1). □

Combining (7.2.11) and (7.2.13) we find

$$\sum_{j=1}^{J} \langle B_j w_j, w_j \rangle \lesssim \Big(\sum_{j=1}^{J} j\Big) |v|_{H^1(\Omega)}^2 \lesssim J^2 |v|_{H^1(\Omega)}^2.$$

In view of (7.2.1) and (7.2.3), we have

$$\sum_{j=1}^{J} \langle B_j w_j, w_j \rangle \lesssim \big(1 + |\ln h_J|^2\big) \langle A_J v, v \rangle,$$

which implies by (7.1.22) that

$$(7.2.14) \qquad \lambda_{\min}(B_{HB} A_J) \gtrsim \big(1 + |\ln h_J|^2\big)^{-1}.$$

We shall need the following lemmas for the estimate of $\lambda_{\max}(B_{HB} A_J)$. The first, a discrete version of Young's inequality on convolutions, is left as an exercise (cf. exercise 7.x.12).

(7.2.15) Lemma. *Let a_j and b_j be nonnegative for $-\infty < j < \infty$. Then we have*

$$\sum_{j=-\infty}^{\infty} \Big(\sum_{k=-\infty}^{\infty} a_{j-k} b_k\Big)^2 \leq \Big(\sum_{k=-\infty}^{\infty} b_k\Big)^2 \Big(\sum_{j=-\infty}^{\infty} a_j^2\Big).$$

(7.2.16) Lemma. *The following estimate holds for* $v_j \in V_j$, $v_k \in V_k$ *and* $1 \le j \le k \le J$.

$$(7.2.17) \qquad \int_\Omega \nabla v_j \cdot \nabla v_k \, dx \lesssim 2^{(j-k)/2} |v_j|_{H^1(\Omega)} \left(h_k^{-1} \|v_k\|_{L^2(\Omega)} \right).$$

Proof. On each triangle $T \in \mathcal{T}_j$ we have

$$\int_T \nabla v_j \cdot \nabla v_k \, dx = \int_{\partial T} \frac{\partial v_j}{\partial n} v_k \, ds$$

$$\lesssim h_j^{-1} |v_j|_{H^1(T)} \int_{\partial T} |v_k| \, ds \qquad\qquad (v_j \in V_j)$$

$$\lesssim \left(h_j^{-1} |v_j|_{H^1(T)} \right) \left(h_k \sum_{p \in \mathcal{V}_k \cap \partial T} |v_k(p)| \right) \qquad\qquad (v_k \in V_k)$$

$$\lesssim \left(h_j^{-1} |v_j|_{H^1(T)} \right) \left[h_k \left(\frac{h_j}{h_k} \right)^{1/2} \left(\sum_{p \in \mathcal{V}_k \cap \partial T} |v_k(p)|^2 \right)^{1/2} \right]$$

$$\text{(Cauchy-Schwarz inequality)}$$

$$\lesssim \left(\frac{h_k}{h_j} \right)^{1/2} |v_j|_{H^1(T)} \left(h_k^{-1} \|v_k\|_{L^2(T)} \right). \qquad\qquad \text{(by 6.2.7)}$$

The estimate (7.2.17) follows by summing up this last estimate over all the triangles $T \in \mathcal{T}_j$, and applying the Cauchy-Schwarz inequality and the relation (7.2.1). $\qquad\square$

(7.2.18) Lemma. (Strengthened Cauchy-Schwarz Inequality)
The following estimate holds for $w_j \in W_j$, $w_k \in W_k$ *and* $1 \le j \le k \le J$.

$$(7.2.19) \qquad \int_\Omega \nabla w_j \cdot \nabla w_k \, dx \lesssim 2^{(j-k)/2} |w_j|_{H^1(\Omega)} |w_k|_{H^1(\Omega)}.$$

Proof. Since $w_k = w_k - \Pi_{k-1} w_k$, the estimate (7.2.19) follows from (7.2.9) and (7.2.17). $\qquad\square$

Given any $v \in V_J$, we have

$$\langle A_J v, v \rangle = \int_\Omega \nabla \left(\sum_{j=1}^J w_j \right) \cdot \nabla \left(\sum_{k=1}^J w_k \right) dx \qquad\qquad \text{(by 7.2.3)}$$

$$\lesssim \sum_{j,k=1}^J 2^{-|j-k|/2} |w_j|_{H^1(\Omega)} |w_k|_{H^1(\Omega)} \qquad\qquad \text{(by 7.2.19)}$$

$$\lesssim \sum_{j=1}^J \left(\sum_{k=1}^J 2^{-|j-k|/2} |w_k|_{H^1(\Omega)} \right) |w_j|_{H^1(\Omega)}$$

$$\lesssim \left[\sum_{j=1}^{J}\left(\sum_{k=1}^{J} 2^{-|j-k|/2}|w_k|_{H^1(\Omega)}\right)^2\right]^{1/2}\left[\sum_{j=1}^{J}|w_j|^2_{H^1(\Omega)}\right]^{1/2}$$

(Cauchy-Schwarz inequality)

$$\lesssim \sum_{i=1}^{J}|w_j|^2_{H^1(\Omega)} \tag{by 7.2.15}$$

$$\approx \sum_{j=1}^{J}\langle B_j w_j, w_j\rangle, \tag{by 7.2.11}$$

which together with (7.1.21) imply that

$$(7.2.20) \qquad\qquad \lambda_{\max}(B_{HB}A_J) \lesssim 1.$$

Combining (7.2.14) and (7.2.20) we have the following theorem on the hierarchical basis preconditioner.

(7.2.21) Theorem. *There exists a positive constant C, independent of J and h_J, such that*

$$\kappa(B_{HB}A_J) = \frac{\lambda_{\max}(B_{HB}A_J)}{\lambda_{\min}(B_{HB}A_J)} \leq C\left(1+|\ln h_J|^2\right).$$

7.3 The BPX Preconditioner

In this section we discuss the preconditioner introduced by Bramble, Pasciak and Xu (cf. (Bramble, Pasciak & Xu 1990)). We will follow the set-up in Sect. 7.2. Again, we want to precondition $A_J : V_J \longrightarrow V'_J$. But this time we take the auxiliary spaces to be $V_1, V_2, \ldots, V_J$, which are connected to V_J by the natural injections $I_j : V_j \longrightarrow V_J$. Condition (7.1.9) is clearly satisfied.

Let the symmetric positive definite operators $B_j : V_j \longrightarrow V'_j$ be defined by

$$(7.3.1) \qquad\qquad \langle B_j v_1, v_2\rangle = \sum_{p\in\mathcal{V}_j} v_1(p)v_2(p).$$

The BPX preconditioner is then given by

$$(7.3.2) \qquad\qquad B_{BPX} = \sum_{j=1}^{J} I_j B_j^{-1} I_j^t.$$

Let $v \in V_J$ and $v = \sum_{j=1}^{J} v_j$, where $v_j \in V_j$, be any decomposition of v. Observe that by (7.2.17) and an inverse estimate (cf. (4.5.12)) we have

$$(7.3.3) \qquad \int_{\Omega} \nabla v_j \cdot \nabla v_k \, dx \lesssim 2^{-|j-k|/2} \Big(h_j^{-1} \|v_j\|_{L^2(\Omega)} \Big) \Big(h_k^{-1} \|v_k\|_{L^2(\Omega)} \Big),$$

for $1 \le j, k \le J$.

We can then obtain, as in the proof of Lemma 7.2.18, the estimate

$$
\begin{aligned}
\langle A_J v, v \rangle &= \int_{\Omega} \nabla \Big(\sum_{j=1}^{J} v_j \Big) \cdot \nabla \Big(\sum_{k=1}^{J} v_k \Big) dx && \text{(by 7.2.3)} \\[2mm]
&\lesssim \sum_{j,k=1}^{J} 2^{-|j-k|/2} \Big(h_j^{-1} \|v_j\|_{L^2(\Omega)} \Big) \Big(h_k^{-1} \|v_k\|_{L^2(\Omega)} \Big) && \text{(by 7.3.3)} \\[2mm]
&\lesssim \sum_{j=1}^{J} h_j^{-2} \|v_j\|_{L^2(\Omega)}^2 && \text{(by 7.2.15)} \\[2mm]
&\approx \sum_{j=1}^{J} \langle B_j v_j, v_j \rangle, && \text{(by 7.3.1 \& 6.2.7)}
\end{aligned}
$$

which implies, in view of (7.1.21),

$$(7.3.4) \qquad \qquad \lambda_{\max}(B_{BPX} A_J) \lesssim 1.$$

Now we turn to finding an estimate for $\lambda_{\min}(B_{BPX} A_J)$. For simplicity, we assume here that Ω is convex, so that we can exploit the full elliptic regularity of the solution of (7.1.24).

Let $P_k : \mathring{H}^1(\Omega) \longrightarrow V_k$ be the Ritz projection operator, i.e.,

$$(7.3.5) \qquad \int_{\Omega} \nabla(P_k \phi) \cdot \nabla v \, dx = \int_{\Omega} \nabla \phi \cdot \nabla v \, dx \quad \forall \, v \in V_k.$$

Given any $v \in V_J$, we define

$$(7.3.6) \qquad \qquad v_j = P_j v - P_{j-1} v \qquad \text{for} \quad 1 \le j \le J,$$

where we take $P_0 v$ to be 0. Clearly $v = \sum_{j=1}^{J} v_j$ and it is easy to see from (7.3.6) (cf. exercise 7.x.14) that

$$(7.3.7) \qquad \qquad \int_{\Omega} \nabla v_j \cdot \nabla v_k \, dx = 0 \qquad \text{for} \quad j \ne k.$$

Moreover, by a duality argument, we have

$$
\begin{aligned}
h_j^{-2} \|v_j\|_{L^2(\Omega)}^2 &= h_j^{-2} \|P_j v - P_{j-1} v\|_{L^2(\Omega)}^2 && \text{(by 7.3.6)} \\[2mm]
&= h_j^{-2} \|P_j v - P_{j-1} P_j v\|_{L^2(\Omega)}^2 && \text{(by 7.x.15)} \\[2mm]
&\lesssim |v_j|_{H^1(\Omega)}^2 && \text{(by 5.4.8 \& 7.3.6)}
\end{aligned}
$$

for $2 \leq j \leq J$. This estimate also holds trivially for $j = 1$ (cf. exercise 7.x.16). It then follows from (7.3.7) that

$$\sum_{j=1}^{J} \langle B_j v_j, v_j \rangle \approx \sum_{j=1}^{J} h_j^{-2} \|v_j\|_{L^2(\Omega)}^2 \lesssim \sum_{j=1}^{J} |v_j|_{H^1(\Omega)}^2 = |v|_{H^1(\Omega)}^2 \,,$$

which implies, by (7.1.22),

$$(7.3.8) \qquad\qquad\qquad \lambda_{\min}(B_{BPX} A_J) \gtrsim 1 \,.$$

(7.3.9) *Remark.* The estimate (7.3.8) also holds for nonconvex polygonal domains (cf. 7.x.17).

Combining (7.3.4) and (7.3.8) we have the following theorem on the BPX preconditioner.

(7.3.10) **Theorem.** *There exists a positive constant C, independent of J and h_J, such that*

$$\kappa(B_{BPX} A_J) = \frac{\lambda_{\max}(B_{BPX} A_J)}{\lambda_{\min}(B_{BPX} A_J)} \leq C \,.$$

7.4 The Two-level Additive Schwarz Preconditioner

Let $\mathcal{T}_h$ be a quasi-uniform triangulation of Ω with mesh size h and $V_h \subset \mathring{H}^1(\Omega)$ be the $\mathcal{P}_1$ finite element space (cf. (3.2.1)) associated with $\mathcal{T}_h$. We define the SPD operator $A_h : V_h \longrightarrow V_h$ by

$$(7.4.1) \qquad \langle A_h v_1, v_2 \rangle = \int_{\Omega} \nabla v_1 \cdot \nabla v_2 \, dx \qquad \forall\, v_1, v_2 \in V_h \,,$$

and $f_h \in V_h'$ by $\langle f_h, v \rangle = F(v)$ for all $v \in V_h$. Then the discrete problem for (7.1.24) can be written as $A_h u_h = f_h$.

In this section we study an overlapping domain decomposition preconditioner for A_h.

Let $\mathcal{T}_H$ be a triangulation of Ω with mesh size H such that $\mathcal{T}_h$ is a subdivision of $\mathcal{T}_H$, $V_H \subset V_h$ be the $\mathcal{P}_1$ finite element space associated with $\mathcal{T}_H$, and $A_H : V_H \longrightarrow V_H$ be defined by

$$(7.4.2) \qquad \langle A_H v_1, v_2 \rangle = \int_{\Omega} \nabla v_1 \cdot \nabla v_2 \, dx \qquad \forall\, v_1, v_2 \in V_H \,.$$

The space V_H and the operator A_H are parts of the construction of the two level additive Schwarz preconditioner.

The other auxiliary spaces are associated with a collection of open subsets $\Omega_1, \Omega_2, \ldots, \Omega_J$ of Ω, whose boundaries are aligned with $\mathcal{T}_h$. We assume that there exist nonnegative C^∞ functions $\theta_1, \theta_2, \ldots \theta_J$ (a *partition of unity*) defined on $\mathbb{R}^2$ with the following properties:

(7.4.3) $\theta_j = 0$ on $\Omega \backslash \Omega_j$.

(7.4.4) $\displaystyle\sum_{j=1}^{J} \theta_j = 1$ on $\overline{\Omega}$.

(7.4.5)   There exists a positive constant $\delta \leq H$ such that

$$\|\nabla \theta_j\|_{L^\infty(\mathbb{R}^2)} \leq C/\delta ,$$

where C is a constant independent of δ, h, H and J.

Furthermore we assume that

(7.4.6)   each point in Ω belongs to at most N_C subdomains.

(7.4.7) Remark. The conditions (7.4.3) and (7.4.4) imply that the subdomains $\Omega_1, \ldots, \Omega_J$ form an overlapping decomposition of Ω. The constant δ in (7.4.5) measures the the amount of overlap between neighboring subdomains.

We associate with each Ω_j the subspace

(7.4.8) $$V_j = \Big\{ v \in V_h : v = 0 \text{ on } \Omega \backslash \Omega_j \Big\},$$

and define the SPD operator $A_j : V_j \longrightarrow V_j'$ by

(7.4.9) $$\langle A_j v_1, v_2 \rangle = \int_{\Omega_j} \nabla v_1 \cdot \nabla v_2 \, dx \qquad \forall\, v_1, v_2 \in V_j .$$

The two-level additive Schwarz preconditioner (cf. (Dryja & Widlund 1987, 1989) and (Nepomnyaschikh 1989)) is then given by

(7.4.10) $$B_{TL} = I_H A_H^{-1} I_H^t + \sum_{j=1}^{J} I_j A_j^{-1} I_j^t ,$$

where $I_H : V_H \longrightarrow V_h$ and $I_j : V_j \longrightarrow V_h$ are natural injections.

(7.4.11) Remark. $\mathcal{T}_H$, $\mathcal{T}_j$ and Ω_j can be constructed in the following way (cf. Fig. 7.1) to satisfy the assumptions (7.4.3)–(7.4.6).
(i)   Construct a coarse triangulation $\mathcal{T}_H$.
(ii)   Divide Ω into *nonoverlapping* subdomains $\tilde{\Omega}_j$, $1 \leq j \leq J$, which are aligned with $\mathcal{T}_H$.
(iii) Subdivide $\mathcal{T}_H$ to obtain $\mathcal{T}_h$.

(iv) Let $\tilde{\Omega}_{j,\delta}$ be the open set obtained by enlarging $\tilde{\Omega}_j$ by a band of width δ such that $\partial\tilde{\Omega}_{j,\delta} \cap \overline{\Omega}$ is aligned with $\mathcal{T}_h$.

(v) Define $\Omega_j = \tilde{\Omega}_{j,\delta} \cap \Omega$.

Since the $\tilde{\Omega}_{j,\delta}$'s form an open cover of $\overline{\Omega}$, the existence of the partition of unity with properties (7.4.3)–(7.4.5) is standard (cf. (Rudin 1991)). Note that in this case the number N_C is determined by the number of $\tilde{\Omega}_j$'s that can share a vertex of $\mathcal{T}_H$, which is in turn determined by the minimum angle of $\mathcal{T}_H$.

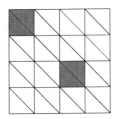 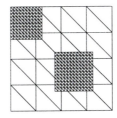

Fig. 7.1. construction of the overlapping domain decomposition

(7.4.12) Lemma. *We have* $V_h = \displaystyle\sum_{j=1}^{J} V_j$.

Proof. Given $v \in V_h$, let $v_j = \Pi_h(\theta_j v)$ for $1 \leq j \leq J$, where $\Pi_h : C(\overline{\Omega}) \longrightarrow V_h$ is the nodal interpolation operator.

Since $\theta_j v = 0$ on $\Omega \backslash \Omega_j$ by (7.4.3) and $\partial \Omega_j$ is aligned with $\mathcal{T}_h$, we also have $v_j = \Pi_h[\theta_j v] = 0$ on $\Omega \backslash \Omega_j$. It follows that $v_j \in V_j$ by (7.4.8).

The v_j's also form a decomposition of v because

$$\sum_{j=1}^{J} v_j = \sum_{j=1}^{J} \Pi_h(\theta_j v) = \Pi_h\left(v \sum_{j=1}^{J} \theta_j\right) = \Pi_h v = v. \quad \square$$

In view of Lemma 7.4.12, condition (7.1.9) is satisfied.

The following lemma will enable us to obtain an upper bound for $\lambda_{\max}(B_{TL}A_h)$.

(7.4.13) Lemma. *Let* $v_H \in V_H$, $v_j \in V_j$ *for* $1 \leq j \leq J$ *and* $v = v_H + \sum_{j=1}^{J} v_j$. *Then the following estimate holds:*

$$\langle A_h v, v \rangle \lesssim \langle A_H v_H, v_H \rangle + \sum_{j=1}^{J} \langle A_j v_j, v_j \rangle.$$

Proof. We have by (7.4.1) and (7.4.2) that

$$\langle A_h v, v \rangle = \int_\Omega \nabla \Big(v_H + \sum_{j=1}^J v_j \Big) \cdot \nabla \Big(v_H + \sum_{k=1}^J v_k \Big) dx$$

(7.4.14)
$$\leq 2 \left[\int_\Omega \nabla v_H \cdot \nabla v_H \, dx + \int_\Omega \nabla \Big(\sum_{j=1}^J v_j \Big) \cdot \nabla \Big(\sum_{k=1}^J v_k \Big) dx \right]$$

$$\lesssim \langle A_H v_H, v_H \rangle + \sum_{j,k=1}^J \int_\Omega \nabla v_j \cdot \nabla v_k \, dx \,.$$

For each $T \in \mathcal{T}_h$, we define $c_{jk}(T) = \begin{cases} 1 & \text{if } T \subseteq \Omega_j \cap \Omega_k, \\ 0 & \text{if } T \nsubseteq \Omega_j \cap \Omega_k. \end{cases}$

Note that

(7.4.15) $$c_{jk}(T) = c_{kj}(T) \,.$$

The second term on the right-hand side of (7.4.14) can now be estimated as follows.

$$\sum_{j,k=1}^J \int_\Omega \nabla v_j \cdot \nabla v_k \, dx = \sum_{j,k=1}^J \sum_{T \in \mathcal{T}_h} c_{jk}(T) \int_T \nabla v_j \cdot \nabla v_k \, dx$$

$$\leq \sum_{j,k=1}^J \sum_{T \in \mathcal{T}_h} \Big(\sqrt{c_{jk}(T)} |v_j|_{H^1(T)} \Big) \Big(\sqrt{c_{jk}(T)} |v_k|_{H^1(T)} \Big)$$

$$\leq \sum_{T \in \mathcal{T}_h} \sum_{j,k=1}^J c_{jk}(T) |v_j|_{H^1(T)}^2 \qquad \text{(Cauchy-Schwarz \& 7.4.15)}$$

$$\leq \sum_{T \in \mathcal{T}_h} \sum_{j=1}^J \left[|v_j|_{H^1(T)}^2 \Big(\sum_{k=1}^J c_{jk}(T) \Big) \right]$$

$$\leq N_C \sum_{j=1}^J \langle A_j v_j, v_j \rangle \,. \qquad \text{(by 7.4.6 \& 7.4.9)}$$

The lemma follows from the last estimate and (7.4.14).   □

Combining (7.4.13) and (7.1.21), we obtain immediately the estimate

(7.4.16) $$\lambda_{\max}(B_{TL} A_h) \lesssim 1 \,.$$

The next lemma will yield a lower bound for $\lambda_{\min}(B_{TL} A_h)$ when combined with (7.1.22).

(7.4.17) Lemma. *Given any $v \in V_h$, there exists a decomposition*

$$v = v_H + \sum_{j=1}^{J} v_j$$

where $v_H \in V_H$, $v_j \in V_j$, such that

$$\langle A_H v_H, v_H \rangle + \sum_{j=1}^{J} \langle A_j v_j, v_j \rangle \lesssim \left(1 + \frac{H}{\delta}\right)^2 \langle A_h v, v \rangle.$$

Proof. Let $\tilde{\mathcal{I}}^H : \overset{\circ}{H}{}^1(\Omega) \longrightarrow V_H$ be the interpolation operator defined in Sect. 4.8. Let $v \in V_h$ and $v_H = \tilde{\mathcal{I}}^H v$. Then we have, by Theorem 4.8.12,

(7.4.18) $$|v_H|_{H^1(\Omega)} \lesssim |v|_{H^1(\Omega)},$$
(7.4.19) $$\|v - v_H\|_{L^2(\Omega)} \lesssim H |v|_{H^1(\Omega)}.$$

Let $w = v - v_H$ and define

(7.4.20) $$v_j = \Pi_h(\theta_j w).$$

It is easy to check that $v = v_H + \sum_{j=1}^{J} v_j$ (cf. exercise 7.x.18).
    It follows from (7.4.2), (7.4.18) and (7.4.1) that

(7.4.21) $$\langle A_H v_H, v_H \rangle = |v_H|^2_{H^1(\Omega)} \lesssim |v|^2_{H^1(\Omega)} = \langle A_h v, v \rangle.$$

We also have, from (7.4.9),

(7.4.22) $$\langle A_j v_j, v_j \rangle = \sum_{\substack{T \in \mathcal{T}_h \\ T \subseteq \Omega_j}} |v_j|^2_{H^1(T)}.$$

Let $T \in \mathcal{T}_h$ and $T \subseteq \Omega_j$, and a, b and c be the vertices of T. A simple calculation (cf. exercise 7.x.19) shows that

(7.4.23) $$|v_j|^2_{H^1(T)} \approx |v_j(b) - v_j(a)|^2 + |v_j(c) - v_j(a)|^2.$$

We have, by (7.4.20),

$$v_j(b) - v_j(a) = \big(\theta_j(b) - \theta_j(a)\big)w(b) + \theta_j(a)\big(w(b) - w(a)\big),$$

and hence we obtain using (7.4.4) and (7.4.5)

(7.4.24)
$$\begin{aligned}
&|v_j(b) - v_j(a)|^2 \\
&\lesssim [\theta_j(b) - \theta_j(a)]^2 [w(b)]^2 + [\theta_j(a)]^2 [w(b) - w(a)]^2 \\
&\lesssim \frac{h^2}{\delta^2}[w(b)]^2 + [w(b) - w(a)]^2.
\end{aligned}$$

Similarly, we find

(7.4.25) $$|v_j(c) - v_j(a)|^2 \lesssim \frac{h^2}{\delta^2}[w(c)]^2 + [w(c) - w(a)]^2 .$$

It follows from (7.4.24), (7.4.25), Lemma 6.2.7 and (7.4.23) that

$$|v_j|^2_{H^1(T)} \lesssim \frac{h^2}{\delta^2}[w^2(b) + w^2(c)] + [w(b) - w(a)]^2 + [w(c) - w(a)]^2$$
$$\lesssim \frac{1}{\delta^2}\|w\|^2_{L^2(T)} + |w|^2_{H^1(T)} ,$$

and hence in view of (7.4.22),

$$\langle A_j v_j, v_j \rangle \lesssim \sum_{\substack{T \in \mathcal{T}_h \\ T \subseteq \Omega_j}} \left[\frac{1}{\delta^2}\|w\|^2_{L^2(T)} + |w|^2_{H^1(T)} \right] .$$

Summing up this last estimate we find

$$\sum_{j=1}^{J} \langle A_j v_j, v_j \rangle \lesssim \sum_{j=1}^{J} \sum_{\substack{T \in \mathcal{T}_h \\ T \subseteq \Omega_j}} \left[\frac{1}{\delta^2}\|w\|^2_{L^2(T)} + |w|^2_{H^1(T)} \right]$$

$$\leq N_C \sum_{T \in \mathcal{T}_h} \left[\frac{1}{\delta^2}\|w\|^2_{L^2(T)} + |w|^2_{H^1(T)} \right] \qquad \text{(by 7.4.6)}$$

(7.4.26) $$\lesssim \frac{1}{\delta^2}\|v - v_H\|^2_{L^2(\Omega)} + |v - v_H|^2_{H^1(\Omega)} \qquad \text{(def. of } w\text{)}$$

$$\lesssim \left(1 + \frac{H^2}{\delta^2}\right)|v|^2_{H^1(\Omega)} . \qquad \text{(by 7.4.18 \& 7.4.19)}$$

$$= \left(1 + \frac{H^2}{\delta^2}\right)\langle A_h v, v \rangle \qquad \text{(by 7.4.1)}$$

The lemma follows from (7.4.21) and (7.4.26). $\qquad\square$

Lemma 7.4.17 and (7.1.22) immediately yield the estimate

(7.4.27) $$\lambda_{\min}(B_{TL}A_H) \gtrsim \left(1 + \frac{H}{\delta}\right)^{-2} .$$

Finally we obtain the following theorem on the two-level additive Schwarz preconditioner by combining (7.4.16) and (7.4.27).

(7.4.28) Theorem. *There exists a positive constant C, independent of H, h, δ and J, such that*

$$\kappa(B_{TL}A_h) = \frac{\lambda_{\max}(B_{TL}A_h)}{\lambda_{\min}(B_{TL}A_h)} \leq C\left(1 + \frac{H}{\delta}\right)^2 .$$

(7.4.29) *Remark.* The two-level additive preconditioner is optimal if H/δ is kept bounded above by a constant. When δ is small with respect to H, the factor $1 + (H/\delta)^2$ becomes significant. Note that the estimates (7.4.24) and (7.4.25) are very conservative, since $\theta_j = 1$ at all the nodes in Ω_j that do not belong to any other subdomains. Under certain shape regularity assumptions on the subdomains Ω_j, the bound for $\kappa(B_{TL}A_h)$ can be improved to (cf. (Dryja & Widlund 1994) and exercise 7.x.21)

$$\kappa(B_{TL}A_h) \leq C\left(1 + \frac{H}{\delta}\right).$$

Furthermore, this bound is sharp (cf. (Brenner 2000)).

(7.4.30) *Remark.* The results in this section remain valid if the exact solves (solution operators) A_H^{-1} and A_j^{-1} in (7.4.10) are replaced by spectrally equivalent inexact solves and the triangulations $\mathcal{T}_h$ and $\mathcal{T}_H$ are only assumed to be regular (with a proper interpretation of H and δ, cf. (Widlund 1999)).

7.5 Nonoverlapping Domain Decomposition Methods

To facilitate the discussion of the Bramble-Pasciak-Schatz (BPS) preconditioner and the Neumann-Neumann preconditioner, we present here a framework for nonoverlapping domain decomposition methods. The notation set up in this section will be followed in the next two sections.

Let Ω be divided into polygonal subdomains $\Omega_1, \ldots, \Omega_J$ (cf. the first figure in Fig. 7.2) such that

(7.5.1) $\qquad \Omega_j \cap \Omega_l = \emptyset \qquad$ if $j \neq l$,

(7.5.2) $\qquad \overline{\Omega} = \bigcup_{j=1}^{J} \overline{\Omega}_j$,

(7.5.3) $\qquad \partial\Omega_j \cap \partial\Omega_l = \emptyset$, a vertex or an edge, if $j \neq l$,

and $\mathcal{T}_h$ be a quasi-uniform triangulation of Ω which is aligned with the boundaries of the subdomains.

We assume that the subdomains satisfy the following shape regularity assumption: There exist reference polygonal domains $D_1, \ldots, D_K$ of unit diameter and a positive number H such that for each subdomain Ω_j there is a reference polygon D_k and a C^1 diffeomorphism $\phi_{j,k} : \overline{D}_k \longrightarrow \overline{\Omega}_j$ which satisfies the estimates

(7.5.4) $\qquad |\nabla\phi_{j,k}(x)| \lesssim H \;\; \forall x \in \overline{D}_k \quad$ and $\quad |\nabla\phi_{j,k}^{-1}(x)| \lesssim H^{-1} \;\; \forall x \in \overline{\Omega}_j$.

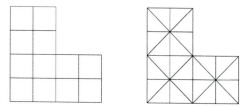

Fig. 7.2. nonoverlapping domain decomposition and coarse grid

(7.5.5) *Remark.* The shape regularity condition (7.5.4) implies that all the estimates involving the subdomains follow from corresponding estimates on the reference domains. It also implies

$$(7.5.6) \qquad \operatorname{diam} \Omega_j \approx H \,.$$

For both the BPS and Neumann-Neumann preconditioner, global communication among the subdomains is provided by a coarse grid space. The coarse grid is constructed by creating a triangulation of each subdomain using the vertices of the subdomain as the nodes. Together they form a triangulation $\mathcal{T}_H$ of Ω with mesh size $\approx H$ (cf. the second figure in Fig. 7.2).

(7.5.7) Definition. *Let* $\Gamma_j = \partial \Omega_j \setminus \partial \Omega$. *The* **interface (skeleton)** *of the domain decomposition* $\Omega_1, \ldots, \Omega_J$ *is* $\Gamma = \bigcup_{j=1}^{J} \Gamma_j$. *The set of nodes of* $\mathcal{T}_h$ *which belong to* Γ_j *(resp.* Γ) *is denoted by* $\Gamma_{j,h}$ *(resp.* Γ_h).

Let the variational bilinear form $a(\cdot, \cdot)$ for the model problem (7.1.24) be defined by

$$(7.5.8) \qquad a(v_1, v_2) = \int_{\Omega} \nabla v_1 \cdot \nabla v_2 \, dx \qquad \forall\, v_1, v_2 \in H^1(\Omega) \,.$$

The discrete problem for (7.1.24) is to find $u_h \in V_h(\Omega)$, the $\mathcal{P}_1$ finite element space in $\overset{\circ}{H}{}^1(\Omega)$ associated with $\mathcal{T}_h$, such that

$$(7.5.9) \qquad a(u_h, v) = F(v) \qquad \forall\, v \in V_h \,.$$

The space $V_h(\Omega)$ can be decomposed, with respect to the interface Γ and the variational form $a(\cdot, \cdot)$, into the direct sum of two subspaces.

(7.5.10) Definition. *The space* $V_h(\Omega \setminus \Gamma)$ *is the subspace of* V_h *whose members vanish at the interface* Γ, *i.e.,*

$$(7.5.11) \qquad V_h(\Omega \setminus \Gamma) = \{v \in V_h : v|_{\Gamma} = 0\} \,.$$

The interface space $V_h(\Gamma) \subset V_h$ *is the orthogonal complement of* $V_h(\Omega \setminus \Gamma)$ *with respect to* $a(\cdot, \cdot)$, *i.e.,*

$$(7.5.12) \qquad V_h(\Gamma) = \{v \in V_h : a(v, w) = 0 \quad \forall\, w \in V_h(\Omega \setminus \Gamma)\} \,.$$

It is clear from Definition 7.5.10 that

$$(7.5.13) \qquad V_h(\Omega) = V_h(\Omega \setminus \Gamma) \oplus V_h(\Gamma).$$

The solution u_h of (7.5.9) can therefore be written as

$$(7.5.14) \qquad u_h = \dot{u}_h + \overline{u}_h,$$

where $\dot{u}_h \in V_h(\Omega \setminus \Gamma)$ and $\overline{u}_h \in V_h(\Gamma)$. Equation (7.5.9) and the orthogonality between $V_h(\Omega \setminus \Gamma)$ and $V_h(\Gamma)$ then imply that

$$(7.5.15) \qquad a(\dot{u}_h, v) = F(v) \qquad \forall\, v \in V_h(\Omega \setminus \Gamma),$$
$$(7.5.16) \qquad a(\overline{u}_h, v) = F(v) \qquad \forall\, v \in V_h(\Gamma).$$

It is easy to see that (cf. 7.x.22) $\dot{u}_{h,j} = \dot{u}_h\big|_{\Omega_j} \in V_h(\Omega_j)$, the $\mathcal{P}_1$ finite element space in $H_0^1(\Omega_j)$ associated with the triangulation induced by $\mathcal{T}_h$, and it satisfies

$$(7.5.17) \qquad a(\dot{u}_{h,j}, v) = F(\tilde{v}) \qquad \forall\, v \in V_h(\Omega_j),$$

where $\tilde{v} \in V_h$ is the trivial extension of v. Therefore $\dot{u}_{h,j}$ for $1 \le j \le J$ can be computed from (7.5.17), which amounts to solving a discrete Dirichlet problem on Ω_j and can be done in parallel.

Let the *Schur complement operator* $S_h : V_h(\Gamma) \longrightarrow V_h(\Gamma)'$ be defined by

$$(7.5.18) \qquad \langle S_h v_1, v_2 \rangle = a(v_1, v_2) \qquad \forall\, v_1, v_2 \in V_h(\Gamma).$$

Then (7.5.16) can be written as

$$(7.5.19) \qquad S_h \overline{u}_h = f_h,$$

where $f_h \in V_h(\Gamma)'$ is defined by $\langle f_h, v \rangle = F(v)$ for all $v \in V_h(\Gamma)$.

It is clear from (7.5.18) that the operator S_h is SPD. The goal of a nonoverlapping domain decomposition method is to provide a good preconditioner for S_h, so that $\overline{u}_h$ can be solved efficiently from (7.5.19) by, for example, the preconditioned conjugate gradient method.

We devote the rest of this section to some basic properties of the interface space $V_h(\Gamma)$, which will be used in the next two sections.

Our first observation is that the functions in $V_h(\Gamma)$ are completely determined by their nodal values on Γ.

(7.5.20) Lemma. *Given any function* $\phi : \Gamma_h \longrightarrow \mathbb{R}$, *there is a unique* $v \in V_h(\Gamma)$ *such that*

$$(7.5.21) \qquad\qquad v(p) = \phi(p) \qquad \forall\, p \in \Gamma_h .$$

Proof. Let $w \in V_h$ be determined by

$$w(p) = \begin{cases} \phi(p) & p \in \Gamma_h, \\ 0 & p \text{ is a node not on } \Gamma. \end{cases}$$

If $v \in V_h(\Gamma)$ satisfies (7.5.21), then $v_* = v - w \in V_h(\Omega \setminus \Gamma)$, and the orthogonality between $V_h(\Omega \setminus \Gamma)$ and $V_h(\Gamma)$ implies that

$$(7.5.22) \qquad a(v_*, \overline{v}) = a(v - w, \overline{v}) = -a(w, \overline{v}) \qquad \forall\, \overline{v} \in V_h(\Omega \setminus \Gamma).$$

Conversely, if $v_* \in V_h(\Omega \setminus \Gamma)$ satisfies (7.5.22), then $v = v_* + w \in V_h(\Gamma)$ and (7.5.21) is satisfied.

Since $a(\cdot, \cdot)$ is an inner product on V_h, equation (7.5.22) has a unique solution in $V_h(\Omega \setminus \Gamma)$. Hence there exists a unique $v \in V_h(\Gamma)$ satisfying (7.5.21). $\qquad\qquad \square$

(7.5.23) Remark. It follows from (7.5.22) that $v \in V_h(\Gamma)$ can be obtained from its nodal values on Γ by solving in parallel a discrete Poisson equation with homogeneous boundary condition on each subdomain.

We have a *minimum energy principle* for functions in $V_h(\Gamma)$.

(7.5.24) Lemma. *Let* $v \in V_h(\Gamma)$ *and* $w \in V_h$ *such that* $v\big|_\Gamma = w\big|_\Gamma$. *Then we have* $a(v, v) \le a(w, w)$.

Proof. Since $w - v \in V_h(\Omega \setminus \Gamma)$, the orthogonality between $V_h(\Gamma)$ and $V_h(\Omega \setminus \Gamma)$ implies that

$$a(w, w) = a((w-v)+v, (w-v)+v) = a(w-v, w-v)+a(v,v) \ge a(v,v). \quad \square$$

There is an important relation between the energy norm of a function in $V_h(\Gamma)$ and a fractional order Sobolev norm of its trace on Γ. (See Chapter 14 for a discussion of general fractional order Sobolev spaces.) We will first give a local version of this relation.

Let D be a bounded polygon and $\hat{V}_{\hat{h}}$ be the $\mathcal{P}_1$ finite element space associated with a quasi-uniform triangulation $\mathcal{T}_{\hat{h}}$ of D. The subspace of $\hat{V}_{\hat{h}}$ whose members vanish on ∂D is denoted by $\hat{V}_{\hat{h}}(D)$. We will denote by $d(\cdot, \cdot)$ the bilinear form

$$d(v_1, v_2) = \int_D \nabla v_1 \cdot \nabla v_2 \, dx .$$

The subspace $\hat{V}_{\hat{h}}(\partial D) \subseteq \hat{V}_{\hat{h}}$ is the orthogonal complement of $\hat{V}_{\hat{h}}(D)$ with respect to $d(\cdot, \cdot)$.

(7.5.25) Definition. *The fractional order Sobolev semi-norm* $|\cdot|_{H^{1/2}(\partial D)}$ *is defined by*

$$|v|^2_{H^{1/2}(\partial D)} = \int_{\partial D} \left[\int_{\partial D} \frac{|v(x) - v(y)|^2}{|x - y|^2} ds(x) \right] ds(y),$$

where ds is the differential of the arc-length. The space $H^{1/2}(\partial D)$ *consists of functions* $v \in L^2(\partial D)$ *such that* $|v|_{H^{1/2}(\partial D)} < \infty$, *and we define*

$$\|v\|^2_{H^{1/2}(\partial D)} = \|v\|^2_{L^2(\partial D)} + |v|^2_{H^{1/2}(\partial D)}.$$

(7.5.26) Lemma. *It holds that* $|v|_{H^{1/2}(\partial D)} \lesssim |v|_{H^1(D)}$ *for all* $v \in H^1(D)$.

Proof. By the trace theorem (cf. (Nečas 1967) and (Arnold, Scott & Vogelius (1988))), we have

$$|v|^2_{H^{1/2}(\partial D)} \leq \|v\|^2_{H^{1/2}(\partial D)} \lesssim \|v\|^2_{H^1(D)}$$

for all $v \in H^1(D)$. It follows that, with $\bar{v} = |D|^{-1} \int_D v(x)\, dx$, we have

$$|v|_{H^{1/2}(\partial D)} = |v - \bar{v}|_{H^{1/2}(\partial D)} \qquad \text{(by 7.x.23)}$$

$$\lesssim \|v - \bar{v}\|^2_{H^1(D)} \lesssim |v|_{H^1(D)}. \qquad \text{(by 4.3.14)}$$

$\square$

(7.5.27) Lemma. *The following relation holds:*

$$|v|_{H^1(D)} \approx |v|_{H^{1/2}(\partial D)} \qquad \forall v \in \hat{V}_{\hat{h}}(\partial D).$$

Proof. Let $v \in \hat{V}_{\hat{h}}(\partial D)$ be arbitrary and $\bar{v} = |\partial D|^{-1} \int_{\partial D} v\, ds$ be the average of v over ∂D. Note that $v - \bar{v}$ also belongs to $\hat{V}_{\hat{h}}(\partial D)$.

By the trace theorem, there exists $w \in H^1(D)$ such that

$$w\big|_{\partial D} = (v - \bar{v})\big|_{\partial D} \quad \text{and} \quad \|w\|_{H^1(D)} \lesssim \|v - \bar{v}\|_{H^{1/2}(\partial D)}.$$

Recall from Section 4.8 that $\tilde{w} = \tilde{\mathcal{I}}^h w \in \hat{V}_{\hat{h}}(D)$, $|\tilde{w}|_{H^1(D)} \lesssim |w|_{H^1(D)}$ and $\tilde{w}\big|_{\partial D} = w\big|_{\partial D}$ (since $w\big|_{\partial D}$ is a piecewise linear function).

Therefore we have

$$|v|_{H^1(D)} = |v - \bar{v}|_{H^1(D)} \leq |\tilde{w}|_{H^1(\Omega_j)} \qquad \text{(by 7.5.24)}$$

$$\lesssim \|v - \bar{v}\|_{H^{1/2}(\partial D)} \approx |v|_{H^{1/2}(\partial D)}. \qquad \text{(by 7.x.24)}$$

The lemma follows from the preceding estimate and Lemma 7.5.26. $\square$

(7.5.28) Remark. The construction of $\tilde{w}$ is known in the literature as a finite element extension theorem (cf. (Widlund 1986)).

It follows from (7.5.4), Lemma 7.5.27 and scaling arguments that

$$(7.5.29) \qquad |v|^2_{H^1(\Omega)} \approx \sum_{j=1}^{J} |v|^2_{H^{1/2}(\partial\Omega_j)} \qquad \forall\, v \in V_h(\Gamma),$$

since there are at most K reference domains.

Finally we prove two lemmas concerning functions in the space

$$\mathcal{L}_{\hat{h}}(\partial D) = \{v \in C(\partial D) : \; v \text{ is piecewise linear with respect}$$
$$\text{to the subdivision of } \partial D \text{ induced by } \mathcal{T}_h\},$$

which is the restriction of $\hat{V}_h$ to ∂D.

(7.5.30) Lemma. *Let* $v \in \mathcal{L}_{\hat{h}}(\partial D)$, E *be an (open) edge of* ∂D, *and* $v_E \in \hat{V}_{\hat{h}}(\partial D)$ *be defined by*

$$v_E(p) = \begin{cases} v(p) & \text{if the node } p \in E, \\ 0 & \text{if the node } p \in \partial D \setminus E. \end{cases}$$

Then the following estimate holds:

$$|v_E|^2_{H^{1/2}(\partial D)} \lesssim |v|^2_{H^{1/2}(\partial D)} + (1 + |\ln \hat{h}|)\|v\|^2_{L^\infty(\partial D)}.$$

Proof. Let a and b be the two endpoints of E, E_a and E_b be the two edges of ∂D neighboring E, and $F = \partial D \setminus \overline{E \cup E_a \cup E_b}$. Then we have

$$
\begin{aligned}
|v_E|^2_{H^{1/2}(\partial D)} &= \int_E \left[\int_E \frac{|v_E(x) - v_E(y)|^2}{|x-y|^2}\, ds(x) \right] ds(y) \\
&\quad + 2\int_E \left[\int_{E_a \cup E_b} \frac{|v_E(y)|^2}{|x-y|^2}\, ds(x) \right] ds(y) \\
&\quad + 2\int_E \left[\int_F \frac{|v_E(y)|^2}{|x-y|^2}\, ds(x) \right] ds(y) \\
&\approx \int_E \left[\int_E \frac{|v_E(x) - v_E(y)|^2}{|x-y|^2}\, ds(x) \right] ds(y) \\
&\quad + \int_E v_E^2(y) \left(\frac{1}{|y-a|} + \frac{1}{|y-b|} \right) ds(y) + \|v_E\|^2_{L^2(E)} \\
&\lesssim |v|^2_{H^{1/2}(\partial D)} + (1 + |\ln \hat{h}|)\|v\|^2_{L^\infty(\partial D)}. \qquad \text{(by 7.x.25 \& 7.x.26)}
\end{aligned}
$$

$\square$

(7.5.31) Lemma. *Let* p *be a node on* ∂D, *and define* $v_p \in \mathcal{L}_{\hat{h}}(\partial D)$ *such that it vanishes at all the nodes except* p *and* $v_p(p) = 1$. *Then we have* $|v_p|_{H^{1/2}(\partial D)} \approx 1$.

Proof. Let a and b be the two nodes next to p, $\hat{h}_1 = |a - p|$ and $\hat{h}_2 = |b - p|$. Then we have (cf. exercise 7.x.27)

$$|v_p|^2_{H^{1/2}(\partial D)} \approx \|\phi\|^2_{L^2(I)} + \int_I \left[\int_I \frac{|\phi(x) - \phi(y)|^2}{|x - y|^2} dx \right] dy$$

$$+ \int_I \phi^2(y) \left(\frac{1}{y + \hat{h}_2} + \frac{1}{\hat{h}_1 - y} \right) dy \,,$$

where $I = [-\hat{h}_2, \hat{h}_1]$ and ϕ is defined by

$$\phi(x) = \begin{cases} 1 - (x/\hat{h}_1) & \text{for } 0 \le x \le \hat{h}_1, \\ 1 + (x/\hat{h}_2) & \text{for } -\hat{h}_2 \le x \le 0. \end{cases}$$

A direct calculation then yields the lemma.   □

7.6 The BPS Preconditioner

In this section we discuss the Bramble-Pasciak-Schatz (BPS) preconditioner (cf. (Bramble, Pasciak & Schatz 1986)) for the Schur complement operator S_h and we follow the notation in Sect. 7.5. There are two types of auxiliary spaces for the BPS preconditioner: those associated with the common edges of the subdomains and one associated with the coarse triangulation $\mathcal{T}_H$.

Let $E_1, \ldots, E_L$ be the (open) edges that are common to two of the subdomains. We define the *edge spaces* to be

(7.6.1) $V_h(E_\ell) = \{v \in V_h(\Gamma) : v = 0 \text{ at all the nodes in } \Gamma_h \backslash E_\ell\}\,.$

The SPD operator $S_\ell : V_h(E_\ell) \longrightarrow V_h(E_\ell)'$ is defined by

(7.6.2) $\langle S_\ell v_1, v_2 \rangle = a(v_1, v_2) \quad \forall\, v_1, v_2 \in V_h(E_\ell)\,.$

The edge space $V_h(E_\ell)$ is connected to $V_h(\Gamma)$ by the natural injections I_ℓ.

(7.6.3) *Remark.* Note that, by the argument in Lemma 7.5.20, the functions in $V_h(E_\ell)$ vanish identically on the subdomains whose boundaries do not contain E_l (cf. 7.x.28). Also, the number of edges on the boundary of each Ω_j is bounded by a constant since there are at most K reference domains.

The coarse grid space $V_H \subseteq \mathring{H}^1(\Omega)$ is the $\mathcal{P}_1$ finite element space associated with $\mathcal{T}_H$, and the SPD operator $A_H : V_H \longrightarrow V_H'$ is defined by

(7.6.4) $\langle A_H v_1, v_2 \rangle = a(v_1, v_2) \quad \forall\, v_1, v_2 \in V_H\,.$

Taking advantage of Lemma 7.5.20, we define the connection operator $I_H : V_H \longrightarrow V_h(\Gamma)$ by

(7.6.5) $(I_H v)|_\Gamma = v|_\Gamma\,.$

The BPS preconditioner is then defined by

$$(7.6.6) \qquad B_{BPS} = I_H A_H^{-1} I_H^t + \sum_{\ell=1}^{L} I_\ell S_\ell^{-1} I_\ell^t.$$

(7.6.7) *Remark.* The inverse operators A_H^{-1} and S_ℓ^{-1} can be replaced by spectrally equivalent approximate inverses (cf. (Bramble, Pasciak & Schatz 1986)).

(7.6.8) Lemma. *The following decomposition holds:*

$$V_h(\Gamma) = I_H V_H \oplus V_h(E_1) \oplus \ldots \oplus V_h(E_L).$$

Proof. Let $v \in V_h(\Gamma)$ and define $v_H = \Pi_H v$, the nodal interpolant of v with respect to $\mathcal{T}_H$. According to Lemma 7.5.20, we can define $v_\ell \in V_h(\Gamma)$ by

$$(7.6.9) \qquad v_\ell(p) = \begin{cases} v(p) - v_H(p) & p \in \Gamma_h \cap E_\ell, \\ 0 & p \in \Gamma_h \backslash E_\ell. \end{cases}$$

It is easy to check (cf. exercise 7.x.29) that

$$(7.6.10) \qquad v = I_H v_H + \sum_{\ell=1}^{L} v_\ell,$$

and it is also the only decomposition for v.   □

It follows from Lemma 7.6.8 that condition (7.1.9) is satisfied.

(7.6.11) Lemma. *The following estimate holds:* $\lambda_{\max}(B_{BPS} S_h) \lesssim 1$.

Proof. Let $v \in V_h(\Gamma)$ be arbitrary. According to (7.1.21), we only need to show that

$$\langle S_h v, v \rangle \lesssim \langle A_H v_H, v_H \rangle + \sum_{\ell=1}^{L} \langle S_l v_\ell, v_\ell \rangle,$$

where $v_H \in V_H$, $v_\ell \in V_h(E_\ell)$ and $v = v_H + \sum_{\ell=1}^{L} v_\ell$. In view of (7.5.8), (7.5.18), (7.6.4) and (7.6.2), it is equivalent to show that

$$(7.6.12) \qquad |v|^2_{H^1(\Omega)} \lesssim |v_H|^2_{H^1(\Omega)} + \sum_{\ell=1}^{L} |v_\ell|^2_{H^1(\Omega)}.$$

First of all, we have, by (7.6.10) and the Cauchy-Schwarz inequality,

$$(7.6.13) \qquad |v|^2_{H^1(\Omega)} \lesssim |I_H v_H|^2_{H^1(\Omega)} + \left| \sum_{\ell=1}^{L} v_\ell \right|^2_{H^1(\Omega)}.$$

Secondly, Lemma 7.5.24, (7.5.8) and (7.6.5) imply that

$$(7.6.14) \qquad |I_H v_H|^2_{H^1(\Omega)} \leq |v_H|^2_{H^1(\Omega)} .$$

Finally, since $a(v_l, v_k) = 0$ unless E_l and E_k share a common subdomain (cf. (7.6.3)), we have (cf. exercise 7.x.30)

$$(7.6.15) \qquad \left| \sum_{\ell=1}^{L} v_\ell \right|^2_{H^1(\Omega)} \lesssim \sum_{\ell=1}^{L} |v_\ell|^2_{H^1(\Omega)} .$$

The estimate (7.6.12) follows from (7.6.13)–(7.6.15). $\qquad\qquad\square$

(7.6.16) Lemma. *The following estimate holds:*

$$\lambda_{\min}(B_{BPS} S_h) \gtrsim \left[1 + \ln(H/h)\right]^{-2} .$$

Proof. Let $v \in V_h(\Gamma)$ be arbitrary. This time we must estimate $|v_H|^2_{H^1(\Omega)} + \sum_{\ell=1}^{L} |v_\ell|^2_{H^1(\Omega)}$ by $|v|^2_{H^1(\Omega)}$, where v_H and v_ℓ are the functions in the decomposition (7.6.10).

First we consider v_H. Recall that $v_H = \Pi_H v$. Let $\mathcal{V}_j$ be the vertices of Ω_j and $\overline{v}_j = |\Omega_j|^{-1} \int_{\Omega_j} v(x) \, dx$. Then we have

$$
\begin{aligned}
|\Pi_H v|^2_{H^1(\Omega_j)} &= |\Pi_H(v - \overline{v}_j)|^2_{H^1(\Omega_j)} \\
&\lesssim \sum_{p \in \mathcal{V}_j} (v - \overline{v}_j)^2(p) &\text{(analog of 7.4.23)} \\
(7.6.17) \qquad &\lesssim \left[1 + \ln(H/h)\right]\left(H^{-2}\|v - \overline{v}_j\|^2_{L^2(\Omega_j)} + |v - \overline{v}_j|^2_{H^1(\Omega_j)}\right) \\
&&\text{(by 4.9.2 \& scaling)} \\
&\lesssim \left[1 + \ln(H/h)\right]|v|^2_{H^1(\Omega_j)} . &\text{(by 4.3.15 \& scaling)}
\end{aligned}
$$

Summing up we obtain

$$(7.6.18) \qquad |v_H|^2_{H^1(\Omega)} \lesssim \left[1 + \ln(H/h)\right]|v|^2_{H^1(\Omega)} .$$

Next we consider v_ℓ defined by (7.6.9). Let E_ℓ be the common edge of Ω_{ℓ_1} and Ω_{ℓ_2}. On each Ω_j, for $j = \ell_1$ or ℓ_2, we have

$$
v_\ell(p) = \begin{cases} [v(p) - \overline{v}_j] - \Pi_H[v - \overline{v}_j](p) & p \in \partial\Omega_{j,h} \cap E_\ell, \\ 0 & p \in \partial\Omega_{j,h} \backslash E_\ell, \end{cases}
$$

where $\partial\Omega_{j,h}$ is the set of nodes of $\mathcal{T}_h$ that belong to $\partial\Omega_j$. It follows that, for $j = \ell_1$ or ℓ_2,

$$|v_\ell|^2_{H^1(\Omega_j)} \approx |v_\ell|^2_{H^{1/2}(\partial\Omega_j)} \qquad\qquad\qquad\qquad \text{(by 7.5.27)}$$

$$\lesssim |v - \bar{v}_j|^2_{H^{1/2}(\Omega_j)} + |\Pi_H(v - \bar{v}_j)|^2_{H^{1/2}(\Omega_j)}$$
$$+ \left[1 + \ln(H/h)\right]\|v - \bar{v}_j\|^2_{L^\infty(\partial\Omega_j)} \qquad \text{(by 7.5.30 \& scaling)}$$

$$\lesssim \left[1 + \ln(H/h)\right]|v|^2_{H^{1/2}(\Omega_j)} \qquad\qquad \text{(by 7.5.26 \& 7.6.17)}$$
$$+ \left[1 + \ln(H/h)\right]^2\left(H^{-2}\|v - \bar{v}_j\|^2_{L^2(\Omega_j)} + |v - \bar{v}_j|^2_{H^1(\Omega_j)}\right)$$
$$\text{(by 4.9.2 \& scaling)}$$

$$\lesssim \left[1 + \ln(H/h)\right]^2|v|^2_{H^1(\Omega_j)}, \qquad\qquad \text{(by 4.3.15 \& scaling)}$$

and hence,

$$|v_\ell|^2_{H^1(\Omega)} \lesssim \left[1 + \ln(H/h)\right]^2\left(|v|^2_{H^1(\Omega_{\ell_1})} + |v|^2_{H^1(\Omega_{\ell_2})}\right).$$

Summing up we find

$$(7.6.19) \qquad \sum_{\ell=1}^L |v_\ell|^2_{H^1(\Omega)} \lesssim \left[1 + \ln(H/h)\right]^2|v|^2_{H^1(\Omega)}.$$

The lemma follows from (7.1.22), (7.6.18) and (7.6.19). $\qquad\qquad\qquad$ □

Combining Lemmas 7.6.11 and 7.6.16 we have the following theorem.

(7.6.20) Theorem. *There exists a positive constant C, independent of h, H and J, such that*

$$\kappa(B_{BPS}S_h) = \frac{\lambda_{\max}(B_{BPS}S_h)}{\lambda_{\min}(B_{BPS}S_h)} \leq C\left(1 + \ln\frac{H}{h}\right)^2.$$

(7.6.21) *Remark.* The bound in Theorem 7.6.20 is sharp (cf. (Brenner \& Sung 2000)).

(7.6.22) *Remark.* The BPS preconditioner can also be applied to the boundary value problem whose variational form is defined by

$$a(v_1, v_2) = \sum_{j=1}^J \rho_j \int_{\Omega_j} \nabla v_1 \cdot \nabla v_2 \, dx \qquad \forall\, v_1, v_2 \in V_h,$$

where the ρ_j's are positive constants. The bound in (7.6.20) remains valid and the constant C is independent of h, H, J and the ρ_j's.

(7.6.23) *Remark.* There are three dimensional generalizations of the BPS preconditioner (cf. (Bramble, Pasciak \& Schatz 1989) and (Smith 1991)).

7.7 The Neumann-Neumann Preconditioner

In this section we discuss the Neumann-Neumann preconditioner (cf. (Dryja & Widlund 1995)) for the Schur complement operator S_h and we follow the notation in Sect. 7.5. The auxiliary spaces for the Neumann-Neumann preconditioner are associated with either a subdomain or a coarse triangulation.

The coarse grid space V_H, the SPD operator $A_H : V_H \longrightarrow V_H$ and the connection operator I_H are defined as in Sect. 7.6.

To define the auxiliary space associated with a subdomain Ω_j, we first introduce the space $V_h(\Omega_j \cup \Gamma_j)$, which is the $\mathcal{P}_1$ finite element space on Ω_j associated with the triangulation induced by $\mathcal{T}_h$ and whose members vanish on $\partial\Omega_j \cap \partial\Omega$. Let the inner product $\hat{a}_j(\cdot,\cdot)$ be defined by

$$(7.7.1) \qquad \hat{a}_j(v_1, v_2) = \int_{\Omega_j} \nabla v_1 \cdot \nabla v_2 \, dx + H^{-2} \int_{\Omega_j} v_1 v_2 \, dx$$

for all $v_1, v_2 \in V_h(\Omega_j \cup \Gamma_j)$.

We can now take the auxiliary space $V_h(\Gamma_j) \subseteq V_h(\Omega_j \cup \Gamma_j)$ to be the $\hat{a}_j(\cdot,\cdot)$-orthogonal complement of the space $V_h(\Omega_j) = \{v \in V_h(\Omega_j \cup \Gamma_j) : v|_{\partial\Omega_j} = 0\}$. It is easy to check (cf. 7.x.32) that a function $v \in V_h(\Gamma_j)$ is completely determined by its nodal values on $\Gamma_{j,h}$. The connection operator $I_j : V_h(\Gamma_j) \longrightarrow V_h(\Gamma)$ is defined by

$$(7.7.2) \qquad (I_j v_j)(p) = \begin{cases} 0 & \text{if } p \in \Gamma_h \setminus \Gamma_{j,h}, \\ v_j(p)/n(p) & \text{if } p \in \Gamma_{j,h}, \end{cases}$$

where $n(p) = $ the number of subdomains sharing the node p, and the SPD linear operator $\hat{S}_j : V_h(\Gamma_j) \longrightarrow V_h(\Gamma_j)'$ is given by

$$(7.7.3) \qquad \langle \hat{S}_j v_1, v_2 \rangle = \hat{a}_j(v_1, v_2) \qquad \forall\, v_1, v_2 \in V_h(\Gamma_j).$$

The Neumann-Neumann preconditioner $B_{NN} : V_h(\Gamma) \longrightarrow V_h(\Gamma)'$ is then defined by

$$(7.7.4) \qquad B_{NN} = I_H A_H^{-1} I_H^t + \sum_{j=1}^{J} I_j \hat{S}_j^{-1} I_j^t.$$

(7.7.5) Remark. We use the weighted integral in the second term of the right-hand side of (7.7.1) so that the bilinear form is invariant under scaling. For subdomains that have at least one edge on $\partial\Omega$, we could also use the inner product

$$a_j(v_1, v_2) = \int_{\Omega_j} \nabla v_1 \cdot \nabla v_2 \, dx.$$

But, for internal subdomains, the bilinear form $a_j(\cdot, \cdot)$ is only semi–definite, since $1 \in V_h(\Omega_j \cup \Gamma_j)$ and $a_j(1,1) = 0$.

In order to show that condition (7.1.9) is satisfied, we introduce the restriction maps $R_j : V_h(\Gamma) \longrightarrow V_h(\Gamma_j)$ defined by

$$(7.7.6) \qquad (R_j v)(p) = v(p) \qquad \forall\, p \in \Gamma_{j,h}\,.$$

The operators R_j and I_j together yield a partition of unity for $V_h(\Gamma)$:

$$(7.7.7) \qquad \sum_{j=1}^{J} I_j R_j v = v \qquad \forall\, v \in V_h(\Gamma)\,,$$

which can be verified as follows. According to Lemma 7.5.20, we only need to check that the two sides of (7.7.7) agree on Γ_h. Let $p \in \Gamma_h$ and $\sigma_p = \{j : p \in \Gamma_j\}$. Then we have by (7.7.2) and (7.7.6)

$$\left(\sum_{j=1}^{J} I_j R_j v \right)(p) = \sum_{j \in \sigma_p} \frac{1}{|\sigma_p|}(R_j v)(p) = \sum_{j \in \sigma_p} \frac{1}{|\sigma_p|} v(p) = v(p)\,.$$

Condition (7.1.9) follows immediately from (7.7.7).

(7.7.8) Lemma. *The following estimate holds:* $\lambda_{\min}(B_{NN} S_h) \gtrsim 1$.

Proof. Let $\tilde{\mathcal{I}}^H : \mathring{H}^1(\Omega) \longrightarrow V_H$ be the interpolation operator defined in Section 4.8. For an arbitrary $v \in V_h(\Gamma)$, we define $v_H = \tilde{\mathcal{I}}^H v$ and $v_j = R_j(v - I_H v_H)$ for $1 \le j \le J$. It follows from (7.7.7) that

$$(7.7.9) \qquad I_H v_H + \sum_{j=1}^{J} I_j v_j = I_H v_H + (v - I_H v_H) = v\,.$$

We have

$$\langle A_H v_H, v_H \rangle = |v_H|^2_{H^1(\Omega)} \lesssim |v|^2_{H^1(\Omega)}\,. \qquad \text{(by 7.6.4 \& 4.8.14)}$$

Since $v_j = v - v_H$ on Γ_j,

$$\langle \hat{S}_j v_j, v_j \rangle = \hat{a}_j(v_j, v_j) \qquad\qquad\qquad \text{(by 7.7.3)}$$
$$\le \hat{a}_j(v - v_H, v - v_H) \qquad\qquad \text{(by 7.x.33)}$$
$$= |v - v_H|^2_{H^1(\Omega)} + H^{-2}\|v - v_H\|^2_{L^2(\Omega)} \qquad \text{(by 7.7.1)}$$
$$\lesssim |v|^2_{H^1(\Omega_j)}\,, \qquad\qquad\qquad\qquad \text{(by 4.8.14)}$$

and hence by (7.5.18)

$$(7.7.10) \qquad \langle A_H v_H, v_H \rangle + \sum_{j=1}^{J} \langle \hat{S}_j v_j, v_j \rangle \lesssim \langle S_h v, v \rangle\,.$$

The lemma follows from (7.1.22), (7.7.9) and (7.7.10). $\qquad\qquad\square$

(7.7.11) Lemma. *The following estimate holds:*

$$\lambda_{\max}(B_{NN} S_h) \lesssim [1 + \ln(H/h)]^2 .$$

Proof. Let $v \in V_h(\Gamma)$ be arbitrary. For any decomposition

$$(7.7.12) \qquad v = I_H v_H + \sum_{j=1}^{J} I_j v_j ,$$

where $v_H \in V_H$ and $v_j \in V_h(\Gamma_j)$ for $1 \le j \le J$, we must show, in view of
(7.1.21), (7.5.18), (7.6.4), (7.7.1) and (7.7.3), that

$$(7.7.13) \qquad |v|^2_{H^1(\Omega)} \lesssim [1 + \ln(H/h)]^2 \Big[|v_H|^2_{H^1(\Omega)}$$

$$+ \sum_{j=1}^{J} \big(|v_j|^2_{H^1(\Omega_j)} + H^{-2} \|v_j\|^2_{L^2(\Omega_j)} \big) \Big] .$$

We have, by (7.7.12), the Cauchy-Schwarz inequality and an argument
analogous to the one in the derivation of (7.6.15),

$$(7.7.14) \qquad |v|^2_{H^1(\Omega)} \lesssim |v_H|^2_{H^1(\Omega)} + \sum_{j=1}^{J} |I_j v_j|^2_{H^1(\Omega)} .$$

Therefore it only remains to estimate $|I_j v_j|^2_{H^1(\Omega)}$.

Note that $I_j v_j$ vanishes at all the subdomains Ω_k such that $\partial\Omega_k \cap \partial\Omega_j = \emptyset$. There are three cases (cf. (7.5.3)) where $\partial\Omega_k \cap \partial\Omega_j \ne \emptyset$:
(1) $\Omega_k = \Omega_j$,
(2) $\partial\Omega_k \cap \partial\Omega_j$ consists of just one vertex,
(3) $\partial\Omega_k \cap \partial\Omega_j$ consists of an open edge E and its endpoints p_1 and p_2.
We will concentrate on the third case, since the analysis for the other two
cases are similar.

Let w_{p_1} (resp. w_{p_2}) be the function in $V_h(\Gamma_k)$ which equals $I_j v_j$ at p_1
(resp. p_2) and vanishes at all the other nodes in Γ_k, and w_E be the function
in $V_h(\Gamma_k)$ which equals $I_j v_j$ at the nodes in E and vanishes at all the other
nodes in Γ_k. Then we can write

$$I_j v_j\big|_{\Omega_k} = w_{p_1} + w_{p_2} + w_E .$$

We have the following estimate on w_{p_j} for $j = 1$ or 2:

$$|w_{p_j}|^2_{H^1(\Omega_k)} \approx |w_{p_j}|^2_{H^{1/2}(\partial\Omega_k)} \qquad\qquad \text{(by 7.5.27)}$$

$$\approx |w_{p_j}(p_j)|^2 \qquad\qquad \text{(by 7.5.31)}$$

$$\lesssim \|v_j\|^2_{L^\infty(\Omega_j)} \qquad\qquad \text{(by 7.7.2)}$$

$$\lesssim [1 + \ln(H/h)] \Big[|v_j|^2_{H^1(\Omega_j)} + H^{-2}\|v_j\|^2_{L^2(\Omega_j)} \Big]. \qquad \text{(by 4.9.2 \& scaling)}$$

For the function w_E, we have

$$
\begin{aligned}
|w_E|^2_{H^1(\Omega_k)} &\approx |w_E|^2_{H^{1/2}(\partial\Omega_k)} & \text{(by 7.5.27)}\\
&\approx |w_E|^2_{H^{1/2}(\partial\Omega_j)} & \text{(by 7.x.34)}\\
&\lesssim |v_j|^2_{H^{1/2}(\partial\Omega_j)} + [1 + \ln(H/h)]\|v_j\|^2_{L^\infty(\partial\Omega_j)} & \text{(by 7.7.2 \& 7.5.30)}\\
&\lesssim [1 + \ln(H/h)]^2 \Big[|v_j|^2_{H^1(\Omega_j)} + H^{-2}\|v_j\|^2_{L^2(\Omega_j)}\Big].
\end{aligned}
$$

$$\text{(by 7.5.26, 4.9.2 \& scaling)}$$

The preceding estimates imply

$$
|I_j v_j|^2_{H^1(\Omega_k)} \lesssim [1 + \ln(H/h)]^2 \Big[|v_j|^2_{H^1(\Omega_j)} + H^{-2}\|v_j\|^2_{L^2(\Omega_j)}\Big]
$$

in case (3). Since the same estimate holds for cases (1) and (2) using similar arguments, we have

$$
(7.7.15) \qquad |I_j v_j|^2_{H^1(\Omega)} \lesssim [1 + \ln(H/h)]^2 \Big[|v_j|^2_{H^1(\Omega_j)} + H^{-2}\|v_j\|^2_{L^2(\Omega_j)}\Big].
$$

The estimate (7.7.13) (and hence the lemma) follows from (7.7.14) and (7.7.15). □

Combining Lemmas 7.7.8 and 7.7.8, we have the following theorem.

(7.7.16) Theorem. *There exists a positive constant C, independent of h, H and J, such that*

$$
\kappa(B_{NN} S_h) = \frac{\lambda_{\max}(B_{NN} S_h)}{\lambda_{\min}(B_{NN} S_h)} \le C \left(1 + \ln\frac{H}{h}\right)^2.
$$

(7.7.17) *Remark.* The bound in (7.7.16) is sharp (cf. (Brenner & Sung 2000)).

(7.7.18) *Remark.* The Neumann-Neumann preconditioner can also be applied to the boundary value problem in (7.6.22). In this case the definition of I_j should be modified as

$$
(I_j v_j)(p) = \begin{cases} 0 & \text{if } p \in \Gamma_h \setminus \Gamma_{j,h}, \\ v_j(p)/[\sum_{k\in\sigma_p} \rho_k^t] & \text{if } p \in \Gamma_{j,h}, \end{cases}
$$

where $\sigma_p = \{k : p \in \Gamma_k\}$ and t is any number greater than or equal to $1/2$. The bound in (7.7.16) remains valid, and the constant C is independent of h, H, J and the ρ_j's.

(7.7.19) *Remark.* The Neumann-Neumann preconditioner can be generalized to three dimensions (cf. (Dryja & Widlund 1995)).

(7.7.20) *Remark.* The balancing domain decomposition preconditioner of Mandel (cf. (Mandel 1993), (Mandel & Brezina 1996), (LeTallec, Mandel & Vidrascu 1998)) is closely related to the Neumann-Neumann preconditioner and it can also be analyzed as an additive Schwarz preconditioner (cf. (Xu & Zou 1998), (Brenner & Sung 1999), (Widlund 1999)).

7.x Exercises

7.x.1 Let $a(\cdot,\cdot)$ be a bilinear form on $V \times V$. Show that there exists a unique $A : V \longrightarrow V'$ such that

$$a(v_1, v_2) = \langle Av_1, v_2 \rangle \qquad \forall v_1, v_2 \in V,$$

and that $a(\cdot,\cdot)$ is an inner product if and only if A is SPD.

7.x.2 Let $\{v_1, \ldots, v_n\}$ be a basis of V and $\alpha_1, \ldots \alpha_n$ be the dual basis of V'. Then the linear operator $A : V \longrightarrow V'$ is represented by a matrix $\underset{\approx}{A}$ with respect to these two bases, where the (i, j) component of $\underset{\approx}{A}$ is $\langle Av_i, v_j \rangle$. Show that A is SPD if and only if the matrix $\underset{\approx}{A}$ is SPD.

7.x.3 Let $V = \mathbb{R}^n$ with canonical basis $e_1, \ldots, e_n$, and $A : V \longrightarrow V'$ be represented by the SPD matrix $\underset{\approx}{A} = (a_{ij})$. Let $V_1 = V_2 = \cdots = V_n = \mathbb{R}$, and $B_j : V_j \longrightarrow V_j'$ be defined by the 1×1 matrix $[a_{jj}]$. Determine the matrix representation of the additive Schwarz preconditioner $B = \sum_{j=1}^n I_j B_j^t I_j^t$, where $I_j : V_j \longrightarrow V$ is defined by $I_j x = x e_j$.

7.x.4 Show that the preconditioner B defined by (7.1.7) is symmetric.

7.x.5 Verify that the v_j's defined in (7.1.18) yield a decomposition of v.

7.x.6 Referring to the operators A and B in Theorem 7.1.20, show that BA is SPD with respect to the inner product $\langle B^{-1}\cdot, \cdot \rangle$.

7.x.7 Let $a(\cdot,\cdot)$ be an inner product on V, and $A : V \longrightarrow V'$ be defined by

$$\langle Av, w \rangle = a(v, w) \qquad \forall v, w \in V.$$

Let V_j be a subspace of V for $0 \le j \le J$, $I_j : V_j \longrightarrow V$ be the natural injection, and $A_j : V_j \longrightarrow V_j'$ be defined by

$$\langle A_j v, w \rangle = a(v, w) \qquad \forall v, w \in V_j.$$

Show that

$$\left(\sum_{j=0}^J I_j A_j^{-1} I_j^t \right) A = \sum_{j=0}^J P_j,$$

where $P_j : V \to V_j$ is the $a(\cdot,\cdot)$-orthogonal projection operator.

7.x.8 Let V, $a(\cdot,\cdot)$, A, V_j and I_j be as in 7.x.7. Let $b_j(\cdot,\cdot)$ be an inner product on V_j, $B_j : V_j \longrightarrow V_j'$ be defined by

$$\langle B_j v, w \rangle = b_j(v, w) \qquad \forall\, v, w \in V_j\,,$$

and B be the additive Schwarz preconditioner given by (7.1.7). For $\phi \in V'$, show that the equation $BAu = B\phi$ can be written as

$$\sum_{j=0}^{J} T_j u = \sum_{j=0}^{J} f_j\,,$$

where $T_j : V \longrightarrow V_j$ and $f_j \in V_j$ are defined by

$$b_j(T_j v, v_j) = a(v, v_j) \qquad \forall\, v \in V\,,\ v_j \in V_j\,,$$
$$b_j(f_j, v_j) = \langle \phi, v \rangle \qquad \forall\, v_j \in V_j\,.$$

7.x.9 Show that $v = \Pi_{j-1} v + (v - \Pi_{j-1} v)$ gives the unique decomposition described in (7.2.4).

7.x.10 Establish (7.2.9) by a direct calculation over the triangles of $\mathcal{T}_{j-1}$.

7.x.11 Show that (referring to the notation in Lemma 7.2.12)

$$\|\Pi_j v - v_T\|_{L^\infty(T)} \lesssim \left(1 + |\ln(h_j/h_J)|\right)^{1/2} |v|_{H^1(T)}$$

by using Lemma 4.9.1, Lemma 4.3.14 and a scaling argument.

7.x.12 Establish the discrete Young's inequality in Lemma 7.2.15.

7.x.13 Carry out the details in the derivation of (7.3.4).

7.x.14 Prove the orthogonality relation (7.3.7).

7.x.15 Show that $P_{j-1} P_j = P_{j-1}$ for the Ritz projection operators defined by (7.3.5).

7.x.16 Use Poincaré's inequality to show that $\|P_1 v\|_{L^2(\Omega)} \lesssim |v|_{H^1(\Omega)}$.

7.x.17 Let $Q_j : \mathring{H}^1(\Omega) \longrightarrow V_j$ be the L^2 orthogonal projection operator and $v_j = Q_j v - Q_{j-1} v$ for $1 \le j \le J$ (with $Q_0 v = 0$). Show that (cf. (Xu 1992) and (Bramble 1995)) $\sum_{j=1}^{J} h_j^{-2} \|v_j\|_{L^2(\Omega)}^2 \approx |v|_{H^1(\Omega)}^2$ for all $v \in V_J$ and derive (7.3.8) for nonconvex domains.

7.x.18 Establish the decomposition in the proof of Lemma 7.4.17.

7.x.19 Use a direct calculation on the standard simplex and a homogeneity (scaling) argument to establish (7.4.23).

7.x.20 Let $B = \sum_{j=1}^{J} I_j A_j^{-1} I_j^t$ be the one-level additive Schwarz preconditioner associated with an overlapping domain decomposition which satisfies (7.4.3)–(7.4.6). Find an upper bound for $\kappa(BA_h)$.

7.x.21 Let $S = (0, 1) \times (0, 1)$ and $B = (0, 1) \times (0, \delta)$ where $0 < \delta < 1$. Show that

$$\|v\|_{L^2(B)}^2 \leq C\delta \|v\|_{H^1(S)}^2 \qquad \forall v \in H^1(S),$$

where the positive constant C is independent of δ. Use this estimate and a scaling argument to establish the improved bound for the two-level additive Schwarz preconditioner stated in (7.4.29) in the case where the subdomains are squares.

7.x.22 Show that $u_{h,j} \in V_h(\Omega_j)$ and (7.5.17) is valid.

7.x.23 Show that the seminorm $|\cdot|_{H^{1/2}(\partial D)}$ is invariant under the addition of a constant.

7.x.24 Let D be a bounded polygon. Prove that

$$\|v\|_{L^2(\partial D)} \lesssim |v|_{H^{1/2}(\partial D)}$$

for all $v \in H^{1/2}(\partial D)$ satisfying $\int_{\partial D} v \, ds = 0$.
(Hint: Use the expression

$$\int_{\partial D} v^2(x) ds(x) = \int_{\partial D} \left[|\partial D|^{-1} \int_{\partial D} (v(x) - v(y)) ds(y) \right]^2 ds(x)$$

and the Cauchy-Schwarz inequality.)

7.x.25 Let $0 = a_0 < a_1 < \cdots < a_n = 1$ be a quasi-uniform partition of the unit interval I so that $a_j - a_{j-1} \approx \rho = 1/n$ for $1 \leq j \leq n$, and $\mathcal{L}_\rho$ be the space of continuous functions on $[0, 1]$ which are piecewise linear with respect to this partition. Given any $v \in \mathcal{L}_\rho$, we define $v_* \in \mathcal{L}_\rho$ by

$$v_*(a_j) = \begin{cases} v(a_j) & 1 \leq j \leq n - 1, \\ 0 & j = 0, n. \end{cases}$$

Prove that $|v_*|_{H^{1/2}(I)} \leq C[|v|_{H^{1/2}(I)} + \|v\|_{L^\infty(I)}]$ for all $v \in \mathcal{L}_\rho$, where C is a positive constant independent of ρ.
(Hint: Calculate the $H^{1/2}(I)$ seminorm of the function in $\mathcal{L}_\rho$ which equals v at a_0 (or a_n) and vanishes at all the other a_j's.)

7.x.26 Following the notation in 7.x.25, show that

$$\int_0^1 v^2(x) \left(\frac{1}{x} + \frac{1}{1-x} \right) dx \leq C(1 + |\ln \rho|) \|v\|_{L^\infty(I)}^2$$

for all $v \in \mathcal{L}_\rho$ such that $v(0) = v(1) = 0$, where C is a positive constant independent of ρ.
(Hint: Break up the integral $\int_0^1 [v^2(x)/x] \, dx$ at the point a_1.)

7.x.27 Prove the norm equivalence in the proof of (7.5.31) and estimate the terms involving $\phi(x)$.

7.x.28 Let $v \in V_h(E_l)$ and $E_l \not\subset \partial\Omega_j$. Prove that $v = 0$ on Ω_j.

7.x.29 Show that decomposition defined in the proof of Lemma 7.6.8 is the unique decomposition of $v \in V_h(\Gamma)$ with respect to the coarse grid space V_H and the edge spaces $V_h(E_l)$ for $1 \leq l \leq L$.

7.x.30 Prove the estimate (7.6.15). (Hint: Consult the proof of (7.4.13).)

7.x.31 Verify the assertion in (7.6.22).

7.x.32 Use an argument similar to the one in the proof of Lemma 7.5.20 to show that a function $v \in V_h(\Gamma_j)$ is completely determined by its nodal values on $\Gamma_{j,h}$.

7.x.33 Show that if $v \in V_h(\Gamma_j)$ and $w \in V_h(\Omega_j \cup \Gamma_j)$ agree on Γ_j, then $\hat{a}_j(v, v) \leq \hat{a}_j(w, w)$.

7.x.34 Show that $|w_E|_{H^{1/2}(\partial\Omega_k)} \approx |w_E|_{H^{1/2}(\partial\Omega_j)}$. (Hint: Consult the proof of Lemma 7.5.30.)

7.x.35 Verify the assertion in (7.7.18).

7.x.36 Given any $p \in \Gamma_h$, let $w_p \in V_h(\Gamma)$ be defined by
$$w_p(q) = \begin{cases} 1 & \text{if } q = p, \\ 0 & \text{if } q \in \Gamma_h \setminus p. \end{cases}$$
Show that $\{w_p : p \in \Gamma_h\}$ is a basis of $V_h(\Gamma)$.

7.x.37 Given any $p \in \Gamma_h$, let $\omega_p \in V_h(\Gamma)'$ be defined by
$$\langle \omega_p, w_q \rangle = \begin{cases} 1 & \text{if } q = p, \\ 0 & \text{if } q \in \Gamma_h \setminus p, \end{cases}$$
where w_q is defined in 7.x.36. Show that $\{\omega_p : p \in \Gamma_h\}$ is a basis of $V_h(\Gamma)'$.

7.x.38 Let $v \in V_h(\Gamma)$ and $S_h v = \sum_{p \in \Gamma_h} \alpha_p \omega_p$ (cf. 7.x.37). Show that $\alpha_p = a(v, v_p)$, where v_p is the natural nodal basis function of V_h associated with the node p. Discuss how $S_h v$ can be computed from the nodal values of v on Γ_h.

7.x.39 Let $v_H \in V_H'$. Show that $A_H^{-1} v_H$ is obtained by solving a discrete Poisson equation on Ω associated with the coarse grid $\mathcal{T}_H$.

7.x.40 Let the (open) edge E_ℓ be the common edge of Ω_{ℓ_1} and Ω_{ℓ_2}, and $v_\ell \in V_h(E_\ell)'$. Show that $S_\ell^{-1} v_\ell$ is obtained by solving a discrete Poisson equation on $\Omega_{\ell_1} \cup \Omega_{\ell_2} \cup E_\ell$ with the homogeneous Dirichlet boundary condition. What is special about the right-hand side of this discrete problem?

7.x.41 Let $v_j \in V_h(\Gamma_j)'$. Show that $\hat{S}_j^{-1} v_j$ is obtained by solving a discrete stabilized Poisson equation on Ω_j, with the Neumann (natural) boundary condition on Γ_j and the homogeneous Dirichlet boundary condition on $\partial\Omega_j \setminus \Gamma_j = \partial\Omega_j \cap \partial\Omega$.

Chapter 8

Max–norm Estimates

The finite element approximation is essentially defined by a mean-square projection of the gradient. Thus, it is natural that error estimates for the gradient of the error directly follow in the L^2 norm. It is interesting to ask whether such a gradient-projection would also be of optimal order in some other norm, for example L^∞. We prove here that this is the case. Although of interest in their own right, such estimates are also crucial in establishing the viability of approximations of nonlinear problems (Douglas & Dupont 1975) as we indicate in Sect. 8.7. Throughout this chapter, we assume that the domain $\Omega \subset \mathbb{R}^d$ is bounded and polyhedral.

8.1 Main Theorem

To begin with, we consider a variational problem with a variational form

$$a(u, v) :=$$

$$\int_\Omega \sum_{i,j=1}^d a_{ij}(x) \frac{\partial u}{\partial x_i}(x) \frac{\partial v}{\partial x_j}(x) + \sum_{i=1}^d b_i(x) \frac{\partial u}{\partial x_i}(x) v(x) + b_0(x) u(x) v(x)\, dx$$

whose leading term is coercive pointwise *a.e.*:

$$(8.1.1) \qquad C_a |\xi|^2 \le \sum_{ij} a_{ij}(x) \xi_i \xi_j \quad \forall 0 \ne \xi \in \mathbb{R}^d, \quad \text{for a.a. } x \in \Omega.$$

We also assume that the coefficients, a_{ij} and b_i are bounded on Ω. For simplicity, we consider only the Dirichlet problem, so let $V = \mathring{H}^1(\Omega)$. Re-garding regularity of the variational problem, we assume that there is a unique solution to

$$(8.1.2) \qquad\qquad a(u, v) = (f, v) \quad \forall v \in V$$

which satisfies

$$(8.1.3) \qquad\qquad \|u\|_{W_p^2(\Omega)} \le C \|f\|_{L^p(\Omega)}$$

for $1 < p < \mu$ for some $\mu > 1$ to be chosen later. We also assume the same regularity for the adjoint problem defined by

$$(8.1.2')\qquad\qquad a(v, u) = (f, v) \quad \forall v \in V.$$

We will also make some independent assumptions on the regularity of the coefficients, a_{ij} and b_i.

We consider spaces, V_h, based on general elements, $\mathcal{E}$, in two and three dimensions. Let us assume, for simplicity, that the elements are conforming, based on a quasi-uniform family of subdivisions, $\mathcal{T}^h$. We will utilize weighted-norm techniques similar to those introduced in Sect. 0.8, following very closely the arguments of (Rannacher & Scott 1982).

We introduce the family of weight functions

$$\sigma_z(x) = \left(|x - z|^2 + \kappa^2 h^2\right)^{\frac{1}{2}},$$

depending on the parameter $\kappa \geq 1$. It is easy to verify that, for any $\lambda \in \mathbb{R}$,

$$(8.1.4)\qquad \max_{T \in \mathcal{T}^h} \left(\sup_{x \in T} \sigma_z^\lambda(x) / \inf_{x \in T} \sigma_z^\lambda(x)\right) \leq C$$

$$(8.1.5)\qquad \left\|\sigma_z^\lambda\right\|_{L^\infty(\Omega)} \leq C \max\left\{1, (\kappa h)^\lambda\right\}$$

$$(8.1.6)\qquad \left|D_x^\beta \sigma_z^\lambda(x)\right| \leq C \sigma_z^{\lambda - |\beta|}(x) \quad \forall x \in \Omega \quad \forall \beta$$

where the constant C depends continuously on λ and is independent of $z \in \Omega$ and h.

We introduce the notation

$$\widetilde{\int_\Omega} \cdots dx := \sum_{T \in \mathcal{T}^h} \int_T \cdots dx.$$

We assume that the finite element spaces, $V_h \subset V$, consist of piecewise polynomials of degree less than $k_{\max}$ and have approximation order $2 \leq k \leq k_{\max}$ in the sense that there is an interpolant (or other projection onto V_h) $\mathcal{I}^h$ such that

$$\int_\Omega \sigma_z^\lambda \left(\psi - \mathcal{I}^h \psi\right)^2 dx + h^2 \int_\Omega \sigma_z^\lambda \left|\nabla \left(\psi - \mathcal{I}^h \psi\right)\right|^2 dx$$

$$\leq C \sum_{r=k}^{k_{\max}} h^{2r} \widetilde{\int_\Omega} \sigma_z^\lambda |\nabla_{\mathcal{E},r} \psi|^2 dx,$$

$$(8.1.7)$$

$$\int_\Omega \sigma_z^\lambda \left(\psi - \mathcal{I}^h \psi\right)^2 dx + h^2 \int_\Omega \sigma_z^\lambda \left|\nabla \left(\psi - \mathcal{I}^h \psi\right)\right|^2 dx$$

$$\leq C h^4 \widetilde{\int_\Omega} \sigma_z^\lambda |\nabla_2 \psi|^2 dx,$$

for any $\psi \in H^1(\Omega)$ satisfying $\psi|_\tau \in H^k(\tau)$ for all $\tau \in \mathcal{T}^h$, and such that

$$(8.1.8) \qquad \nabla_{\mathcal{E},r} (\psi_h|_\tau) = 0 \quad \forall \tau \in \mathcal{T}^h \quad \forall \psi_h \in V_h \quad \forall k \leq r \leq k_{\max},$$

where ∇_j denotes the vector of all partial derivatives of order j and $\nabla_{\mathcal{E},r}$ denotes a vector of partial derivatives of order r depending on the element used. Moreover, we assume an inverse estimate of the form

$$(8.1.9) \qquad \int_\Omega \sigma_z^\lambda |\nabla_j \psi_h|^2 \, dx \leq C h^{-2j} \int_\Omega \sigma_z^\lambda \psi_h^2 \, dx \quad \forall \psi_h \in V_h, \ j \geq 1.$$

It is assumed that the constants C in (8.1.7) and (8.1.9) depend continuously on λ. It is easily verified that the above assumptions hold for all the elements studied in Chapter 3 (cf. exercise 8.x.6). For elements based on triangles and tetrahedra, for which the displacement functions consist of all polynomials of degree less than k, take $\nabla_{\mathcal{E},r} = \nabla_k$ for $r = k$ (and nothing otherwise). For the tensor-product elements, we also have $\nabla_{\mathcal{E},r}$ to be null unless $r = k$; for $= k$ one takes $\nabla_{\mathcal{E},r}$ to consist only of the derivatives D^α for $\alpha_i = k$ for some i (and $\alpha_j = 0$ for $i \neq j$) (cf. Theorem 4.6.11). For the serendipity elements, see Sect. 4.6 for possible choices of $\nabla_{\mathcal{E},r}$.

As usual, let $u_h \in V_h$ solve

$$(8.1.10) \qquad a(u_h, v) = (f, v) \quad \forall v \in V_h.$$

The main result of the chapter is the following.

(8.1.11) Theorem. *Let the finite element spaces satisfy (8.1.7), (8.1.8) and (8.1.9). We assume (8.1.1) holds, and that a_{ij} and b_j are all in $L^\infty(\Omega)$. Suppose that $d \leq 3$, $a_{ij} \in W_p^1(\Omega)$ for $p > 2$ if $d = 2$ and $p \geq 12/5$ if $d = 3$, for all $i, j = 1, \ldots, d$, and that (8.1.3) holds for $\mu > d$. Then there is an $h_0 > 0$ and $C < \infty$ such that*

$$\|u_h\|_{W_\infty^1(\Omega)} \leq C \|u\|_{W_\infty^1(\Omega)}$$

for $0 < h < h_0$.

The theorem can also be proved with ∞ replaced by p, with $2 < p < \infty$ (cf. exercise 8.x.1). The following corollary is easily derived (cf. exercise 8.x.2).

(8.1.12) Corollary. *Under the assumptions of Theorem 8.1.11*

$$\|u - u_h\|_{W_\infty^1(\Omega)} \leq C h^{k-1} \|u\|_{W_\infty^k(\Omega)}.$$

(8.1.13) Remark. We have not attempted to prove a result with minimal smoothness requirements, but it is interesting to note that our regularity assumption (8.1.3) with $\mu > d$ implies that the solution, u, to (8.1.2) satisfies $u \in W_\infty^1(\Omega)$, and that this would not hold for $\mu \leq d$.

The verification of (8.1.7) and (8.1.9) can be done for most of the elements studied in Chapters 3 and 4. The inverse estimate (8.1.9) follows from Lemma 4.5.3 and (8.1.4) provided the mesh is quasi-uniform (cf. exercise 8.x.15). The bound (8.1.7) also follows from (8.1.4) in view of the results of Sects. 4.4 and 4.6 for most simplicial and tensor-product elements (cf. exercise 8.x.16).

8.2 Reduction to Weighted Estimates

For $z \in K^z$, let $\delta^z \in C_0^\infty(K^z)$ satisfy

(8.2.1)
$$\int_\Omega \delta^z(x) P(x) \, dx = P(z), \quad \forall P \in \mathcal{P}_{k_{\max}}$$
$$\|\nabla_k \delta^z\|_{L^\infty(\Omega)} \leq C_k h^{-d-k}, \quad k = 0, 1, \ldots,$$

where the constant C_k depends only on the triangulation and ∇_k denotes the vector of all k-th order derivatives (see exercise 8.x.7).

Note that, by construction,

(8.2.2)
$$\nu \cdot \nabla v(z) = (\nu \cdot \nabla v, \delta^z) \quad \forall v \in V_h$$

for any direction vector, ν.

Define $g^z \in V$ by solving the adjoint variational problem

$$a(v, g^z) = (-\nu \cdot \nabla \delta^z, v) \quad \forall v \in V.$$

Note that both δ^z and g^z depend on h, but we suppress this to simplify the notation. Let g_h^z denote the finite element approximation to g^z,

$$a(v, g_h^z) = (-\nu \cdot \nabla \delta^z, v) \quad \forall v \in V_h.$$

Then the definitions of u, u_h, g^z and g_h^z, together with integration by parts, yield

(8.2.3)
$$\begin{aligned}
\nu \cdot \nabla u_h(z) &= (\nu \cdot \nabla u_h, \delta^z) = (u_h, -\nu \cdot \nabla \delta^z) \\
&= a(u_h, g^z) = a(u_h, g_h^z) = a(u, g_h^z) \\
&= a(u, g^z) - a(u, g^z - g_h^z) \\
&= (-\nu \cdot \nabla \delta^z, u) - a(u, g^z - g_h^z) \\
&= (\delta^z, \nu \cdot \nabla u) - a(u, g^z - g_h^z).
\end{aligned}$$

From Hölder's inequality, we have

$$|a(g^z - g_h^z, u)| \leq C \|g^z - g_h^z\|_{W_1^1(\Omega)} \|u\|_{W_\infty^1(\Omega)}$$
$$\leq C \left(\int_\Omega \sigma_z^{d+\lambda} |\nabla (g^z - g_h^z)|^2 \, dx \right)^{\frac{1}{2}} \left(\int_\Omega \sigma_z^{-d-\lambda} \, dx \right)^{\frac{1}{2}} \|u\|_{W_\infty^1(\Omega)}.$$

Letting

$$(8.2.4) \qquad M_h := \sup_{z \in \Omega} \left(\int_\Omega \sigma_z^{d+\lambda} \left(|\nabla (g^z - g_h^z)|^2 + (g^z - g_h^z)^2 \right) dx \right)^{\frac{1}{2}}$$

we have

$$(8.2.5) \qquad |a(g^z - g_h^z, u)| \le C M_h \left(\int_\Omega \sigma_z^{-d-\lambda} dx \right)^{\frac{1}{2}} \|u\|_{W_\infty^1(\Omega)}$$

$$\le C M_h \lambda^{-1/2} (\kappa h)^{-\lambda/2} \|u\|_{W_\infty^1(\Omega)} .$$

Thus, we need to show that the following holds.

(8.2.6) Lemma. *For appropriate $\lambda > 0$ and κ sufficiently large, there is an $h_0 > 0$ such that*

$$\sup_{z \in \Omega} \int_\Omega \sigma_z^{d+\lambda} \left(|\nabla (g^z - g_h^z)|^2 + (g^z - g_h^z)^2 \right) dx \le C h^\lambda$$

for all $0 < h < h_0$.

The proof will be given in a series of propositions in the next section. As a corollary, we immediately obtain the following.

(8.2.7) Corollary. *There is a $C < \infty$ and an $h_0 > 0$ such that*

$$\|g^z - g_h^z\|_{W_1^1(\Omega)} \le C$$

for all $0 < h < h_0$.

8.3 Proof of Lemma 8.2.6

We begin by proving a general estimate for the finite element error in weighted norms. It is similar to those proved in Chapter 0, but slightly more technical due to the particular nature of the weight, σ_z.

(8.3.1) Proposition. *Suppose that $d = 2$ or 3 and (8.1.3) holds for some $\mu > d$. Let the finite element spaces satisfy (8.1.7), (8.1.8) and (8.1.9). Let w solve (8.1.2) and w_h solve (8.1.10). For any $\lambda > 0$ and $\kappa > 1$, there is a $C < \infty$ such that*

$$\int_\Omega \sigma_z^{d+\lambda} |\nabla (w - w_h)|^2 dx \le C \Bigg(\int_\Omega \sigma_z^{d+\lambda-2} (w - w_h)^2 dx$$

$$+ \int_\Omega \sigma_z^{d+\lambda-2} (w - \mathcal{I}^h w)^2 + \sigma_z^{d+\lambda} |\nabla (w - \mathcal{I}^h w)|^2 dx \Bigg),$$

for all $z \in \Omega$ and all h.

Proof. Let $e := w - w_h$, $\tilde{e} := \mathcal{I}^h w - w_h$ and set $\psi = \sigma_z^{d+\lambda}\tilde{e}$. Note that $a(e, v) = 0$ for all $v \in V_h$. From (8.1.1),

$$C_a \int_\Omega \sigma_z^{d+\lambda} |\nabla e|^2 \, dx$$

$$\leq \int_\Omega \sigma_z^{d+\lambda} \sum_{ij} a_{ij} e_{,i} e_{,j} \, dx$$

$$= a(e, \sigma_z^{d+\lambda} e) - \int_\Omega \sum_{ij} a_{ij} e_{,i} e \left(\sigma_z^{d+\lambda}\right)_{,j} \, dx$$

(8.3.2)
$$- \int_\Omega \sum_i b_i e_{,i} \sigma_z^{d+\lambda} e + b_0 \sigma_z^{d+\lambda} e^2 \, dx$$

$$= a(e, \sigma_z^{d+\lambda}(w - \mathcal{I}^h w) + \psi)$$

$$- \int_\Omega \sum_{ij} a_{ij} e_{,i} e \left(\sigma_z^{d+\lambda}\right)_{,j} + \sum_i b_i e_{,i} \sigma_z^{d+\lambda} e + b_0 \sigma_z^{d+\lambda} e^2 \, dx$$

$$= a(e, \sigma_z^{d+\lambda}(w - \mathcal{I}^h w)) + a(e, \psi - \mathcal{I}^h \psi)$$

$$- \int_\Omega \sum_{ij} a_{ij} e_{,i} e \left(\sigma_z^{d+\lambda}\right)_{,j} + \sum_i b_i e_{,i} \sigma_z^{d+\lambda} e + b_0 \sigma_z^{d+\lambda} e^2 \, dx$$

where $v_{,j}$ denotes the partial derivative with respect to the j-th coordinate. The last two terms in (8.3.2) may be estimated as follows:

$$\left| \int_\Omega \sum_{ij} a_{ij} e_{,i} e \left(\sigma_z^{d+\lambda}\right)_{,j} + \sum_i b_i e_{,i} \sigma_z^{d+\lambda} e + b_0 \sigma_z^{d+\lambda} e^2 \, dx \right|$$

$$\leq C \int_\Omega |\nabla e| \, |e| \, \sigma_z^{d+\lambda-1} + \sigma_z^{d+\lambda} e^2 \, dx \qquad \text{(by 8.1.5 and 8.1.6)}$$

$$\leq C \left(\int_\Omega \sigma_z^{d+\lambda} |\nabla e|^2 \, dx \right)^{\frac{1}{2}} \left(\int_\Omega \sigma_z^{d+\lambda-2} e^2 \, dx \right)^{\frac{1}{2}}$$

$$+ C \int_\Omega \sigma_z^{d+\lambda} e^2 \, dx \qquad \text{(Schwarz' inequality)}$$

$$\leq \frac{C_a}{4} \int_\Omega \sigma_z^{d+\lambda} |\nabla e|^2 \, dx + C' \int_\Omega \sigma_z^{d+\lambda-2} e^2 \, dx. \quad \text{(by 8.1.5 and 0.9.5)}$$

Similarly, the first term in (8.3.2) may be estimated:

$$|a(e, \sigma_z^{d+\lambda}(w - \mathcal{I}^h w))|$$

$$\leq C \int_\Omega |\nabla e| \left(\sigma_z^{d+\lambda} \left| \nabla \left(w - \mathcal{I}^h w\right) \right| + \sigma_z^{d+\lambda-1} \left| w - \mathcal{I}^h w \right| \right) \, dx$$

$$+ C \int_\Omega |e| \sigma_z^{d+\lambda} \left| w - \mathcal{I}^h w \right| \, dx \qquad \text{(by 8.1.5 and 8.1.6)}$$

$$\leq C \left(\int_\Omega \sigma_z^{d+\lambda} \left(|\nabla e|^2 + e^2 \right) \, dx \right)^{\frac{1}{2}} \qquad \text{(Schwarz' inequality and 8.1.5)}$$

$$\times \left(\int_\Omega \sigma_z^{d+\lambda-2} \left(w - \mathcal{I}^h w \right)^2 + \sigma_z^{d+\lambda} \left| \nabla \left(w - \mathcal{I}^h w \right) \right|^2 dx \right)^{\frac{1}{2}}$$

$$\leq \frac{C_a}{4} \int_\Omega \sigma_z^{d+\lambda} \left(|\nabla e|^2 + e^2 \right) dx \qquad \text{(by 0.9.5)}$$

$$+ C \int_\Omega \sigma_z^{d+\lambda-2} \left(w - \mathcal{I}^h w \right)^2 + \sigma_z^{d+\lambda} \left| \nabla \left(w - \mathcal{I}^h w \right) \right|^2 dx.$$

The second term in (8.3.2) may be estimated similarly:

$$\left| a(e, \psi - \mathcal{I}^h \psi) \right|$$

$$\leq C \int_\Omega |\nabla e| \left(\left| \nabla \left(\psi - \mathcal{I}^h \psi \right) \right| + \left| \psi - \mathcal{I}^h \psi \right| \right) + |e| \left| \psi - \mathcal{I}^h \psi \right| dx$$

$$\leq C \left(\int_\Omega \sigma_z^{d+\lambda} \left(|\nabla e|^2 + e^2 \right) dx \right)^{\frac{1}{2}} \qquad \text{(Schwarz' inequality)}$$

$$\times \left(\int_\Omega \sigma_z^{-d-\lambda} \left(\left| \nabla \left(\psi - \mathcal{I}^h \psi \right) \right|^2 + \left(\psi - \mathcal{I}^h \psi \right)^2 \right) dx \right)^{\frac{1}{2}}$$

$$\leq \frac{C_a}{4} \int_\Omega \sigma_z^{d+\lambda} \left(|\nabla e|^2 + e^2 \right) dx$$

$$+ C \int_\Omega \sigma_z^{-d-\lambda} \left(\left| \nabla \left(\psi - \mathcal{I}^h \psi \right) \right|^2 + \left(\psi - \mathcal{I}^h \psi \right)^2 \right) dx. \qquad \text{(by 0.9.5)}$$

Inserting these estimates in (8.3.2) and using (8.1.5) yields

$$(8.3.3) \qquad \begin{aligned} \frac{C_a}{4} \int_\Omega \sigma_z^{d+\lambda} |\nabla e|^2 \, dx &\leq C \Bigg(\int_\Omega \sigma_z^{d+\lambda-2} e^2 \, dx \\ &+ \int_\Omega \sigma_z^{-d-\lambda} \left(\left| \nabla \left(\psi - \mathcal{I}^h \psi \right) \right|^2 + \left(\psi - \mathcal{I}^h \psi \right)^2 \right) dx \\ &+ \int_\Omega \sigma_z^{d+\lambda-2} \left(w - \mathcal{I}^h w \right)^2 + \sigma_z^{d+\lambda} \left| \nabla \left(w - \mathcal{I}^h w \right) \right|^2 dx \Bigg). \end{aligned}$$

We see that

$$\int_\Omega \sigma_z^{-d-\lambda} \left(\left| \nabla \left(\psi - \mathcal{I}^h \psi \right) \right|^2 + \left(\psi - \mathcal{I}^h \psi \right)^2 \right) dx$$

$$\leq \sum_{r=k}^{k_{\max}} h^{2r-2} \int_\Omega \sigma_z^{-d-\lambda} |\widetilde{\nabla_{\mathcal{E},r} \psi}|^2 \, dx \qquad \text{(from 8.1.7)}$$

$$\leq C \sum_{r=k}^{k_{\max}} h^{2r-2} \sum_{j=0}^{r-1} \int_\Omega \sigma_z^{d+\lambda-2(r-j)} |\widetilde{\nabla_j \tilde{e}}|^2 \, dx \qquad \text{(from 8.1.8 \& 8.1.6)}$$

$$\leq C \sum_{r=k}^{k_{\max}} \sum_{j=0}^{r-1} h^{2r-2j-2} \int_\Omega \sigma_z^{d+\lambda-2(r-j)} \tilde{e}^2 \, dx \qquad \text{(from 8.1.9)}$$

$$\leq C \int_\Omega \sigma_z^{d+\lambda-2} \tilde{e}^2 \, dx. \qquad \text{(from 8.1.5)}$$

Substituting this into (8.3.3) completes the proof of Proposition 8.3.1.

We note that the previous estimate may be written

$$(8.3.4) \qquad \int_\Omega \sigma_z^{-d-\lambda} \left| \nabla \left(\psi - \mathcal{I}^h \psi \right) \right|^2 \, dx \; \leq \; C \int_\Omega \sigma_z^{-d-\lambda-2} \psi^2 \, dx.$$

Such an estimate is often called a *superapproximation* estimate (cf. Nitsche & Schatz 1974) since the right hand side involves a weaker norm (and no negative power of h).   □

To estimate the term involving $(w - w_h)^2$ on the right-hand side of (8.3.1), we use the following.

(8.3.5) Proposition. *Suppose that $d = 2$ or 3, (8.1.3) holds for $\mu > d$ and $0 < \lambda < 2(1 - \frac{d}{\mu})$. Let the finite element spaces satisfy (8.1.7), (8.1.8) and (8.1.9). Let w solve (8.1.2) and w_h solve (8.1.10). Let $\epsilon > 0$. There is a $\kappa_1 < \infty$ such that*

$$\int_\Omega \sigma_z^{d+\lambda-2} \left(w - w_h \right)^2 \, dx \leq \epsilon \int_\Omega \sigma_z^{d+\lambda} \left| \nabla \left(w - w_h \right) \right|^2 \, dx$$

for all $\kappa \geq \kappa_1$ and all h.

Proof. We use a duality argument. Let $v \in V$ solve

$$(8.3.6) \qquad\qquad a(\phi, v) = (\sigma_z^{d+\lambda-2} e, \phi) \quad \forall \phi \in V.$$

Thus,

$$
\begin{aligned}
\int_\Omega \sigma_z^{d+\lambda-2} e^2 \, dx &= a(e, v) & \text{(8.3.6 with } \phi = e) \\
&= a(e, v - \mathcal{I}^h v) & \text{(8.1.2 \& 8.1.10)} \\
&\leq C \left(\int_\Omega \sigma_z^{d+\lambda} \left(|\nabla e|^2 + e^2 \right) \, dx \right)^{\frac{1}{2}} & \text{(Schwarz' inequality)} \\
&\quad \times \left(\int_\Omega \sigma_z^{-d-\lambda} \left(\left| \nabla \left(v - \mathcal{I}^h v \right) \right|^2 + \left(v - \mathcal{I}^h v \right)^2 \right) \, dx \right)^{\frac{1}{2}} \\
&\leq \epsilon \int_\Omega \sigma_z^{d+\lambda} \left(|\nabla e|^2 + e^2 \right) \, dx & \text{(by 0.9.5)} \\
&\quad + \frac{C}{\epsilon} \int_\Omega \sigma_z^{-d-\lambda} \left(\left| \nabla \left(v - \mathcal{I}^h v \right) \right|^2 + \left(v - \mathcal{I}^h v \right)^2 \right) \, dx \\
&\leq \epsilon \int_\Omega \sigma_z^{d+\lambda} \left(|\nabla e|^2 + e^2 \right) \, dx + \frac{Ch^2}{\epsilon} \int_\Omega \sigma_z^{-d-\lambda} |\nabla_2 v|^2 \, dx. & \text{(by 8.1.7)}
\end{aligned}
$$

In the next section, we will prove the following.

(8.3.7) Lemma. *Suppose that $d = 2$ or 3, (8.1.3) holds for $\mu > d$ and λ satisfies $0 < \lambda < 2(1 - \frac{d}{\mu})$. Then the solution to*

$$a(\phi, v) = (f, \phi) \quad \forall \phi \in V,$$

for $f \in \mathring{H}^1(\Omega)$, satisfies

$$\int_\Omega \sigma^{-d-\lambda} |\nabla_2 v|^2 \, dx \le C \lambda^{-1} \zeta^{-2} \int_\Omega \sigma^{4-d-\lambda} |\nabla f|^2 \, dx$$

where

$$\sigma(x) := \left(|x - x^0|^2 + \zeta^2 \right)^{\frac{1}{2}}$$

and $x^0 \in \Omega$ is arbitrary.

Applying Lemma 8.3.7 with $f = \sigma_z^{d+\lambda-2} e$, we find

$$\int_\Omega \sigma_z^{-d-\lambda} |\nabla_2 v|^2 \, dx \le C \lambda^{-1} (\kappa h)^{-2} \int_\Omega \sigma_z^{4-d-\lambda} \left| \nabla \left(\sigma_z^{d+\lambda-2} e \right) \right|^2 \, dx$$

$$\le C \lambda^{-1} (\kappa h)^{-2} \left(\int_\Omega \sigma_z^{d+\lambda} |\nabla e|^2 \, dx + \int_\Omega \sigma_z^{d+\lambda-2} e^2 \, dx \right).$$

Therefore,

$$\int_\Omega \sigma_z^{d+\lambda-2} e^2 \, dx \le \epsilon \int_\Omega \sigma_z^{d+\lambda} \left(|\nabla e|^2 + e^2 \right) \, dx$$

$$+ \frac{C}{\lambda \epsilon \kappa^2} \left(\int_\Omega \sigma_z^{d+\lambda} |\nabla e|^2 \, dx + \int_\Omega \sigma_z^{d+\lambda-2} e^2 \, dx \right).$$

For any fixed ϵ and λ, we can pick κ_1 large enough that

$$(8.3.8) \qquad \int_\Omega \sigma_z^{d+\lambda-2} e^2 \, dx \le 2\epsilon \int_\Omega \sigma_z^{d+\lambda} |\nabla e|^2 \, dx$$

for all $\kappa \ge \kappa_1$. Renaming ϵ completes the proof of (8.3.5). $\square$

Combining the two Propositions yields

$$(8.3.9) \qquad \begin{aligned} \int_\Omega \sigma_z^{d+\lambda} \left| \nabla (w - w_h) \right|^2 + \sigma_z^{d+\lambda-2} \left(w - w_h \right)^2 \, dx \\ \le C \int_\Omega \sigma_z^{d+\lambda-2} \left(w - \mathcal{I}^h w \right)^2 + \sigma_z^{d+\lambda} \left| \nabla \left(w - \mathcal{I}^h w \right) \right|^2 \, dx \end{aligned}$$

for all $z \in \Omega$. To complete the proof of (8.2.6), we apply (8.3.9) to $w = g^z$. Thus, we need to show that

$$(8.3.10) \qquad \int_\Omega \sigma_z^{d+\lambda} |\nabla_2 g^z|^2 \, dx \le C h^{\lambda-2}.$$

This is a consequence of the following *a priori* estimate.

(8.3.11) Lemma. *Suppose that $d \leq 3$, $0 < \lambda < 1$, and $a_{ij} \in W_p^1(\Omega)$ for $p > 2$ if $d = 2$ and $p \geq 12/5$ if $d = 3$, for all $i, j = 1, \ldots, d$. Suppose that (8.1.3) holds for $\mu > \frac{2d}{4-\lambda}$. Then the solution to*

$$a(\phi, v) = (\nu \cdot \nabla f, \phi) \quad \forall \phi \in V,$$

for $f \in \overset{\circ}{H}{}^1(\Omega)$, satisfies

$$\int_\Omega \sigma^{d+\lambda} |\nabla_2 v|^2 \, dx \leq C \int_\Omega \sigma^{d+\lambda} |\nabla f|^2 \, dx + C\zeta^{-2} \int_\Omega \sigma^{d+\lambda} f^2 \, dx$$

where σ is the weight defined in (8.3.7).

8.4 Proofs of Lemmas 8.3.7 and 8.3.11

We begin with Lemma 8.3.7. We have

$$\int_\Omega \sigma^{-d-\lambda} |\nabla_2 v|^2 \, dx$$

$$\leq \left(\int_\Omega \sigma^{-(d+\lambda)p'} \, dx \right)^{1/p'} \|\nabla_2 v\|_{L^{2p}(\Omega)}^2 \qquad \text{(Hölder's inequality)}$$

$$\leq C\zeta^{-d-\lambda+d/p'} \|v\|_{W_{2p}^2(\Omega)}^2$$

$$= C\zeta^{-\lambda-d/p} \|v\|_{W_{2p}^2(\Omega)}^2$$

$$\leq C\zeta^{-\lambda-d/p} \|f\|_{L^{2p}(\Omega)}^2 \qquad \text{(by 8.1.3)}$$

$$\leq C\zeta^{-\lambda-d/p} \|\nabla f\|_{L^{2pd/(d+2p)}(\Omega)}^2. \qquad \text{(exercise 8.x.12)}$$

Provided that $p < d/(d-2)$ then $2pd/(d+2p) < 2$ so that we may again apply Hölder's inequality, with $q = (d+2p)/pd$ to bound

$$\|\nabla f\|_{L^{2pd/(d+2p)}(\Omega)}^{\frac{2}{q}} = \int_\Omega |\nabla f|^{\frac{2}{q}} \, dx = \int_\Omega \sigma^{\frac{-(4-d-\lambda)}{q}} \left(\sigma^{\frac{4-d-\lambda}{q}} |\nabla f|^{\frac{2}{q}} \right) dx$$

$$\leq \left(\int_\Omega \sigma^{-(4-d-\lambda)\frac{q'}{q}} \, dx \right)^{\frac{1}{q'}} \left(\int_\Omega \sigma^{4-d-\lambda} |\nabla f|^2 \, dx \right)^{\frac{1}{q}}$$

$$= \left(\int_\Omega \sigma^{\frac{-(4-d-\lambda)}{q-1}} \, dx \right)^{\frac{q-1}{q}} \left(\int_\Omega \sigma^{4-d-\lambda} |\nabla f|^2 \, dx \right)^{\frac{1}{q}}.$$

Using the definition of σ, we have

$$\left(\int_\Omega \sigma^{\frac{-(4-d-\lambda)}{q-1}} \, dx \right)^{q-1} \leq C\zeta^{-(4-d-\lambda)+d(q-1)}$$

$$= C\zeta^{-2+\lambda+d/p}$$

provided $p > d/(2-\lambda)$. Because of our assumption on λ, $d/(2-\lambda) < \mu$, so we can pick p satisfying the above constraints. Combining previous estimates, we prove Lemma 8.3.7.

Now consider Lemma 8.3.11. Expanding the expression $\nabla_2\left(\sigma^{(d+\lambda)/2}v\right)$, we find

$$(8.4.1) \quad \sigma^{d+\lambda}|\nabla_2 v|^2 \leq \left|\nabla_2\left(\sigma^{(d+\lambda)/2}v\right)\right|^2 + C\left(\sigma^{d+\lambda-2}|\nabla v|^2 + \sigma^{d+\lambda-4}v^2\right).$$

In order to use (8.1.3), we consider the equation satisfied by $\sigma^{(d+\lambda)/2}v$:

$$a(\phi, \sigma^{(d+\lambda)/2}v) = a(\sigma^{(d+\lambda)/2}\phi, v) + \left(\sum_{i,j=1}^{d} a_{ij}\phi_{,i}\left(\sigma^{(d+\lambda)/2}\right)_{,j}, v\right)$$

$$- \left(\sum_{i,j=1}^{d} a_{ij}\left(\sigma^{(d+\lambda)/2}\right)_{,j}v_{,i}, \phi\right) - \left(\sum_{i=1}^{d} b_i\left(\sigma^{(d+\lambda)/2}\right)_{,i}v, \phi\right)$$

$$= a(\sigma^{(d+\lambda)/2}\phi, v) - \left(\sum_{i,j=1}^{d}\left(a_{ij}\left(\sigma^{(d+\lambda)/2}\right)_{,j}v\right)_{,i}, \phi\right)$$

$$- \left(\sum_{i,j=1}^{d} a_{ij}\left(\sigma^{(d+\lambda)/2}\right)_{,j}v_{,i}, \phi\right) - \left(\sum_{i=1}^{d} b_i\left(\sigma^{(d+\lambda)/2}\right)_{,i}v, \phi\right)$$

$$= (\nu\cdot\nabla f, \sigma^{(d+\lambda)/2}\phi) - \left(\sum_{i,j=1}^{d}\left(a_{ij}\left(\sigma^{(d+\lambda)/2}\right)_{,j}v\right)_{,i}, \phi\right)$$

$$- \left(\sum_{i,j=1}^{d} a_{ij}\left(\sigma^{(d+\lambda)/2}\right)_{,j}v_{,i}, \phi\right) - \left(\sum_{i=1}^{d} b_i\left(\sigma^{(d+\lambda)/2}\right)_{,i}v, \phi\right)$$

$$= (F, \phi),$$

where the second step is integration by parts and F is defined by

$$F = \sigma^{(d+\lambda)/2}\nu\cdot\nabla f - \sum_{i,j=1}^{d}\left(a_{ij}\left(\sigma^{(d+\lambda)/2}\right)_{,j}v\right)_{,i}$$

$$- a_{ij}\left(\sigma^{(d+\lambda)/2}\right)_{,j}v_{,i} - b_i\left(\sigma^{(d+\lambda)/2}\right)_{,i}v.$$

Therefore, $\left(\int_\Omega\left|\nabla_2\left(\sigma^{(d+\lambda)/2}v\right)\right|^2 dx\right)^{\frac{1}{2}}$ is bounded by a constant times $\|F\|_{L^2(\Omega)}$. Integrating (8.4.1) and using (8.1.6), we thus find

$$\int_\Omega \sigma^{d+\lambda}|\nabla_2 v|^2 dx \leq C\left(\int_\Omega \sigma^{d+\lambda}|\nabla f|^2 dx + \int_\Omega \sigma^{d+\lambda-2}|\nabla v|^2 dx\right.$$

$$\left. + \int_\Omega \sigma^{d+\lambda-4}v^2 dx\right) + \int_\Omega A(x)^2\sigma^{d+\lambda-2}v^2 dx$$

where $A(x) = \max_{ij}|a_{ij,i}(x)|$. To bound the last integral, we apply Hölder's and Sobolev's (exercise 8.x.12) inequalities:

$$\int_\Omega A(x)^2\sigma^{d+\lambda-2}v^2\,dx \leq \left(\int_\Omega A(x)^{2r}\,dx\right)^{1/r}\left(\int_\Omega \left|\sigma^{(d+\lambda-2)/2}v\right|^{2r'}dx\right)^{1/r'}$$

$$\leq \left(\int_\Omega A(x)^{2r}\,dx\right)^{1/r}\int_\Omega \left|\nabla\left(\sigma^{(d+\lambda-2)/2}v\right)\right|^2\,dx$$

provided that $r' \leq 2d/d - 2$ ($r' < \infty$ if $d = 2$). Thus, we may take $r > 1$ if $d = 2$ or $r \geq 2d/(d+2)$ for $d \geq 3$. We therefore require $A \in L^p(\Omega)$ for $p > 2$ if $d = 2$ and $p \geq 4d/(d+2)$ for $d \geq 3$. It is interesting to note that $p \geq 4d/(d+2)$ does not necessarily imply that $p > d$, the condition for continuity of the coefficients, a_{ij}, for $d \geq 3$. Expanding the last term and inserting the resulting estimate in the previous one yields

$$\int_\Omega \sigma^{d+\lambda}|\nabla_2 v|^2\,dx \leq C\left(\int_\Omega \sigma^{d+\lambda}|\nabla f|^2\,dx + \int_\Omega \sigma^{d+\lambda-2}|\nabla v|^2\,dx\right.$$
$$\left. + \int_\Omega \sigma^{d+\lambda-4}v^2\,dx\right).$$

Recalling (8.3.2) and using the convention that $v_{,0} = v$ we have

$$C_a\int_\Omega \sigma_z^{d+\lambda-2}|\nabla v|^2\,dx \leq \int_\Omega \sigma_z^{d+\lambda-2}\sum_{i,j=1}^d a_{ij}v_{,i}v_{,j}\,dx$$

$$= a(v, \sigma_z^{d+\lambda-2}v) - \int_\Omega \sum_{i,j=1}^d a_{ij}v_{,i}v\left(\sigma_z^{d+\lambda-2}\right)_{,j}$$

$$+ \sum_{i=0}^d b_i v_{,i}v\sigma_z^{d+\lambda-2}\,dx$$

$$(8.4.2) \qquad = \left(\nu\cdot\nabla f, \sigma_z^{d+\lambda-2}v\right) - \int_\Omega \sum_{i,j=1}^d a_{ij}v_{,i}v\left(\sigma_z^{d+\lambda-2}\right)_{,j}$$

$$+ \sum_{i=0}^d b_i v_{,i}v\sigma_z^{d+\lambda-2}\,dx$$

$$= \left((\nu\cdot\nabla f)\sigma_z^{d+\lambda-2}, v\right) - \int_\Omega \sum_{i,j=1}^d a_{ij}v_{,i}v\left(\sigma_z^{d+\lambda-2}\right)_{,j}$$

$$+ \sum_{i=0}^d b_i v_{,i}v\sigma_z^{d+\lambda-2}\,dx.$$

Estimating as before (see 8.x.9), we find

$$(8.4.3) \quad \int_\Omega \sigma_z^{d+\lambda-2} |\nabla v|^2 \, dx \leq C \left(\int_\Omega \sigma^{d+\lambda} |\nabla f|^2 \, dx + \int_\Omega \sigma^{d+\lambda-4} v^2 \, dx \right).$$

Hölder's inequality implies that

$$\int_\Omega \sigma^{d+\lambda-4} v^2 \, dx \leq \left(\int_\Omega \sigma^{(d+\lambda-4)P'} \, dx \right)^{1/P'} \left(\int_\Omega v^{2P} \, dx \right)^{1/P}$$

$$\leq C\zeta^{(d+\lambda-4)+d/P'} \left(\int_\Omega v^{2P} \, dx \right)^{1/P}$$

provided that $P' > d/(4 - \lambda - d)$. The proof is completed by a duality argument.

Let w solve (8.1.2) with right-hand side given by $\text{sign}(v)|v|^{2P-1}$. Then

$$
\begin{aligned}
\|v\|_{L^{2P}(\Omega)}^{2P} &= (\text{sign}(v)|v|^{2P-1}, v) = a(w, v) &&\text{(definition of } w\text{)} \\
&= (\nu \cdot \nabla f, w) &&\text{(definition of } v\text{)} \\
&= -(f, \nu \cdot \nabla w) &&\text{(integration by parts)} \\
&\leq \|f\|_{L^r(\Omega)} \|w\|_{W_{r'}^1(\Omega)} &&\text{(Hölder's inequality)} \\
&\leq C\|f\|_{L^r(\Omega)} \|w\|_{W_{2P/2P-1}^2(\Omega)} &&\text{(Sobolev's inequality 4.x.11)} \\
&\leq C\|f\|_{L^r(\Omega)} \||v|^{2P-1}\|_{L^{2P/2P-1}(\Omega)} &&\text{(by 8.1.3)} \\
&= C\|f\|_{L^r(\Omega)} \|v\|_{L^{2P}(\Omega)}^{2P-1}
\end{aligned}
$$

so that

$$\|v\|_{L^{2P}(\Omega)} \leq C\|f\|_{L^r(\Omega)}.$$

Sobolev's inequality requires that

$$\frac{1}{r'} = \frac{2P-1}{2P} - \frac{1}{d} = 1 - \frac{1}{2P} - \frac{1}{d}$$

so that $r = 2Pd/(2P + d)$. The condition $P' > d/(4 - \lambda - d)$ translates to

$$r < \frac{2d}{2d - 2 + \lambda}.$$

Note that we must have $2P/(2P - 1) = (1 - 1/r + 1/d)^{-1} < \mu$ in order to apply (8.1.3). This condition, which translates to

$$r > \left(1 - \frac{1}{\mu} + \frac{1}{d} \right)^{-1},$$

becomes vacuous if $\mu > d$. If $\mu > \frac{2d}{4-\lambda}$, then there is an open interval of r's that can satisfy both conditions. Therefore, Hölder's inequality implies

$$\int_\Omega \sigma^{d+\lambda-4} v^2 \, dx \le C \zeta^{2d(1-\frac{1}{r})+\lambda-2} \|f\|_{L^r(\Omega)}^2$$

$$\le C \zeta^{2d(1-\frac{1}{r})+\lambda-2} \int_\Omega \sigma_z^{d+\lambda} f^2 \, dx \left(\int_\Omega \sigma_z^{-(d+\lambda)\frac{r}{2-r}} \, dx \right)^{\frac{2-r}{r}}$$

$$\le C \zeta^{2d(1-\frac{1}{r})+\lambda-2} \int_\Omega \sigma_z^{d+\lambda} f^2 \, dx \left(C \zeta^{-(d+\lambda)+d\frac{2-r}{r}} \right)$$

$$\le C \zeta^{-2} \int_\Omega \sigma_z^{d+\lambda} f^2 \, dx.$$

Substituting this into an earlier expression completes the proof of Lemma 8.3.11.   □

(8.4.4) *Remark.* The proofs of the lemmas use Hölder's inequality repeatedly with what may appear to be "magic" indices at first reading. This technique of proof might be called "index engineering." There is a basic scientific principle underlying the inequalities that indicates whether the approach will work or not. For example, in (8.3.7), both sides of the expression have the same "units." The weight σ has the units of length (L) and $|\nabla_2 v|$ is essentially like f. Suppose that f has units F. Then the left-hand side has units of $L^{-\lambda} F^2$ since dx has units of L^d. The units of $|\nabla f|^2$ are $L^{-2} F^2$, because differentiation involves dividing by a length. Noting that ζ has units of L completes the verification. This comparison can be seen more precisely if we scale the x variable by L, and compare the powers of L on each side (they will be equal, cf. exercise 8.x.10).

8.5 L^p Estimates (Regular Coefficients)

The proof that Theorem 8.1.11 holds with p finite, at least for $2 \le p \le \infty$, is given in exercise 8.x.1. Such an estimate can be extended by the following duality argument for $1 < p \le 2$. Let $a_{ij} \in C^0(\overline{\Omega})$. From (Simader 1972), we have (for $\frac{1}{p} + \frac{1}{q} = 1$)

$$\frac{1}{C} \|u_h\|_{W_p^1(\Omega)} \le \sup_{0 \neq v \in \mathring{W}_q^1(\Omega)} \frac{a(u_h, v)}{\|v\|_{W_q^1(\Omega)}} + \|u_h\|_{L^p(\Omega)}$$

$$= \sup_{0 \neq v \in \mathring{W}_q^1(\Omega)} \frac{a(u_h, v_h)}{\|v\|_{W_q^1(\Omega)}} + \|u_h\|_{L^p(\Omega)}$$

$$= \sup_{0 \neq v \in \mathring{W}_q^1(\Omega)} \frac{a(u, v_h)}{\|v\|_{W_q^1(\Omega)}} + \|u_h\|_{L^p(\Omega)},$$

where v_h represents the projection of v with respect to $a(\cdot, \cdot)$. Applying exercise 8.x.1 and Hölder's inequality, we find

$$\|u_h\|_{W_p^1(\Omega)} \le C\left(\|u\|_{W_p^1(\Omega)} + \|u_h\|_{L^p(\Omega)}\right).$$

A bound for $\|u_h\|_{L^p(\Omega)}$ can be obtained easily by duality, modifying the proof of Theorem 5.4.8 only slightly. Consider the variational problem: find $w \in V$ such that

$$(8.5.1) \qquad a(v,w) = (f,v) \quad \forall v \in V.$$

This has the property that for $1 < q \le \infty$

$$(8.5.2) \qquad \|w\|_{W_q^1(\Omega)} \le C\|f\|_{L^q(\Omega)}$$

as follows from (8.1.3), provided $\mu > d$ (see exercise 8.x.14):

$$
\begin{aligned}
\|w\|_{W_q^1(\Omega)} &\le C\|w\|_{W_r^2(\Omega)} && \text{(Sobolev's inequality 4.x.11)} \\
&\le C\|f\|_{L^r(\Omega)} && \text{(by 8.1.3)} \\
&\le C\|f\|_{L^q(\Omega)}. && \text{(Hölder's inequality)}
\end{aligned}
$$

Applying this to $f = |u_h|^{p-1}\,\mathrm{sign}(u_h)$ and with $\frac{1}{p} + \frac{1}{p'} = 1$ we find

$$
\begin{aligned}
\|u_h\|_{L^p(\Omega)}^p &= a(u_h, w) && \text{(from 8.5.1)} \\
&= a(u_h, w_h) && (w_h \text{ Galerkin projection of } w) \\
&= a(u, w_h) && (u_h \text{ Galerkin projection of } u) \\
&\le C\|u\|_{W_p^1(\Omega)}\|w_h\|_{W_{p'}^1(\Omega)} && \text{(Hölder's inequality)} \\
&\le C\|u\|_{W_p^1(\Omega)}\|w\|_{W_{p'}^1(\Omega)} && \text{(exercise 8.x.1)} \\
&\le C\|u\|_{W_p^1(\Omega)}\left\||u_h|^{p-1}\,\mathrm{sign}(u_h)\right\|_{L^{p'}(\Omega)} && \text{(from 8.5.2)} \\
&= C\|u\|_{W_p^1(\Omega)}\|u_h\|_{L^p(\Omega)}^{p-1}.
\end{aligned}
$$

Therefore,

$$\|u_h\|_{L^p(\Omega)} \le C\|u\|_{W_p^1(\Omega)} \quad \forall\, 1 < p \le \infty$$

as the case $2 \le p \le \infty$ was treated earlier. Combining previous estimates proves the following.

(8.5.3) Theorem. *Under the conditions of Theorem 8.1.11 and assuming $a_{ij} \in C^0(\overline{\Omega})$, there is an $h_0 > 0$ and $C < \infty$ such that*

$$\|u_h\|_{W_p^1(\Omega)} \le C\|u\|_{W_p^1(\Omega)} \quad \forall\, 1 < p \le \infty$$

for $0 < h < h_0$.

Of course, it follows that

$$(8.5.4) \qquad \|u - u_h\|_{W_p^1(\Omega)} \le C \inf_{v \in V_h} \|u - v\|_{W_p^1(\Omega)} \quad \forall\, 1 < p \le \infty$$

by applying the previous result to $u_h - v$ and using the triangle inequality (cf. exercise 8.x.2). We leave to the reader (exercise 8.x.13) the proof of the following using duality techniques:

$$(8.5.5) \qquad \|u - u_h\|_{L^p(\Omega)} \le Ch \, \|u - u_h\|_{W_p^1(\Omega)} \quad \forall \, \mu' < p < \infty$$

for $0 < h < h_0$, where $\frac{1}{\mu} + \frac{1}{\mu'} = 1$ (cf. (8.1.3)).

The case $p = \infty$ is special due, in part, to the lack of regularity in (8.1.3) for $p = 1$. The best order of approximation by piecewise linears, at least in two dimensions (Scott 1976 and Haverkamp 1984), is

$$\|u - u_h\|_{L^\infty(\Omega)} \le Ch|\log h| \, \|u - u_h\|_{W_\infty^1(\Omega)} \, .$$

For higher-degree approximation, one can show (Scott 1976) that

$$\|u - u_h\|_{L^\infty(\Omega)} \le Ch \, \|u - u_h\|_{W_\infty^1(\Omega)} \, .$$

Rather than derive more L^p estimates in the case of regular coefficients, we now show that some estimates can be derived for very rough coefficients.

8.6 L^p Estimates (Irregular Coefficients)

The estimates in the previous section were based on the results from earlier sections which yielded estimates for the gradient of the error in maximum norm. In particular, it was necessary to assume some conditions on the coefficients in the bilinear form that might not always hold. In this section we give a more general result. For simplicity, we restrict to the symmetric case, that is, we assume that

$$(8.6.1) \qquad a(u, v) := \int_\Omega \sum_{i,j=1}^d a_{ij}(x) \frac{\partial u}{\partial x_i}(x) \frac{\partial v}{\partial x_j}(x) \, dx$$

but we only assume that a_{ij} are bounded, measurable coefficients.

(8.6.2) Proposition. *Suppose that $a(\cdot, \cdot)$ is as given in (8.6.1), where a_{ij} are bounded, measurable coefficients such that (8.1.1) holds. Then there are constants $\alpha < \infty$, $h_0 > 0$ and $\epsilon > 0$ such that for all $0 < h \le h_0$ and $u_h \in V_h$*

$$|u_h|_{W_p^1(\Omega)} \le \alpha \sup_{0 \neq v_h \in V_h} \frac{a(u_h, v_h)}{|v_h|_{W_q^1(\Omega)}},$$

whenever $|2 - p| \le \epsilon$, where q is the dual index to p, $\frac{1}{p} + \frac{1}{q} = 1$.

Before proving this result, let us relate it to estimates for the finite element projection. Let $P_h u$ be the projection with respect to the bilinear form $a(u,v)$ of an element $u \in V$ onto V_h, i.e., $P_h u$ is the unique element of V_h which satisfies

$$a(u - P_h u, v) = 0 \quad \forall\, v \in V_h.$$

(8.6.3) Corollary. *Under the assumptions of Proposition 8.6.2, the projection P_h is stable in $W_p^1(\Omega)$, i.e., there is a positive constant C, independent of h and u, such that*

$$\|P_h u\|_{W_p^1(\Omega)} \leq C \, \|u\|_{W_p^1(\Omega)}, \quad |2 - p| \leq \epsilon.$$

This follows from Proposition 8.6.2 because

$$\|P_h u\|_{W_p^1(\Omega)} \leq C \, |P_h u|_{W_p^1(\Omega)} \qquad \text{(Poincaré's inequality 5.3.5)}$$

$$\leq C \sup_{0 \neq v_h \in V_h} \frac{a(P_h u, v_h)}{|v_h|_{W_q^1(\Omega)}} \qquad \text{(by 8.6.2)}$$

$$= C \sup_{0 \neq v_h \in V_h} \frac{a(u, v_h)}{|v_h|_{W_q^1(\Omega)}} \qquad \text{(definition of } P_h)$$

$$\leq C \, |u|_{W_p^1(\Omega)}. \qquad \text{(Hölder's inequality)}$$

For simplicity, we use the following definitions for the remainder of the proof. Define the L^p norm on vector (d component) functions, $\mathbf{F}$, via

$$\|\mathbf{F}\|_{L^p(\Omega)^d} := \left(\int_\Omega |\mathbf{F}(x)|^p \, dx \right)^{1/p}$$

where $|\mathbf{F}(x)|$ denotes the Euclidean length of $\mathbf{F}(x)$. Note that with this definition of vector norm, Hölder's inequality takes a convenient form, namely

$$\left| \int_\Omega \mathbf{F}(x) \cdot \mathbf{G}(x) \, dx \right| \leq \|\mathbf{F}\|_{L^p(\Omega)^d} \|\mathbf{G}\|_{L^q(\Omega)^d} \quad \text{where} \quad \frac{1}{p} + \frac{1}{q} = 1.$$

We also use the following definition of Sobolev semi-norm which is equivalent with our earlier Definition 1.3.7, namely

$$|v|_{W_p^1(\Omega)} := \|\nabla v\|_{L^p(\Omega)^d}.$$

The proof of Proposition 8.6.2 is based on the ideas of (Meyers 1963). First, we establish the corresponding inequality in the case that the bilinear form in question is much simpler. Consider the projection $P_h^* : V \to V_h$ defined by

$$\langle \nabla(P_h^* u - u), \nabla v_h \rangle = 0 \quad \forall v_h \in V_h,$$

where $\langle \cdot, \cdot \rangle$ denotes the vector-L^2 inner-product on Ω. The results in the previous section imply that

$$(8.6.4) \qquad |P_h^* u|_{W_p^1(\Omega)} \leq C^* \, |u|_{W_p^1(\Omega)}, \quad 1 < p \leq \infty.$$

From (Simader 1972) or (Meyers 1963) we have

$$(8.6.5) \qquad |w|_{W_p^1(\Omega)} \leq c_p \sup_{0 \neq v \in \mathring{W}_q^1(\Omega)} \frac{\langle \nabla w, \nabla v \rangle}{|v|_{W_q^1(\Omega)}} \quad \forall w \in \mathring{W}_p^1(\Omega),$$

where the constant c_p was observed to be log-convex (hence continuous) as a function of $1/p$ by (Meyers 1963). Obviously, $c_2 = 1$. Thus,

$$
\begin{aligned}
|u_h|_{W_p^1(\Omega)} &\leq c_p \sup_{0 \neq v \in \mathring{W}_q^1(\Omega)} \frac{\langle \nabla u_h, \nabla v \rangle}{|v|_{W_q^1(\Omega)}} \\
&= c_p \sup_{0 \neq v \in \mathring{W}_q^1(\Omega)} \frac{\langle \nabla u_h, \nabla P_h^* v \rangle}{|v|_{W_q^1(\Omega)}} \\
&\leq c_p C_p \sup_{0 \neq v \in \mathring{W}_q^1(\Omega)} \frac{\langle \nabla u_h, \nabla P_h^* v \rangle}{|P_h^* v|_{W_q^1(\Omega)}} \\
&= c_p C_p \sup_{0 \neq v_h \in V_h} \frac{\langle \nabla u_h, \nabla v_h \rangle}{|v_h|_{W_q^1(\Omega)}}
\end{aligned}
$$

(8.6.6)

for all $1 < p < \infty$, where

$$C_p := \|P_h^*\|_{\mathring{W}_p^1(\Omega) \to \mathring{W}_p^1(\Omega)} = \sup_{0 \neq v \in \mathring{W}_p^1(\Omega)} \frac{|P_h^* v|_{W_p^1(\Omega)}}{|v|_{W_p^1(\Omega)}} \leq C^*.$$

Obviously, $C_2 = 1$. We now wish to show that $C_p = \|P_h^*\|_{\mathring{W}_p^1(\Omega) \to \mathring{W}_p^1(\Omega)}$ is continuous as a function of p, uniformly in h.

We view P_h^* as inducing a mapping, $\mathcal{P}_h^* : \nabla u \longrightarrow \nabla u_h$, of $L^p(\Omega)^d$ to itself. More precisely, given $\mathbf{F} \in L^p(\Omega)^d$, let $u_h(\mathbf{F}) \in V_h$ solve

$$\langle \nabla u_h, \nabla v_h \rangle = \langle \mathbf{F}, \nabla v_h \rangle \quad \forall v_h \in V_h.$$

Then $\mathcal{P}_h^*(\mathbf{F}) := \nabla u_h$. Similarly, we can let $u \in \mathring{W}_p^1(\Omega)$ solve

$$\langle \nabla u, \nabla v \rangle = \langle \mathbf{F}, \nabla v \rangle \quad \forall v \in \mathring{W}_q^1(\Omega),$$

and we have

$$
\begin{aligned}
|u|_{W_p^1(\Omega)} &\leq c_p \sup_{0 \neq v \in \mathring{W}_q^1(\Omega)} \frac{\langle \nabla u, \nabla v \rangle}{|v|_{W_q^1(\Omega)}} && \text{(by 8.6.5)} \\
&= c_p \sup_{0 \neq v \in \mathring{W}_q^1(\Omega)} \frac{\langle \mathbf{F}, \nabla v \rangle}{|v|_{W_q^1(\Omega)}} && \text{(definition of } u\text{)} \\
&\leq c_p \|\mathbf{F}\|_{L^p(\Omega)^d} \, . && \text{(Hölder's inequality)}
\end{aligned}
$$

Therefore,

$$\|\mathcal{P}_h^* \mathbf{F}\|_{L^p(\Omega)^d} \leq C^* c_p \|\mathbf{F}\|_{L^p(\Omega)^d} \, .$$

Using operator interpolation (see Chapter 14), we conclude that for $P > 2$

$$\|\mathcal{P}_h^* \mathbf{F}\|_{[L^2(\Omega)^d, L^P(\Omega)^d]_\theta} \leq (C^* c_P)^\theta \|\mathbf{F}\|_{[L^2(\Omega)^d, L^P(\Omega)^d]_\theta}.$$

Whether using the real interpolation method (with appropriate second index) or the complex interpolation method (Bergh & Löfstrom 1976),

$$c(p)^{-1} \|\mathbf{F}\|_{L^{2(1-\theta) + P\theta}(\Omega)^d}$$
$$\leq \|\mathbf{F}\|_{[L^2(\Omega)^d, L^P(\Omega)^d]_\theta}$$
$$\leq c(p) \|\mathbf{F}\|_{L^{2(1-\theta) + P\theta}(\Omega)^d}$$

where $c(2) = 1$ and $c(p)$ is a smooth function of p near $p = 2$. Therefore,

$$\|P_h^*\|_{\mathring{W}_p^1(\Omega) \to \mathring{W}_p^1(\Omega)} \leq c(p)^2 (C^* c_P)^\theta, \quad p = 2(1-\theta) + P\theta.$$

A similar inequality holds for $P < 2$. By taking θ small, we can make $(C^* c_P)^\theta$ as close to one as we like. Thus, for all $\delta > 0$ there exists $\epsilon > 0$ such that

(8.6.7) $\qquad |u_h|_{W_p^1(\Omega)} \leq (1+\delta) \sup_{0 \neq v_h \in V_h} \dfrac{\langle \nabla u_h, \nabla v_h \rangle}{|v_h|_{W_q^1(\Omega)}} \quad \forall |2 - p| \leq \epsilon$

where $\frac{1}{p} + \frac{1}{q} = 1$ and δ and ϵ are independent of h

We now consider the general case via a perturbation argument. Let M be a constant such that

(8.6.8) $\qquad \displaystyle\sum_{i,j=1}^d a_{ij}(x)\xi_i\xi_j \leq M|\xi|^2 \quad$ for almost all $x \in \Omega$.

Define a bilinear form $\mathcal{B} : W_p^1(\Omega) \times W_q^1(\Omega) \to \mathbb{R}$ by

$$\mathcal{B}(u,v) := \langle \nabla u, \nabla v \rangle - \frac{1}{M} a(u,v),$$
$$= \int_\Omega \sum_{i,j=1}^d B_{ij}(x) \frac{\partial u}{\partial x_i} \frac{\partial v}{\partial x_j} \, dx.$$

It follows from (8.1.1) and (8.6.8) that the eigenvalues of $(B_{ij}(x))$ are in $[0, 1 - \frac{C_a}{M}]$ for almost all $x \in \Omega$ (see exercise 8.x.17). Note that $C_a/M < 1$. Therefore, exercise 8.x.17 and Hölder's inequality imply that

(8.6.9)
$$|\mathcal{B}(u,v)| \leq \left(1 - \frac{C_a}{M}\right) \int_\Omega |\nabla u(x)| \, |\nabla v(x)| \, dx$$
$$\leq \left(1 - \frac{C_a}{M}\right) |u|_{W_p^1(\Omega)} |v|_{W_q^1(\Omega)}.$$

Using the identity

$$\langle \nabla u, \nabla v \rangle \;=\; \mathcal{B}(u,v) + \frac{1}{M} a(u,v),$$

together with estimates (8.6.7) and (8.6.9) yields

$$\left(\frac{1}{1+\delta} - \left(1 - \frac{C_a}{M}\right) \right) |u_h|_{W_p^1(\Omega)} \le \frac{1}{M} \sup_{0 \neq v_h \in V_h} \frac{a(u_h, v_h)}{|v_h|_{W_q^1(\Omega)}}.$$

Let $\delta = \left(\frac{1}{2}M - C_a\right)/\left(M - C_a\right)$ and choose ϵ to be as given in (8.6.7) for this choice of δ. This completes the proof of Proposition 8.6.2. Note that ϵ and α depend only on the constants C_a in (8.1.1), C^* in (8.6.4) and M in (8.6.8). $\qquad\square$

8.7 A Nonlinear Example

We consider a very simple model problem in two dimensions to show how the refined estimates of the previous section can obtain basic existence results for nonlinear problems. Let

$$(8.7.1) \qquad\qquad a(u,v;w) := \int_\Omega A(w)\nabla u \cdot \nabla v \, dx$$

where the function $A : \mathbb{R} \to \mathbb{R}$ satisfies

$$(8.7.2) \qquad\qquad 0 < C_a \le A(s) \quad \forall s \in \mathbb{R}.$$

Further, we assume only that $A(\cdot)$ is bounded on bounded subsets of $\mathbb{R}$. We seek u such that

$$(8.7.3) \qquad\qquad a(u,v;u) = F(v) \quad \forall v \in V$$

with, say, $V = \overset{\circ}{H}^1(\Omega)$.

Note that $a(u,u;u)$ is not actually defined for arbitrary $u \in H^1(\Omega)$, so the variational formulation of such a problem requires some elaboration. We will not dwell on such issues, but instead we simply show how one can insure the existence of a solution to the discrete problem

$$(8.7.4) \qquad\qquad a(u_h,v;u_h) = F(v) \quad \forall v \in V_h$$

under suitable conditions. Let

$$(8.7.5) \qquad\qquad V_h^{K,p} := \left\{ v \in V_h \;:\; \|v\|_{W_p^1(\Omega)} \le K \right\}$$

for a given $K > 0$. Define the simple iteration map, $T_h : V_h \to V_h$ via

$$(8.7.6) \qquad\qquad a(T_h u_h, v; u_h) = F(v) \quad \forall v \in V_h.$$

This is always well defined, since $u_h \in V_h$ implies $A(u_h) \in L^\infty(\Omega)$.

(8.7.7) Theorem. *Suppose that A is bounded on bounded sets and (8.7.2) holds. There exists $K > 0$, $p > 2$, $h_0 > 0$ and $\delta > 0$ such that for any F such that $\|F\|_{W_p^{-1}} \leq \delta$, T_h maps $V_h^{K,p}$ into itself for all $0 < h \leq h_0$.*

Proof. For any $u_h \in V_h^{K,p}$, $A(u_h)$ satisfies the conditions of Proposition 8.6.2 with the constant in (8.6.8) bounded by

$$(8.7.8) \qquad\qquad M = \sup\{A(s) \ : \ |s| \leq c_p K\}$$

where c_p is the constant in Sobolev's inequality:

$$\|v\|_{L^\infty(\Omega)} \leq c_p \|v\|_{W_p^1(\Omega)} \quad \forall v \in W_p^1(\Omega)$$

because $v \in V_h^{K,p}$ implies $\|v\|_{L^\infty(\Omega)} \leq c_p K$ and hence $\|A(v)\|_{L^\infty(\Omega)} \leq M$. For sufficiently small K (e.g., $K = C/c_p$) there is a $p > 2$ such that the inequality in Proposition 8.6.2 holds. Then

$$
\begin{aligned}
\|T_h u_h\|_{W_p^1(\Omega)} &\leq \alpha \sup_{0 \neq v_h \in V_h} \frac{a(T_h u_h, v_h; u_h)}{|v_h|_{W_q^1(\Omega)}} \qquad \text{(from 8.6.2)} \\
&= \alpha \sup_{0 \neq v_h \in V_h} \frac{F(v_h)}{|v_h|_{W_q^1(\Omega)}} \\
&\leq C\|F\|_{W_p^{-1}}.
\end{aligned}
$$

Choose $\delta = K/C$. $\qquad\qquad\qquad\qquad\qquad\qquad\qquad\qquad\qquad\qquad\qquad\square$

(8.7.9) Corollary. *Suppose A is continuous and (8.7.2) holds. Let $K > 0$, $p > 2$, $h_0 > 0$ and $\delta > 0$ be as in Theorem 8.7.7. For $\|F\|_{W_p^{-1}} \leq \delta$, (8.7.4) has a solution u_h which satisfies*

$$\|u_h\|_{W_p^1(\Omega)} \leq K$$

for $0 < h \leq h_0$.

Proof. Apply the Brouwer fixed-point theorem (cf. Dugundji 1966).

Not only does the corollary guarantee a solution to the discrete problem which remains uniformly bounded as $h \to 0$, it also provides a stability result because the family of problems (8.7.4) are all uniformly continuous (and coercive) independent of h. This allows us to establish convergence estimates as follows:

$$
\begin{aligned}
a(u - u_h, v; u_h) &= a(u, v; u_h) - F(v) \qquad\quad \text{(from 8.7.4)} \\
&= a(u, v; u_h) - a(u, v; u) \qquad \text{(from 8.7.3)} \\
&= \int_\Omega (A(u) - A(u_h)) \nabla u \cdot \nabla v \, dx
\end{aligned}
$$

for any $v \in V_h$. Assuming $u \in W_p^1(\Omega)$ and $A \in W_\infty^1(I)$ for any bounded interval $I \subset \mathbb{R}$, we find for any $v \in V_h$

$$a(u-u_h, u - u_h; u_h) = a(u - u_h, u - v; u_h) + a(u - u_h, v - u_h; u_h)$$

$$= a(u - u_h, u - v; u_h) + \int_\Omega (A(u) - A(u_h)) \nabla u \cdot \nabla (v - u_h)\, dx$$

$$\leq M \|u - u_h\|_{H^1(\Omega)} \|u - v\|_{H^1(\Omega)} \qquad \text{(from 8.7.8)}$$

$$+ \int_\Omega |A(u) - A(u_h)| |\nabla u \cdot \nabla (v - u_h)|\, dx$$

$$\leq M \|u - u_h\|_{H^1(\Omega)} \|u - v\|_{H^1(\Omega)}$$

$$+ \|A\|_{W_\infty^1(I)} \int_\Omega |u - u_h| |\nabla u \cdot \nabla (v - u_h)|\, dx$$

$$\leq M \|u - u_h\|_{H^1(\Omega)} \|u - v\|_{H^1(\Omega)}$$

$$+ \|A\|_{W_\infty^1(I)} \| |u - u_h| |\nabla u| \|_{L^2(\Omega)} \|v - u_h\|_{H^1(\Omega)}$$

$$\leq M \|u - u_h\|_{H^1(\Omega)} \|u - v\|_{H^1(\Omega)}$$

$$+ \|A\|_{W_\infty^1(I)} \|u - u_h\|_{L^q(\Omega)} \|u\|_{W_p^1(\Omega)} \|v - u_h\|_{H^1(\Omega)}$$

$$\leq \|u - u_h\|_{H^1(\Omega)} \Big(M \|u - v\|_{H^1(\Omega)} \qquad \text{(Sobolev's inequality 4.x.11)}$$

$$+ c_p' \|A\|_{W_\infty^1(I)} \|u\|_{W_p^1(\Omega)} \|v - u_h\|_{H^1(\Omega)} \Big)$$

where $I = [0, \max\{M, \|u\|_{L^\infty(\Omega)}\}]$, $q = 2p/(p-2)$ and we have used Hölder's inequality repeatedly. Therefore,

$$\text{(8.7.10)} \qquad \|u - u_h\|_{H^1(\Omega)} \leq \frac{1}{C_a} \Big(M \|u - v\|_{H^1(\Omega)}$$

$$+ c_p' \|A\|_{W_\infty^1(I)} \|u\|_{W_p^1(\Omega)} \|v - u_h\|_{H^1(\Omega)} \Big).$$

From the triangle inequality, we find

$$(1 - \gamma) \|u - u_h\|_{H^1(\Omega)} \leq \left(\frac{M}{C_a} + \gamma \right) \|u - v\|_{H^1(\Omega)}$$

where

$$\gamma := \frac{c_p'}{C_a} \|A\|_{W_\infty^1(I)} \|u\|_{W_p^1(\Omega)}.$$

Thus, if $\gamma < 1$ we obtain a result similar to Ceá's Theorem:

$$\text{(8.7.11)} \qquad \|u - u_h\|_{H^1(\Omega)} \leq \frac{M + \gamma C_a}{(1 - \gamma) C_a} \|u - v\|_{H^1(\Omega)} \qquad \forall v \in V_h.$$

Using only slightly more complicated techniques than used above (cf. Meyers 1963, Douglas & Dupont 1975), one can show the existence of a solution $u \in W_p^1(\Omega)$ to (8.7.3) under the conditions of Theorem 8.7.7 for sufficiently small $\delta > 0$, using a map $T : \mathring{W}_p^1(\Omega) \to \mathring{W}_p^1(\Omega)$ defined by

$$a(Tu, v; u) = F(v) \qquad \forall v \in \mathring{W}_q^1(\Omega).$$

Such a result follows more simply from the fact that both T and T_h are Lipschitz continuous, provided A is Lipschitz continuous.

We show that T_h is Lipschitz continuous, as this also demonstrates both the uniqueness of the solution and the convergence of the fixed point iteration,

$$(8.7.12) \qquad\qquad u_h^{n+1} := T_h u_h^n \,,$$

to u_h as $n \to \infty$. For $v, w, \phi \in V_h$, we find using the techniques leading to (8.7.10) that

$$|a(T_h v - T_h w, \phi; v)| = \left| \int_\Omega (A(w) - A(v)) \, \nabla T_h w \cdot \nabla \phi \, dx \right|$$
$$\leq c_p' \|A\|_{W_\infty^1 (0,M)} \|w - v\|_{H^1(\Omega)} \|T_h v\|_{W_p^1(\Omega)} \|\phi\|_{H^1(\Omega)} \,.$$

Choosing $\phi = T_h v - T_h w$ yields

$$(8.7.13) \qquad \|T_h v - T_h w\|_{H^1(\Omega)} \leq \frac{c_p' \|A\|_{W_\infty^1 (0,M)} K}{C_a} \|w - v\|_{H^1(\Omega)}$$

for all $v, w \in V_h^{K,p}$. A similar result can be proved for T. Note that K can be made arbitrarily small by choosing δ appropriately. We collect these results in the following.

(8.7.14) Theorem. *Suppose that A is Lipschitz continuous and (8.7.2) holds. There exist $\delta > 0$, $h_0 > 0$ and $p > 2$ such that (8.7.3) and (8.7.4), for all $0 < h \leq h_0$, have unique solutions for arbitrary F such that $\|F\|_{W_p^{-1}} \leq \delta$. Moreover, u_h can be approximated to arbitrary accuracy via the simple iteration (8.7.12), which involves solving only linear equations at each step. Finally, the error $u - u_h$ satisfies (8.7.11).*

Using more sophisticated techniques and further assumptions on the coefficient A, (Douglas & Dupont 1975) prove stronger results of a global character and consider more efficient iterative techniques such as Newton's method. For another application of the results of the previous section to a different type of nonlinear problem, see (Saavedra & Scott 1991).

8.x Exercises

8.x.1 Prove Theorem 8.1.11 with ∞ replaced by p, for $2 < p < \infty$. (Hint: use Hölder's inequality in such a way to leave a weighted L^p norm of ∇u in (8.2.5), integrate the p-th power of (8.2.3) and then apply Fubini, cf. Rannacher & Scott 1982.)

8.x.2 Prove (8.1.12). (Hint: write $u - u_h = u - \mathcal{I}^h u + (\mathcal{I}^h u - u_h)$ and apply (8.1.11) to $\mathcal{I}^h u - u$ playing the role of u.)

8.x.3 Prove (8.1.4). How does the constant depend on κ and h?

8.x.4 Prove (8.1.5). How does the constant depend on κ and h?

8.x.5 Prove (8.1.6). How does the constant depend on κ and h? (Hint: consider first the case $\lambda = 1$ and then use induction.)

8.x.6 Prove (8.1.7). How does the constant depend on κ and h? (Hint: use (8.1.4).)

8.x.7 Prove (8.2.1). How does the constant depend on k? (Hint: see (Scott 1976).)

8.x.8 Prove that (8.3.10) follows from Lemma 8.3.11. (Hint: use (8.1.5).)

8.x.9 Derive (8.4.3) from (8.4.2) in detail. (Hint: use Schwarz' inequality and 0.9.5, splitting the weight functions appropriately.)

8.x.10 Do the change of variables suggested in Remark (8.4.4) and verify that the powers of L on each side of the inequality are the same.

8.x.11 Prove Sobolev's embedding $\mathring{W}_1^1(\Omega) \subset L^2(\Omega)$ in two dimensions. (Hint: write

$$u(x_1, x_2)^2 = \int_{-\infty}^{x_1} \frac{\partial u}{\partial x_1}(t, x_2)\, dt \int_{-\infty}^{x_2} \frac{\partial u}{\partial x_2}(x_1, s)\, ds,$$

integrate with respect to x and use Fubini.)

8.x.12 Prove Sobolev's inequality $\|u\|_{L^p(\Omega)} \le c_p \|u\|_{W_q^1(\Omega)}$ for $u \in \mathring{W}_q^1(\Omega)$, where $1 + \frac{2}{p} = \frac{2}{q}$, in two dimensions. How does c_p behave as $p \to \infty$? (Hint: write

$$\|u\|_{L^p(\Omega)}^p = \left\| |u|^{p/2} \right\|_{L^2(\Omega)}^2 \le C \left\| |u|^{p/2} \right\|_{W_1^1(\Omega)}^2$$

using exercise 8.x.11. Expand $\nabla \left(|u|^{p/2} \right)$ and apply Hölder's inequality.)

8.x.13 Prove (8.5.5) under the conditions of Theorem 8.1.11. (Hint: solve (8.5.1) with $f = |u - u_h|^{p-1} \operatorname{sign}(u - u_h)$ and follow the subsequent estimates.)

8.x.14 What value for r should be chosen in the proof of (8.5.2)? (Hint: let q and r be related by Sobolev's inequality 4.x.11.)

8.x.15 Prove (8.1.7) for Lagrange, Hermite and Argyris elements in two and three dimensions on a quasi-uniform family of triangulations and for tensor-product and serendipity elements on a quasi-uniform rectangular subdivision. (Hint: see the proof of Theorem 4.4.20 and use (8.1.4).)

8.x.16 Prove (8.1.9) for all the piecewise-polynomial finite-elements introduced in Chapter 3 on a quasi-uniform subdivision. (Hint: see the proof of Theorem 4.5.11 and use (8.1.4).)

8.x.17 Let $A = (a_{ij})$ be a $d \times d$ symmetric matrix such that

$$C_a |\xi|^2 \le \sum_{i,j=1}^d a_{ij} \xi_i \xi_j \le M |\xi|^2 \quad \forall \xi \in \mathbb{R}^d$$

where $C_a > 0$. Prove that $AX = \lambda X$ for some vector $X \ne 0$ implies $C_a \le \lambda \le M$. Moreover, show that C_a (resp. M) can be taken to be the minimum (resp. maximum) of such λ. Finally, show that

$$\sum_{i,j=1}^d a_{ij} \xi_i \nu_j \le M |\xi| |\nu| \quad \forall \xi, \nu \in \mathbb{R}^d.$$

8.x.18 Prove Theorem 8.1.11 for the adjoint problem (8.1.2′), with u_h defined by $a(v, u_h) = (f, v)$ for $v \in V_h$. (Hint: consider the case when

$$a(u, v) := \int_\Omega \sum_{i,j=1}^d a_{ij} \frac{\partial u}{\partial x_i} \frac{\partial v}{\partial x_j} + \sum_{i=1}^d b_i u \frac{\partial v}{\partial x_i} + b_0 uv \, dx$$

and make appropriate changes to the arguments.)

8.x.19 Suppose $G_h^z \in V_h$ is the discrete Green's function defined by

$$a(G_h^z, v) = v(z) \quad \forall v \in V_h$$

for some space V_h of piecewise polynomials of degree m in a two-dimensional domain Ω. Prove that

$$G_h^z(z) \le C |\log h|.$$

(Hint: since $G_h^z \in V_h$, we have $G_h^z(z) = a(G_h^z, G_h^z)$. Apply the discrete Sobolev inequality (4.9.2) to prove that

$$G_h^z(z) \le C |\log h|^{1/2} a(G_h^z, G_h^z)^{1/2}.$$

See also (Bramble 1966).)

8.x.20 Suppose $G_h^z \in V_h$ is the discrete Green's function defined by

$$a(G_h^z, v) = v(z) \quad \forall v \in V_h$$

for some space V_h of piecewise polynomials of degree m satisfying Dirichlet boundary conditions on $\partial \Omega$, in a two-dimensional domain Ω. Suppose also that the distance from z to $\partial \Omega$ is $O(h)$. Prove that

$$G_h^z(z) \leq C.$$

(Hint: see exercise 8.x.19. Since $G_h^z = 0$ on $\partial\Omega$, we have

$$G_h^z(z) \leq Ch\|G_h^z\|_{W_\infty^1(\Omega)}.$$

Use an inverse inequality instead of (4.9.2) to complete the proof. See also (Draganescu, Dupont & Scott 2002).)

Chapter 9

Adaptive Meshes

In Sect. 0.8, we demonstrated the possibility of dramatic improvements in approximation power resulting from adaptive meshes. In current computer simulations, meshes are often adapted to the solution either using *a priori* information regarding the problem being solved or *a posteriori* after an initial attempt at solution (Babuška et al. 1983 & 1986). The resulting meshes tend to be strongly graded in many important cases, no longer being simply modeled as quasi-uniform. Here we present some basic estimates that show that such meshes can be effective in approximating difficult problems. For further references, see (Eriksson, Estep, Hansbo and Johnson 1995), (Verfürth 1996), (Ainsworth and Oden 2000), (Becker and Rannacher 2001), (Babuška and Strouboulis 2001) and (Dörfler and Nochetto 2001).

First of all, we need to see that the resulting finite element (Galerkin) method does indeed provide the appropriate approximation on strongly graded meshes. The approximation theory results of Chapter 4 are sufficiently localized to guarantee good approximation from non-degenerate meshes, but it must be shown that the Galerkin method, which is based on global information, will actually provide localized approximation.

Once we know that graded meshes can give local benefit, we turn to the question of predicting where mesh refinement is needed. Error estimators provide such a guide. We give a brief introduction to some key ideas in one simple case.

Finally, the value of the localized approximation will be lost if the resulting linear equations are ill-conditioned. We show that under mild restrictions, this does not happen.

Throughout, we assume Ω is a bounded, polyhedral domain in $\mathbb{R}^n$, $n \geq 2$ (for the one-dimensional case, see Sect. 0.8). We consider a general elliptic equation as in Sect. 5.7, where the finite element spaces satisfy the approximation properties of Theorem 4.4.4. For simplicity, we assume that the variational form is

$$(9.0.1) \qquad a(v, w) = \int_\Omega \alpha(x) \nabla v \cdot \nabla w \, dx$$

where $\alpha(x)$ denotes a function in $L^\infty(\Omega)$ such that $\alpha(x) \geq \alpha_0 > 0$ for all

$x \in \Omega$. However, the coefficient α need not be smooth. We take V as in (5.1.3).

9.1 A priori Estimates

We restrict our attention to a non-degenerate family (see Definition 4.4.13) of simplicial meshes, $\mathcal{T}^h$. Suppose $h(x)$ is a function that measures the local mesh size near the point x. In particular, we will assume that h is a piecewise linear function satisfying

$$(9.1.1) \qquad h(x) = \max_{K \in \mathrm{star}(x)} \mathrm{diam}\,(K)$$

for each vertex x, where $\mathrm{star}(x)$ denotes the union of simplices, $K \in \mathcal{T}^h$, meeting at x. Note that for all simplices, $K \in \mathcal{T}^h$,

$$(9.1.2) \qquad h(x) \geq \mathrm{diam}\,(K) \quad \forall x \in K,$$

since this holds at each vertex of K and h is linear between them. Furthermore,

$$(9.1.3) \qquad h|_K \leq C \mathrm{diam}\,(K)$$

because a non-degenerate mesh is locally quasi-uniform, in two or higher dimensions (exercise 9.x.1). Correspondingly, we assume that α has comparable values on each element, namely

$$(9.1.4) \qquad \max\{\alpha(x) \,:\, x \in K\} \leq C \min\{\alpha(x) \,:\, x \in K\} \quad \forall K \in \mathcal{T}^h.$$

Note that (9.1.4) does not preclude large jumps in α, it just implies that the mesh has been chosen to match such jumps. For example, it could be possible to have α piecewise constant and the constant $C = 1$ in (9.1.4), independent of the size of the different constant values of α.

We begin by deriving a basic estimate analogous to Theorem 5.7.6. Suppose that V_h is a space of piecewise polynomials of degree less than k and let $\mathcal{I}^h$ be a corresponding interpolant. From Theorem 4.4.4 and (9.1.1), it follows that

$$\left\| u - \mathcal{I}^h u \right\|_{H^1(K)}^2 \leq C \sum_{|\beta|=k} \int_K h(x)^{2k-2} \left| D^\beta u \right|^2 dx \quad \forall K \in \mathcal{T}^h.$$

Define a natural "energy" (semi-)norm by

$$(9.1.5) \qquad \|v\|_E := \left(\int_\Omega \alpha(x)|\nabla v(x)|^2\, dx \right)^{1/2}.$$

From (2.5.10) and (9.1.4), it follows that

$$(9.1.6) \qquad \|u - u_h\|_E \leq C \left(\sum_{|\beta|=k} \int_\Omega \alpha(x) h(x)^{2k-2} \left| D^\beta u \right|^2 dx \right)^{1/2}.$$

This result says that the energy error is always reduced by adapting the mesh to the solution.

We summarize the above result in the following theorem.

(9.1.7) Theorem. *For any non-degenerate mesh, we have*

$$\|u - u_h\|_E \leq C \left\| \sqrt{\alpha} h^{k-1} \left| \nabla_k u \right| \right\|_{L^2(\Omega)}$$

where $|\nabla_k u| (x) := \left(\sum_{|\beta|=k} \left| D^\beta u(x) \right|^2 \right)^{1/2}$ *and* C *is independent of* α *and* h.

We next derive an L^2 estimate. Choosing w as was done in the proof of Theorem 5.4.8, we can develop a standard duality argument. The key new ingredient is to multiply and divide by h, e.g.,

$$a(e, w - \mathcal{I}^h w) = \int_\Omega \alpha h \nabla(u - u_h) \nabla(w - \mathcal{I}^h w)/h \, dx$$

$$\leq \left(\int_\Omega \alpha^{2-\lambda} \left(h \nabla(u - u_h) \right)^2 dx \right)^{1/2} \left(\int_\Omega \alpha^\lambda \left| \nabla(w - \mathcal{I}^h w) \right|^2 h^{-2} dx \right)^{1/2}$$

where λ is an arbitrary parameter satisfying $0 \leq \lambda \leq 2$. From the results of Sect. 4.4 we have

$$\int_\Omega \alpha^\lambda \left| \nabla(w - \mathcal{I}^h w)(x) \right|^2 h^{-2} \, dx \leq C \int_\Omega \alpha^\lambda |\nabla_2 w|^2 \, dx.$$

Assuming (5.4.7) holds, we can bound the second derivatives of w in terms of $u - u_h$. The choice of λ is open to us. However, full use of this parameter depends on available estimates such as (5.4.7). For example, we can prove the following result.

(9.1.8) Theorem. *Suppose that (5.4.7) holds for the boundary value problem with variational form (9.0.1). For any non-degenerate mesh, we have*

$$(9.1.9) \qquad \|u - u_h\|_{L^2(\Omega)} \leq C \left(\int_\Omega \alpha(x)^2 h(x)^2 \left| \nabla(u - u_h)(x) \right|^2 dx \right)^{1/2}$$

where C *depends on* α.

This says that the L^2 error can be estimated in terms of a weighted integral of the squared derivative error, where the weight is given by the mesh function (9.1.1). It is also possible to estimate "weighted energy" norms such as the right-hand side of (9.1.9). Since the estimates are similar to Sect. 0.9, and Chapter 7, we only summarize a typical result.

(9.1.10) Theorem. *There is a constant $\kappa > 0$ such that if the mesh size variation, $\||\nabla h|\|_{L^\infty(\Omega)} < \kappa$, then*

$$\|u - u_h\|_{L^2(\Omega)} \leq C \||\sqrt{\alpha}h^2|\nabla_2 u|\|_{L^2(\Omega)}$$

where C is independent of the mesh.

Recall that the condition that the derivative of h be small does not preclude strong mesh gradings, e.g., a geometrically graded mesh, as depicted in Fig. 9.1. However, automatic mesh generators may violate this condition.

9.2 Error Estimators

One successful error estimator is based on the *residual*. We will consider such an estimator in a very simple case here. We assume, for simplicity, that our variational problem is of the form (9.0.1) with α piecewise smooth, but not necessarily continuous. We will consider the Dirichlet problem on a polyhedral domain Ω in the n dimensions, so that $V = \mathring{H}^1(\Omega)$. Moreover, we assume that the right-hand side for the variational problem is also a piecewise smooth function, f.

Let V_h be piecewise polynomials of degree less than k on a mesh $\mathcal{T}^h$, and assume that the discontinuities of α and f fall on mesh faces (edges in two dimensions) in $\mathcal{T}^h$. That is, both α and f are smooth on each $T \in \mathcal{T}^h$. However, we will otherwise only assume that $\mathcal{T}^h$ is non-degenerate, since we will want to allow significant local mesh refinement.

(9.2.1) Lemma. *Let $u_h \in V_h$ be the standard Galerkin approximation, and let $e_h := u - u_h$. Then e_h satisfies the residual equation*

$$(9.2.2) \qquad a(e_h, v) = R(v) \quad \forall v \in V$$

where $R \in V'$ is the residual which can be computed by

$$
\begin{aligned}
R(v) := &\sum_T \int_T (f + \nabla \cdot (\alpha \cdot \nabla u_h)) v \, dx \\
&+ \sum_e \int_e [\alpha \mathbf{n}_e \cdot \nabla u_h]_{\mathbf{n}_e} v \, ds \quad \forall v \in V
\end{aligned}
$$

$$(9.2.3)$$

where $\mathbf{n}_e$ denotes a unit normal to e.

One way to interpret this lemma is to say that R can be defined equivalently by either (9.2.2) or (9.2.3), with the other being a consequence of the definition. That is, (9.2.2) and (9.2.3) are equivalent. Although (9.2.3) can be viewed as just a re-writing of (9.2.2), it gives an expression of the error in terms of a right-hand side $R \in V'$.

The key point is that R has two parts. One is the *absolutely continuous* part R_A which is an $L^1(\Omega)$ function defined on each element T by

$$(9.2.4) \qquad R_A|_T := (f + \nabla \cdot (\alpha \nabla u_h))\,|_T.$$

The other term in the definition of R is the "jump" term

$$(9.2.5) \qquad R_J(v) := \sum_e \int_e [\alpha \mathbf{n}_e \cdot \nabla u_h]_{\mathbf{n}_e}\, v\, ds \quad \forall v \in V$$

where $[\phi]_\mathbf{n}$ denotes the jump in ϕ (across the face in question). More precisely,

$$[\phi]_\mathbf{n}(x) := \lim_{\epsilon \to 0} \phi(x + \epsilon \mathbf{n}) - \phi(x - \epsilon \mathbf{n})$$

so that the expression in (9.2.5) is independent of the choice of normal $\mathbf{n}$ on each face.

Proof. The relations (9.2.2)–(9.2.3) are derived simply by integrating by parts on each T, and the resulting boundary terms are collected in the term R_J. If $\mathcal{A}$ is the differential operator formally associated with the form (9.0.1), namely, $\mathcal{A}v := -\nabla \cdot (\alpha \nabla v)$, then we see that $R_A = \mathcal{A}(u - u_h) = \mathcal{A}e_h$ on each T. Note that we used the fact that ∇u_h is a polynomial on each T, as well as our assumptions about α and f. □

Inserting $v = e_h$ in (9.2.2), we see that

$$(9.2.6) \qquad \alpha_0 \, |e_h|^2_{H^1(\Omega)} \leq |R(e_h)| \leq \|R\|_{H^{-1}(\Omega)} \, \|e_h\|_{H^1(\Omega)}\,.$$

Therefore

$$(9.2.7) \qquad \|e_h\|_{H^1(\Omega)} \leq C\|R\|_{H^{-1}(\Omega)}.$$

Thus we find that the error may be estimated simply by computing the $H^{-1}(\Omega)$ norm of the residual, and we note that the residual is something that involves only the data of the problem (f and α) and u_h. Although all of these are explicitly available, there are two complications. First of all, it is difficult to compute a negative norm explicitly. Instead, we will estimate it. Moreover, since R has two radically different parts, one an integrable function, the other consisting of "interface Delta functions," it is challenging to estimate the combination of the two terms.

The residual has special properties. In particular, the fundamental orthogonality implies that

$$(9.2.8) \qquad R(v) := a(e_h, v) = 0 \quad \forall v \in V_h.$$

Suppose we have an interpolant $\mathcal{I}^h$ as defined in Sect. 4.8 that satisfies

$$(9.2.9) \qquad \left\| v - \mathcal{I}^h v \right\|_{L^2(T)} \le \gamma_0 h_T |v|_{H^1(\widehat{T})}$$

for all $T \in \mathcal{T}^h$ and

$$(9.2.10) \qquad \left\| v - \mathcal{I}^h v \right\|_{L^2(e)} \le \gamma_0 h_e^{1/2} |v|_{H^1(\widehat{T}_e)}$$

for some constant γ_0 and for all faces e in $\mathcal{T}^h$. For each interior face e, let T_e denote the union of the two elements sharing that face. From now on, we will drop the subscript "e" when referring to a normal $\mathbf{n}$ to e. Then

$$
\begin{aligned}
|R(v)| ={} & |R(v - \mathcal{I}^h v)| \\
={} & \left| \int_\Omega R_A(v - \mathcal{I}^h v)\, dx + R_J(v - \mathcal{I}^h v) \right| \\
={} & \left| \sum_T \int_T R_A(v - \mathcal{I}^h v)\, dx \right. \\
& \left. + \sum_e \int_e [\alpha \mathbf{n} \cdot \nabla u_h]_{\mathbf{n}}(v - \mathcal{I}^h v)\, ds \right| \\
\le{} & \sum_T \|R_A\|_{L^2(T)} \|v - \mathcal{I}^h v\|_{L^2(T)} \\
& + \sum_e \| [\alpha \mathbf{n} \cdot \nabla u_h]_{\mathbf{n}} \|_{L^2(e)} \|v - \mathcal{I}^h v\|_{L^2(e)} \\
\le{} & \sum_T \|R_A\|_{L^2(T)} \gamma_0 h_T |v|_{H^1(\widehat{T})} \qquad \text{(see 9.2.10)} \\
& + \sum_e \| [\alpha \mathbf{n} \cdot \nabla u_h]_{\mathbf{n}} \|_{L^2(e)} \gamma_0 h_e^{1/2} |v|_{H^1(\widehat{T}_e)} \\
\le{} & \gamma \left(\sum_T \|R_A\|_{L^2(T)}^2 h_T^2 \right. \\
& \left. + \sum_e \| [\alpha \mathbf{n} \cdot \nabla u_h]_{\mathbf{n}} \|_{L^2(e)}^2 h_e \right)^{1/2} |v|_{H^1(\Omega)}
\end{aligned}
$$

where h_e (resp. h_T) is a measure of the size of e (resp. T), and $\widehat{T}$ (resp. $\widehat{T}_e$) denotes the neighborhood of elements touching T (resp. T_e). For this reason, we define the *local error indicator* $\mathcal{E}_e$ by

$$(9.2.11) \qquad
\begin{aligned}
\mathcal{E}_e(u_h)^2 :={} & \sum_{T \subset T_e} h_T^2 \|f + \nabla \cdot (\alpha \nabla u_h)\|_{L^2(T)}^2 \\
& + h_e \| [\alpha \mathbf{n} \cdot \nabla u_h]_{\mathbf{n}} \|_{L^2(e)}^2.
\end{aligned}
$$

where a natural choice for h_K (with $K = T$ or e) is the measure of K raised to the power $1/\dim(K)$. With this definition, the previous inequalities can be summarized as

(9.2.12) $$|R(v)| \leq \gamma \left(\sum_e \mathcal{E}_e(u_h)^2 \right)^{1/2} |v|_{H^1(\Omega)}$$

which in view of (9.2.6) implies that

(9.2.13) $$|e_h|_{H^1(\Omega)} \leq \frac{\gamma}{\alpha_0} \left(\sum_e \mathcal{E}_e(u_h)^2 \right)^{1/2}$$

where γ is a constant only related to interpolation error.

Note that we have chosen to define the local error estimator based on an "edge" (or "face") point of view. This turns out to be more convenient in describing lower bound estimates (see Sect. 9.3). However, one could also take an "element" point of view, viz.,

(9.2.14) $$\mathcal{E}_T(u_h)^2 := h_T^2 \|R_A\|_{L^2(T)}^2 + \sum_{e \subset \partial T} h_e \| [\alpha \mathbf{n} \cdot \nabla u_h]_{\mathbf{n}} \|_{L^2(e)}^2$$

which would provide similar results.

Returning to (9.2.13), we take $\gamma = c_0 \gamma_0$ where c_0 depends only on the constant in (9.1.3) and the maximum number of neighbors of an element K in $\mathcal{T}^h$. Thus c_0 depends only on the non-degeneracy of the mesh $\mathcal{T}^h$.

Summarizing these arguments, we have proved the following.

(9.2.15) Theorem. *Suppose that the coefficient α in (9.0.1) and the right-hand side f are piecewise smooth on the non-degenerate mesh family $\mathcal{T}^h$. Under the assumptions (9.2.9) and (9.2.10), the upper bound (9.2.13) holds, where γ depends only on the non-degeneracy of the mesh $\mathcal{T}^h$ and α_0 is a lower bound for α on Ω.*

9.3 Local Error Estimates

In the previous section, we established an estimate for the global error $|e_h|_{H^1(\Omega)}$ in terms of locally defined error estimators (9.2.11). It is reasonable to ask about the correlation between the local estimator (9.2.11) and the local *error*. It is possible to establish an inequality in one direction under fairly general conditions. First of all, note that for any $T \in \mathcal{T}^h$, (9.2.2) implies

(9.3.1) $$a(e_h, v) = (R_A, v) \quad \forall v \in \overset{\circ}{H}{}^1(T).$$

Therefore

$$\left| \int_T R_A v \, dx \right| \leq \alpha_1 |e_h|_{H^1(T)} |v|_{H^1(T)}$$

and thus

$$\alpha_1 |e_h|_{H^1(T)} \geq \sup_{0 \neq v \in \mathring{H}^1(T)} \frac{\left| \int_T v R_A \, dx \right|}{|v|_{H^1(T)}}.$$

We need some way to control the information content of α and f in order to derive a rigorous estimate. For this reason, we assume that α and f are piecewise polynomials (not necessarily continuous) of degree at most $r - k + 2$ and r, respectively. Thus R_A is a piecewise polynomials of degree at most r. It can be shown by a homogeneity argument (see exercise 9.x.5 for a proof) that for any polynomial P of degree r

$$(9.3.2) \qquad \sup_{v \in \mathring{H}^1(T)} \frac{\int_T v P \, dx}{|v|_{H^1(T)}} \geq c_r h_T \|P\|_{L^2(T)}$$

where c_r depends only on the degree r and the chunkiness of the mesh. Applying this with $P = R_A$ proves

$$(9.3.3) \qquad \alpha_1 |e_h|_{H^1(T)} \geq c_1 h_T \left(\int_T R_A^2 \, dx \right)^{1/2}.$$

We can obtain a similar bound involving the jump terms. Again, make the previous assumptions that imply R_A is a piecewise polynomial of degree at most r. Note that for any face e in $\mathcal{T}^h$, (9.2.2) implies

$$a(e_h, v) = \int_e [\alpha \mathbf{n} \cdot \nabla u_h]_\mathbf{n} v \, ds \quad \forall v \in V_e$$

where T_e denotes the union of the two elements, T_e^+ and T_e^-, sharing e, and

$$V_e = \left\{ v \in \mathring{H}^1(T_e) : \int_{T_e^+} v P \, dx = \int_{T_e^-} v P \, dx = 0 \quad \forall P \in \mathcal{P}_r \right\}.$$

Then

$$\left| \int_e [\alpha \mathbf{n} \cdot \nabla u_h]_\mathbf{n} v \, ds \right| \leq \alpha_1 |e_h|_{H^1(T_e)} |v|_{H^1(T_e)} \quad \forall v \in V_e$$

and thus

$$\alpha_1 |e_h|_{H^1(T_e)} \geq \sup_{0 \neq v \in V_e} \frac{\left| \int_e [\alpha \mathbf{n} \cdot \nabla u_h]_\mathbf{n} v \, ds \right|}{|v|_{H^1(T_e)}}.$$

Since $[\alpha \mathbf{n} \cdot \nabla u_h]_\mathbf{n}$ is a polynomial of degree at most $r + 1$ on e, it can be shown (cf. exercise 9.x.7) that

$$(9.3.4) \qquad \sup_{0 \neq v \in \mathring{H}^1(T_e)} \frac{\left| \int_e [\alpha \mathbf{n} \cdot \nabla u_h]_\mathbf{n} v \, ds \right|}{|v|_{H^1(T_e)}} \geq c_r' h_e^{1/2} \| [\alpha \mathbf{n} \cdot \nabla u_h]_\mathbf{n} \|_{L^2(e)}$$

where $c_r' > 0$ depends only on r and the non-degeneracy constant for $\mathcal{T}^h$. Combining these estimates proves that

$$(9.3.5) \qquad \alpha_1 |e_h|_{H^1(T_e)} \geq c \mathcal{E}_e(u_h)$$

where $c > 0$ depends only on the non-degeneracy constant for $\mathcal{T}^h$.

Collecting these arguments, we have proved the following.

(9.3.6) Theorem. *Suppose that α and f are piecewise polynomials (not necessarily continuous) of degree at most $r - k + 2$ and r, respectively, on a non-degenerate mesh $\mathcal{T}^h$. Then the lower bound (9.3.5) holds for all faces e in $\mathcal{T}^h$, where c depends only on r and the non-degeneracy of the mesh $\mathcal{T}^h$ and where α_1 is an upper bound for α on Ω.*

One corollary of this theorem is the reverse inequality to (9.2.13), namely, a global lower bound. Squaring the lower bound (9.3.5) and summing over all elements yields the following.

(9.3.7) Theorem. *Suppose that α and f are as in Theorem 9.3.6. Then*

$$\alpha_1 |e_h|_{H^1(\Omega)} \geq c \left(\sum_{e \in \mathcal{T}^h} \mathcal{E}_e(u_h)^2 \right)^{1/2}$$

where $c > 0$ and α_1 are as in Theorem 9.3.6.

The reverse inequality to (9.3.5), i.e., a local *upper* bound is not true in general. However, the message of the local lower bound (9.3.5) is that one should refine the mesh wherever the local error indicator $\mathcal{E}_e(u_h)$ is big. Unfortunately, we cannot be sure that where it is small that the error will necessarily be small. Distant effects may pollute the error and make it large even if the error indicator $\mathcal{E}_e(u_h)$ is small nearby.

9.4 Estimators for Linear Forms and Other Norms

Suppose that, instead of wanting to estimate the energy norm of the error, we want to estimate just some (continuous) linear functional L of the error. For example, this might be some integral of the solution that we are particularly interested in. In this case, we want to know that $L(u_h)$ approximates $L(u)$ well, and we do not particularly care about anything else. This means that we just want to know that $L(e_h)$ is small.

To estimate $L(e_h)$, we introduce a dual function $\phi_L \in V$ by solving

(9.4.1) $$a(v, \phi_L) = L(v) \quad \forall v \in V.$$

Note that the norm of ϕ_L can be estimated from $L(\phi_L)$:

(9.4.2) $$a(\phi_L, \phi_L) = L(\phi_L),$$

and the (positive) number $L(\phi_L)$ depends only on L. Then

$$L(e_h) = a(e_h, \phi_L) = R(\phi_L)$$

and the estimate (9.2.12) implies

$$|L(e_h)| \leq \gamma \Big(\sum_e \mathcal{E}_e(u_h)^2 \Big)^{1/2} |\phi_L|_{H^1(\Omega)} \,.$$

In view of (9.4.2) we have proved the following.

(9.4.3) Theorem. *Suppose that L is a continuous linear functional on V and that ϕ_L is defined by (9.4.1)*

$$|L(e_h)| \leq C \Big(\sum_e \mathcal{E}_e(u_h)^2 \Big)^{1/2} \sqrt{L(\phi_L)}$$

where C is a constant related only to interpolation error and the coefficient α.

The local, lower-bound error estimate (9.3.5) also provides an estimate in other norms. For example, under the conditions of Theorem 9.3.6, Hölder's inequality implies that

$$h_e^{-n/2} \mathcal{E}_e(u_h) \leq C h_e^{-n/2} |e_h|_{H^1(T_e)} \leq C |e_h|_{W_\infty^1(T_e)} \,.$$

Note that $h_e^{-n/2} \mathcal{E}_e(u_h)$ is equivalent (exercise 9.x.12) to

$$(9.4.4) \quad \mathcal{E}_e^\infty(u_h) := \max_{T \subset T_e} h_T \| f + \nabla \cdot (\alpha \nabla u_h) \|_{L^\infty(T)} + \| [\alpha \mathbf{n} \cdot \nabla u_h] \|_{L^\infty(e)}$$

and hence we have proved the following.

(9.4.5) Theorem. *Under the conditions of Theorem 9.3.6, the estimator $\mathcal{E}_e^\infty(u_h)$ in (9.4.4) satisfies*

$$\mathcal{E}_e^\infty(u_h) \leq C |e_h|_{W_\infty^1(T_e)} \quad \forall e \in \mathcal{T}^h .$$

To get an upper bound for the error, we modify the derivation of (9.2.11). Let g be the smoothed derivative Green's function (for a given direction ν) introduced following (8.2.2). Then

$$
\begin{aligned}
|a(e_h, g)| &= |R(g)| \\
&= |R(g - \mathcal{I}^h g)| \\
&= \Big| \sum_T \int_T R_A (g - \mathcal{I}^h g) \, dx \\
&\quad + \sum_e \int_e [\alpha \mathbf{n} \cdot \nabla u_h](g - \mathcal{I}^h g) \, ds \Big| \\
&\leq \sum_T \| R_A \|_{L^\infty(T)} \| g - \mathcal{I}^h g \|_{L^1(T)}
\end{aligned}
$$

$$+ \sum_e \| [\alpha \mathbf{n} \cdot \nabla u_h] \|_{L^\infty(e)} \| g - \mathcal{I}^h g \|_{L^1(e)}$$

$$\leq C \sum_e \mathcal{E}_e^\infty(u_h) h_e \int_{T_e} |\nabla_2 g(x)| \, dx$$

and therefore we have (recall (8.2.3))

$$|(\nu \cdot \nabla u, \delta^z) - \nu \cdot \nabla u_h(z)| \leq C \sum_e \mathcal{E}_e^\infty(u_h) h_e \int_{T_e} |\nabla_2 g(x)| \, dx.$$

Let us define a piecewise constant function $\mathcal{E}$ on each $T \in \mathcal{T}^h$ by

$$(9.4.6) \qquad \mathcal{E}(x) = \max_{e \subset T} h_e \mathcal{E}_e^\infty(u_h) \quad \forall x \in T$$

for all $T \in \mathcal{T}^h$. Then the above estimates imply that

$$|(\nu \cdot \nabla u, \delta^z) - \nu \cdot \nabla u_h(z)| \leq C \int_\Omega \mathcal{E}(x) |\nabla_2 g(x)| \, dx.$$

Applying (8.3.11) we find

$$\left(\int_\Omega |\nabla_2 g(x)| \mathcal{E}(x) \, dx \right)^2 \leq \int_\Omega \sigma_z^{n+\lambda} |\nabla_2 g(x)|^2 \, dx \int_\Omega \sigma_z^{-n-\lambda} \mathcal{E}(x)^2 \, dx$$

$$\leq C \left(\int_\Omega \sigma_z^{n+\lambda} |\nabla \delta^z(x)|^2 \, dx \right.$$

$$\left. + h_z^{-2} \int_\Omega \sigma_z^{n+\lambda} |\delta^z(x)|^2 \, dx \right) \int_\Omega \sigma_z^{-n-\lambda} \mathcal{E}(x)^2 \, dx.$$

Recalling the definition of σ_z in Sect. 8.1 and the definition of δ^z in Sect. 8.2, we see (exercise 9.x.15) that the following holds.

(9.4.7) Theorem. *Under the assumptions of Lemma 8.3.11, we find*

$$|(\nu \cdot \nabla u, \delta^z) - \nu \cdot \nabla u_h(z)| \leq C \left(h_z^{\lambda-2} \int_\Omega \sigma_z^{-n-\lambda} \mathcal{E}(x)^2 \, dx \right)^{1/2}.$$

Note that we estimate how well $\nu \cdot \nabla u_h(z)$ approximates an explicit average (using the kernel δ^z) of $\nu \cdot \nabla u(z)$. If u is smooth enough, $(\nu \cdot \nabla u, \delta^z)$ will be very close to $\nu \cdot \nabla u(z)$, but in general we have no way of estimating this. Since u_h is defined on a mesh of finite size, we can only expect $\nu \cdot \nabla u_h(z)$ to approximate some appropriate average of $\nu \cdot \nabla u(z)$ on the scale of the mesh. By comparing with this particular average, which is naturally associated with u_h, we avoid introducing further approximation conditions on u, such as (9.4.8) which appears subsequently in the context of L^1 error estimators.

We can interpret the estimate in the theorem in different ways. In particular, if $\mathcal{E}_T^\infty(u_h) = \epsilon$ for all $T \in \mathcal{T}^h$ and $h(x) = h$ is constant, then $\mathcal{E}(x) = \epsilon h$ for all x. In this case, the above expression simplifies to

$$|(\nu \cdot \nabla u, \delta^z) - \nu \cdot \nabla u_h(z)| \leq C\epsilon$$

as expected. Given the simple nature of $\mathcal{E}$ and σ_z we can re-write this as the following

$$|(\nu \cdot \nabla u, \delta^z) - \nu \cdot \nabla u_h(z)|$$
$$\leq C \Big(\sum_T h_z^{\lambda-2} h_T^{2+n} \mathcal{E}_T^\infty(u_h)^2 (h_z + \mathrm{dist}(z,T))^{-(n+\lambda)}\Big)^{1/2}$$
$$\leq C \left(\sum_T \left(\frac{h_T}{h_z}\right)^{2+n} \mathcal{E}_T^\infty(u_h)^2 \left(1 + \frac{\mathrm{dist}(z,T)}{h_z}\right)^{-(n+\lambda)}\right)^{1/2}$$

where $\mathcal{E}_T^\infty(u_h)$ is defined analogous to (9.2.14).

See (Nochetto 1995) for results with less restrictive assumptions than those of (8.3.11). In (Nochetto & Liao 2002), results with weighted error estimators suitable for non-convex domains are given.

Estimates for the error in W_1^1 are a bit more complex. Under the conditions of Theorem 9.3.6 we find, using (4.5.4), that

$$h_T^{n/2} \mathcal{E}_T(u_h) \leq C h_T^{n/2} |e_h|_{H^1(\widehat{T})}$$
$$\leq C h_T^{n/2} \left(|u - \mathcal{I}^h u|_{H^1(\widehat{T})} + |\mathcal{I}^h u - u_h|_{H^1(\widehat{T})}\right)$$
$$\leq C \left(h_T^{n/2} |u - \mathcal{I}^h u|_{H^1(\widehat{T})} + |\mathcal{I}^h u - u_h|_{W_1^1(\widehat{T})}\right)$$

We make a *saturation assumption* that

(9.4.8) $$|u - \mathcal{I}^h u|_{H^1(\widehat{T})} \leq C h_T^{-n/2} |u - \mathcal{I}^h u|_{W_1^1(\widehat{T})}.$$

Then we have

$$h_T^{n/2} \mathcal{E}_T(u_h) \leq C \left(|u - \mathcal{I}^h u|_{W_1^1(\widehat{T})} + |\mathcal{I}^h u - u_h|_{W_1^1(\widehat{T})}\right)$$
$$\leq C \left(|u - \mathcal{I}^h u|_{W_1^1(\widehat{T})} + |e_h|_{W_1^1(\widehat{T})}\right).$$

Observe that we can write $u - \mathcal{I}^h u = e_h - \mathcal{I}^h e_h$. Now suppose that the interpolant $\mathcal{I}^h$ is bounded in W_1^1 (see Sect. 4.8), namely, that

(9.4.9) $$\|\mathcal{I}^h v\|_{W_1^1(\widehat{T})} \leq C \|v\|_{W_1^1(S_{\widehat{T}})}.$$

Then (9.4.9) implies that

(9.4.10) $$h_T^{n/2} \mathcal{E}_T(u_h) \leq C \|e_h\|_{W_1^1(S_{\widehat{T}})}.$$

Note that $h_T^{n/2} \mathcal{E}_T(u_h)$ is equivalent (exercise 9.x.13) to

$$(9.4.11) \quad \mathcal{E}_T^1(u_h) := h_T \|f + \nabla \cdot (a\nabla u_h)\|_{L^1(T)} + \max_{e \subset \partial T} \|\, [a\mathbf{n} \cdot \nabla u_h]\,\|_{L^1(e)}$$

and hence we have proved the following.

(9.4.12) Theorem. *Under the assumptions* (9.4.8) *and* (9.4.9),

$$\mathcal{E}_T^1(u_h) \leq C|e_h|_{W_1^1(S_{\widehat{T}})}.$$

9.5 Conditioning of Finite Element Equations

So far, we have addressed only the question of quality of approximation using highly refined meshes. But what if large mesh variation caused the corresponding linear systems (cf. Sect. 0.2) to degenerate in some sense? In the remaining sections of the chapter, we show that this need not happen when the mesh is refined locally, provided certain restrictions on the mesh are met and a natural scaling of the basis functions is used. Much of the material is drawn from (Bank & Scott 1989).

The convergence properties of iterative methods, such as the conjugate-gradient method, for solving such linear systems can be estimated (cf. Luenberger 1973) in terms of the condition number of the system. The sensitivity of the solution to perturbations in the right hand side can be estimated using the condition number, and error bounds for direct methods, such as Gaussian elimination, also imply a degradation of performance for an ill-conditioned system (cf. Isaacson and Keller 1966). Thus without further justification, it would not be a remedy simply to use a standard direct method for an ill-conditioned system. Fortunately, the condition number of linear systems for finite element methods need not degrade unacceptably as the mesh is refined.

A particular setting that we have in mind is the refinement of meshes (perhaps adaptively) to resolve singularities arising at angular points on the domain boundary or at points of discontinuity of the coefficients of the differential equation. It might seem, naïvely, that there would be large ratios of eigenvalues of the linear system (which would imply a large condition number) resulting from large mesh ratios. However, we show that this is not the case if a natural scaling of the finite element basis functions is used and the mesh is *non-degenerate* (Definition 4.4.13).

On a regular mesh of size h, the condition number of the finite element equations for a second-order elliptic boundary value problem can easily be seen to be $\mathcal{O}(h^{-2})$ using inverse estimates (see Sect. 4.6). Also, the number,

N, of degrees of freedom in this case is $\mathcal{O}(h^{-n})$. Thus, the condition number can be expressed in terms of the number of degrees of freedom as $\mathcal{O}(N^{2/n})$. In the case that $n \geq 3$, we shall show that the condition number is bounded by $\mathcal{O}(N^{2/n})$ for non-degenerate meshes. In the case $n = 2$, estimates for the condition number increase slightly by a logarithmic factor depending essentially on the ratio of the largest and smallest mesh sizes.

We consider a variational problem with form as in (9.0.1). We now write $\mathcal{T}^h = \mathcal{T}_N$ to focus on the number of elements as the key parameter rather than the mesh size. Let V_N denote the corresponding finite element space. Let us write the variational equation (9.0.1) as a matrix equation utilizing a particular basis for V_N. Specifically, suppose that $\{\psi_i \ : \ i = 1, \cdots, N\}$ is a given basis for V_N, and define a matrix, $\mathbf{A}$, and a vector, $\mathbf{F}$, via

$$\mathbf{A}_{ij} := a(\psi_i, \psi_j) \qquad \text{and} \qquad \mathbf{F}_i := f(\psi_i) \quad \forall i, j = 1, \cdots, N.$$

Let u^N denote the solution of the standard variational problem (2.5.7) with the form in (9.0.1) on V_N. Then this problem is equivalent to solving

$$\mathbf{AU} = \mathbf{F}$$

where $u^N = \sum_{i=1}^{N} u_i \psi_i$ and $\mathbf{U} = (u_i)$. We now give conditions on V_N and the basis $\{\psi_i \ : \ i = 1, \cdots, N\}$ that will be used to guarantee that the condition number of $\mathbf{A}$ is well behaved.

For all V_N studied in Chapter 4, we note that following holds, namely, that $\{\psi_i \ : \ i = 1, \cdots, N\}$ is a *local* basis:

$$(9.5.1) \qquad \max_{1 \leq i \leq N} \text{cardinality} \{T \in \mathcal{T}_N \ : \ \text{supp}(\psi_i) \cap T \neq \emptyset\} \leq \alpha_5$$

on a non-degenerate mesh.

The main assumption concerning the scaling of the basis is that for all $T \in \mathcal{T}_N$

$$(9.5.2) \qquad C^{-1} h_T^{n-2} \|v\|_{L^\infty(T)}^2 \ \leq \sum_{\text{supp}(\psi_i) \cap T \neq \emptyset} v_i^2 \ \leq \ C h_T^{n-2} \|v\|_{L^\infty(T)}^2$$

where $C < \infty$, $v = \sum_{i=1}^{N} v_i \psi_i$ and (v_i) is arbitrary.

We assume the domain, Ω, is Lipschitz ruling out "slit" domains. When $n \geq 3$, we thus have Sobolev's inequality (cf. exercise 8.x.11)

$$(9.5.3) \qquad \|v\|_{L^{2n/n-2}(\Omega)} \ \leq \ C_S \|v\|_{H^1(\Omega)} \quad \forall v \in H^1(\Omega).$$

In two dimensions ($n = 2$), since we assume that Ω is bounded, the Sobolev imbedding $H^1(\Omega) \subset L^p(\Omega)$ holds for all $p < \infty$. Moreover, it has a norm, $\sigma(p)$, that is bounded by a constant times the norm of the Sobolev imbedding $H_0^1(B) \subset L^p(B)$ for a sufficiently large ball, B, namely, $\sigma(p) \leq C_S \sqrt{p}$ (cf. Gilbarg & Trudinger 1983, especially the proof of Theorem 7.15). Thus for $n = 2$ we have the following Sobolev inequality:

$$(9.5.4) \qquad \|v\|_{L^p(\Omega)} \le C_S \sqrt{p} \, \|v\|_{H^1(\Omega)} \qquad \forall v \in H^1(\Omega), \, p < \infty.$$

In practice we have only a finite number of (finite) triangulations to deal with, and any finite family is non-degenerate. However, all constants discussed below will be bounded in terms of the parameter, ρ, in Definition 4.2.16.

(9.5.5) Example. Let V_N denote the Lagrange space of C^0 piecewise polynomials of degree k on the mesh $\mathcal{T}_N$ that are contained in the subspace V. We denote by $\{\phi_i : i = 1, \cdots, N\}$ the standard Lagrangian nodal basis for V_N consisting of functions that equal one at precisely one nodal point in the triangulation. We also introduce a scaled basis that is of interest in three (and higher) dimensions. Define a new basis $\{\psi_i : i = 1, \cdots, N\}$ by

$$\psi_i := h(z_i)^{(2-n)/2} \phi_i$$

where h is the function defined in (9.1.1), $\{z_i : i = 1, \cdots, N\}$ denotes the set of nodal points and n is the dimension of Ω. Note that this basis does not differ from the original one if $n = 2$.

Our choice of scaling yields (9.5.2), with C depending only on ρ and k, in view of (9.1.1) and (9.1.3). $\qquad \square$

(9.5.6) Example. To obtain (9.5.2) for Hermite elements in two dimensions, one chooses the basis functions corresponding to derivative nodes to have the corresponding derivative of order $\mathcal{O}\left(h(z_i)^{-1}\right)$, with the remaining basis functions scaled as in the Lagrangian case. The other assumptions for this element follow as in the Lagrangian case. $\qquad \square$

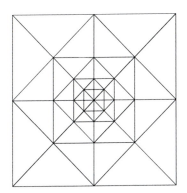 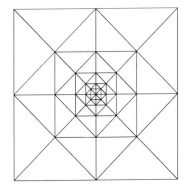

Fig. 9.1. two highly graded meshes with similar triangles

(9.5.7) Example. Let us show that our assumption of non-degeneracy does not exclude radical mesh refinements, as we have already observed in (9.1.1) and (9.1.3) that it does imply local quasi-uniformity. Let Ω_0 denote the square of side 1 centered at the origin, i.e.,

$$\Omega_0 = \left\{ (x,y) \in \mathbb{R}^2 : |x| < \frac{1}{2}, |y| < \frac{1}{2} \right\}.$$

Let $\mathcal{T}_{N_0}$ denote the triangulation of Ω_0 generated by its diagonals and the two axes, i.e., consisting of eight isosceles, right triangles (each having two sides of length $1/2$). We subdivide to construct $\mathcal{T}_{N_1}$ by adding the edges of the square, Ω_1, of side $1/2$ centered at the origin together with eight more edges running parallel with the diagonals. We obtain 24 similar triangles in this way. Also note that $\mathcal{T}_{N_1}$ restricted to the square Ω_1 is a triangulation similar to $\mathcal{T}_{N_0}$. Thus we may repeat the process above to this part of the domain alone to define a triangulation $\mathcal{T}_{N_2}$ consisting of isosceles, right triangles. Continuing in this way, we obtain a sequence of triangulations, $\mathcal{T}_{N_i}$, consisting of similar triangles. Fig. 9.1 shows the cases $i = 3, 4$. The ratio of largest to smallest side length is 2^i yet only $16i + 8$ triangles are used. (There are $8i + 1$ interior vertices in $\mathcal{T}_{N_i}$, so $N_i = 8i + 1$ in the case of Lagrange piecewise linear approximation of the Dirichlet problem.) Such a geometric refinement is far more severe than is often used to resolve boundary or interface singularities, but it shows that the assumption of non-degeneracy in Definition 4.4.13 need not restrict mesh refinement. $\quad\square$

9.6 Bounds on the Condition Number

We now give bounds on the condition number of the matrix $\mathbf{A} := (a(\psi_i, \psi_j))$, where $\{\psi_i : i = 1, \cdots, N\}$ is the (scaled) basis for V_N specified by our assumptions (and defined explicitly in the previous examples). Applications of these results to convergence rates for the conjugate-method for solving $\mathbf{AX} = \mathbf{F}$ will be given in Sect. 9.7. We begin with the general case $n \geq 3$.

(9.6.1) Theorem. *Suppose the basis* $\{\psi_i : i = 1, \cdots, N\}$ *satisfies* (9.5.2). *Then the* l_2*-condition number,* $\kappa_2(\mathbf{A})$*, of* $\mathbf{A}$ *is bounded by*

$$\kappa_2(\mathbf{A}) \leq CN^{2/n}.$$

Proof. First note that if we set $u = \sum_i u_i \psi_i$ then

(9.6.2) $$a(u, u) = \mathbf{U}^t \mathbf{A} \mathbf{U}$$

where $\mathbf{U} = (u_i)$, because $a(\cdot, \cdot)$ is bilinear. Observe that

$$
\begin{aligned}
a(u, u) &\leq C \|u\|_{H^1(\Omega)}^2 && \text{(continuity)} \\
&= C \sum_{T \in \mathcal{T}_N} \|u\|_{H^1(T)}^2 && (\mathcal{T}_N \text{ is a subdivision})
\end{aligned}
$$

$$\leq C \sum_{T \in \mathcal{T}_N} h_T^{n-2} \|u\|_{L^\infty(T)}^2 \qquad \text{(by 4.5.3)}$$

$$\leq C \sum_{T \in \mathcal{T}_N} \sum_{\text{supp}(\psi_i) \cap T \neq \emptyset} u_i^2 \qquad \text{(by 9.5.2)}$$

$$\leq C \mathbf{U}^t \mathbf{U}. \qquad \text{(by 9.5.1)}$$

Here h_T denotes the diameter of T. A complementary inequality can be derived as follows:

$$\mathbf{U}^t \mathbf{U} \leq \sum_{T \in \mathcal{T}_N} \sum_{\text{supp}(\psi_i) \cap T \neq \emptyset} u_i^2$$

$$\leq C \sum_{T \in \mathcal{T}_N} h_T^{n-2} \|u\|_{L^\infty(T)}^2 \qquad \text{(by 9.5.2)}$$

$$\leq C \sum_{T \in \mathcal{T}_N} \|u\|_{L^{2n/n-2}(T)}^2 \qquad \text{(by 4.5.3)}$$

$$\leq C \left(\sum_{T \in \mathcal{T}_N} 1 \right)^{2/n} \|u\|_{L^{2n/n-2}(\Omega)}^2 \qquad \text{(Hölder's inequality)}$$

$$\leq C N^{2/n} \|u\|_{L^{2n/n-2}(\Omega)}^2$$

$$\leq C N^{2/n} \|u\|_{H^1(\Omega)}^2 \qquad \text{(Sobolev's inequality)}$$

$$\leq C N^{2/n} a(u,u). \qquad \text{(coercivity)}$$

Using these estimates we show

$$C^{-1} N^{-2/n} \mathbf{U}^t \mathbf{U} \leq \mathbf{U}^t \mathbf{A} \mathbf{U} \leq C \mathbf{U}^t \mathbf{U}$$

where $C < \infty$. This proves that

$$C^{-1} N^{-2/n} \leq \lambda_{\min}(\mathbf{A}) \qquad \text{and} \qquad \lambda_{\max}(\mathbf{A}) \leq C$$

where $\lambda_{\min}(\mathbf{A})$ and $\lambda_{\max}(\mathbf{A})$ denote, respectively, the smallest and largest eigenvalues of $\mathbf{A}$. Recall (cf. Isaacson and Keller 1966) that the ℓ_2-condition number, $\kappa_2(\mathbf{A})$, of $\mathbf{A}$ satisfies

$$\kappa_2(\mathbf{A}) = \lambda_{\max}(\mathbf{A})/\lambda_{\min}(\mathbf{A}).$$

Thus the previous two estimates yield the stated result. □

A similar result can be given in two dimensions ($n = 2$) as follows.

(9.6.3) Theorem. *Suppose the basis $\{\psi_i : i = 1, \cdots, N\}$ satisfies (9.5.2). Then the l_2-condition number, $\kappa_2(\mathbf{A})$, of $\mathbf{A}$ is bounded by*

$$\kappa_2(\mathbf{A}) \leq C N \left(1 + \left| \log \left(N \, h_{\min}(N)^2 \right) \right| \right)$$

where $h_{\min} = \min \{ h_T : T \in \mathcal{T}_N \}$.

Proof. As in the proof of Theorem 9.6.1, it is sufficient to prove that

$$C^{-1}\left(N\left(1+\left|\log\left(N\,h_{\min}(N)^2\right)\right|\right)\right)^{-1}\mathbf{U}^t\mathbf{U} \leq \mathbf{U}^t\mathbf{A}\mathbf{U} \leq C\mathbf{U}^t\mathbf{U}$$

where $C < \infty$. The proof of these inequalities is quite similar to the case $n \geq 3$. For $u \in V_N$, we again write $u = \sum_i u_i \psi_i$ and recall from (9.6.2) that $a(u, u) = \mathbf{U}^t \mathbf{A} \mathbf{U}$. Then the same argument as in the proof of Theorem 9.6.1 yields

$$a(u, u) \leq C\,\mathbf{U}^t\mathbf{U}.$$

For the remaining inequality, we have (for $p > 2$)

$$\mathbf{U}^t\mathbf{U} \leq \sum_{T\in\mathcal{T}_N}\ \sum_{\mathrm{supp}(\psi_i)\cap T\neq\emptyset} u_i^2$$

$$\leq C\sum_{T\in\mathcal{T}_N}\|u\|_{L^\infty(T)}^2 \qquad\qquad\text{(by 9.5.2)}$$

$$\leq C\sum_{T\in\mathcal{T}_N}h_T^{-4/p}\|u\|_{L^p(T)}^2 \qquad\qquad\text{(by 4.5.3)}$$

$$\leq C\left(\sum_{T\in\mathcal{T}_N}h_T^{-4/(p-2)}\right)^{(p-2)/p}\|u\|_{L^p(\Omega)}^2 \qquad\text{(Hölder's inequality)}$$

$$\leq C\left(\sum_{T\in\mathcal{T}_N}h_T^{-4/(p-2)}\right)^{(p-2)/p}p\,\|u\|_{H^1(\Omega)}^2 \qquad\text{(Sobolev's inequality)}$$

$$\leq C\left(\sum_{T\in\mathcal{T}_N}h_T^{-4/(p-2)}\right)^{(p-2)/p}p\,a(u, u). \qquad\text{(coercivity)}$$

A crude estimate yields

$$\left(\sum_{T\in\mathcal{T}_N}h_T^{-4/(p-2)}\right)^{(p-2)/p} \leq h_{\min}(N)^{-4/p}\,(CN)^{(p-2)/p}$$

$$= C^{1-2/p}\left(N\,h_{\min}(N)^2\right)^{-2/p}N.$$

Thus the estimate above can be simplified to

$$\mathbf{U}^t\mathbf{U} \leq C\left(p\left(N\,h_{\min}(N)^2\right)^{-2/p}\right)Na(u, u).$$

Choosing $p = \max\{2, \left|\log\left(N\,h_{\min}(N)^2\right)\right|\}$ in this estimate yields the stated result. $\qquad\square$

(9.6.4) Remark. In (Bank & Scott 1989), it is shown that the estimate of Theorem 9.6.3 is sharp for the special mesh introduced in Example 9.5.7.

9.7 Applications to the Conjugate-Gradient Method

The conjugate-gradient method for solving a linear system of the form $\mathbf{AU} = \mathbf{F}$ is an iterative method whose convergence properties can be estimated in terms of the condition number of $\mathbf{A}$ (cf. Luenberger 1973). Specifically, define

$$\|\mathbf{U}\|_A := \left(\mathbf{U}^t \mathbf{AU}\right)^{1/2},$$

and let $\mathbf{U}^{(k)}$ denote the sequence of vectors generated by the conjugate-gradient method starting with $\mathbf{U}^{(0)} = \mathbf{0}$. Then

$$\|\mathbf{U} - \mathbf{U}^{(k)}\|_A \leq C \exp\left(-2k/\sqrt{\kappa_2(\mathbf{A})}\right) \|\mathbf{U}\|_A$$

where $\mathbf{U}$ denotes the solution to $\mathbf{AU} = \mathbf{F}$. This can be easily interpreted in terms of norms on V. Define the energy norm on V by

(9.7.1) $$\|u\|_a := \sqrt{a(u,u)}.$$

Then (9.6.2) implies that, for $v = \sum_i y_i \psi_i$,

$$\|v\|_a = \|\mathbf{Y}\|_A$$

where $\mathbf{Y} = (y_i)$. Let $u_N = \sum_i u_i \psi_i$ and $u_N^{(k)} = \sum_i u_i^{(k)} \psi_i$, where $(u_i) = \mathbf{U}$ and $(u_i^{(k)}) = \mathbf{U}^{(k)}$. Then the above estimate may be written

$$\|u - u^{(k)}\|_a \leq C \exp\left(-2k/\sqrt{\kappa_2(\mathbf{A})}\right) \|u\|_a.$$

This estimate says that to reduce the relative error $\|u - u^{(k)}\|_a / \|u\|_a$ to $\mathcal{O}(\epsilon)$ requires at most $k = \mathcal{O}\left(\sqrt{\kappa_2(\mathbf{A})} \, |\log \epsilon|\right)$ iterations.

Suppose that we only require $\epsilon = \mathcal{O}(N^{-q})$ for some $q < \infty$. In $n \geq 3$ dimensions, the above estimate says that this order of accuracy will be achieved after only $\mathcal{O}(N^{1/n} \log N)$ iterations. In two dimensions the above estimate becomes slightly more complicated. In typical applications, even with very severe refinements, we have $h_{\min} = \mathcal{O}(h_{\max}^p) = \mathcal{O}(N^{-p/2})$ for some $p < \infty$, as we shall now assume. In this case, the estimate above says that $\mathcal{O}(\epsilon)$ accuracy will be achieved after only $\mathcal{O}\left(N^{1/2}(\log N)^2\right)$ iterations.

Each conjugate-gradient iteration requires $\mathcal{O}(N)$ operations. Thus the final work estimates for the conjugate-gradient method on refined meshes as described previously would be

$$\mathcal{O}\left(N^{1+\frac{1}{n}} \log N\right)$$

in $n \geq 3$ dimensions. The work estimates for multigrid method are $\mathcal{O}(N)$ operations. Thus it would appear that the relative benefit of multigrid over conjugate-gradient iteration decreases as n increases. However, typically N might also increase with n as well. Comparisons with direct methods are given in (Bank & Scott 1989).

9.x Exercises

9.x.1 Prove that a non-degenerate mesh is locally quasi-uniform, in two or higher dimensions. (Hint: neighboring elements are all connected to each other via a sequence of elements with common faces.)

9.x.2 Prove Theorem 9.1.10.

9.x.3 Compute $\|\nabla h\|_{L^\infty(\Omega)}$ for the mesh in Fig. 9.1. How does this depend on the level, i?

9.x.4 Prove that the condition number of the matrix A in Sect. 9.6 satisfies $\kappa(A) = \mathcal{O}\left(h^{-2}\right)$ on a regular mesh of size h.

9.x.5 Let T be the right-triangle $\{(x,y) \; : \; x, y > 0; \, x + y < 1\}$, and let $V_T = \overset{\circ}{H}^1(T)$. Prove that for any polynomial P of degree r

$$\sup_{v \in V_T} \frac{\int_T vP \, dx}{|v|_{H^1(T)}} \geq c_r \|P\|_{L^2(T)}$$

where c_r depends only on the degree r. (Hint: it suffices to take V_T consisting of polynomials of fixed degree; use equivalence of norms on finite-dimensional spaces, and see (12.5.2).)

9.x.6 Prove the result in exercise 9.x.5 in the case $r = 0$ ($P = $ constant) by explicit construction. (Hint: let v be a "bubble" function, e.g., a cubic polynomial that is postive in the interior of T and zero on ∂T.)

9.x.7 Let T be the right-triangle $\{(x,y) \; : \; x, y > 0; \, x + y < 1\}$, and let e denote one edge. Let V_e denote functions vanishing on $\partial T \backslash e$ that are orthogonal on T to polynomials of degree r. Prove that for any polynomial P of degree r

$$\sup_{v \in V_e} \frac{\int_e vP \, ds}{|v|_{H^1(T)}} \geq c_r \|P\|_{L^2(e)}$$

where c_r depends only on the degree r. (Hint: it suffices to take V_e consisting of polynomials of fixed degree; use equivalence of norms on finite-dimensional spaces, and see (12.5.2).)

9.x.8 Prove the result in exercise 9.x.7 in the case $r = 0$ ($P = $ constant) by explicit construction. (Hint: consider the cubic finite element with the usual Lagrange nodes on the edges and the remaining nodal value being the integral over the triangle. Construct a cubic in V_e that is postive in the interior of e.)

9.x.9 Formulate and prove the results in exercises 9.x.5 and 9.x.7 in the case of $d \geq 3$ dimensions (for T a simplex).

9.x.10 Prove Theorem 9.2.15 for α and f piecewise polynomials of a fixed degree r, with the constant c depending only on r and the chunkiness of $\mathcal{T}^h$. (Hint: use exercises 9.x.5 and 9.x.7.)

9.x.11 Prove that, if $|h|_{W^1_\infty(\Omega)}$ is sufficiently small and (5.4.7) holds, then

$$\|u - u_h\|_{L^2(\Omega)} \le C \left\| \sqrt{\alpha} h^2 \nabla^2 u \right\|_{L^2(\Omega)}.$$

(Hint: see Sect. 0.9.)

9.x.12 Prove that $h_T^{-n/2} \mathcal{E}_T(u_h)$ and $\mathcal{E}_T^\infty(u_h)$ are equivalent for α and f piecewise polynomials of degree r, that is

$$\frac{1}{c_r} \mathcal{E}_T^\infty(u_h) \le h_T^{-n/2} \mathcal{E}_T(u_h) \le c_r \mathcal{E}_T^\infty(u_h) \quad \forall T$$

where c_r depends only on the degree r. (Hint: use inverse estimates and Hölder's inequality.)

9.x.13 Prove that $h_T^{n/2} \mathcal{E}_T(u_h)$ and $\mathcal{E}_T^1(u_h)$ are equivalent for α and f piecewise polynomials of degree r, that is

$$\frac{1}{c_r} \mathcal{E}_T^1(u_h) \le h_T^{n/2} \mathcal{E}_T(u_h) \le c_r \mathcal{E}_T^1(u_h) \quad \forall T$$

where c_r depends only on the degree r. (Hint: use inverse estimates and Hölder's inequality.)

9.x.14 Show that it is necessary to restrict the space in (9.3.2) that P lies in by showing that

$$\inf_{P \in L^2(T)} \sup_{v \in V_T} \frac{\int_T vP \, dx}{\|P\|_{L^2(T)} |v|_{H^1(T)}} = 0$$

where T and V_T are as in exercise 9.x.5. (Hint: let $\mathcal{T}^h$ be a triangulation of T and take P to be orthogonal to constants on each triangle in $\mathcal{T}^h$. Approximate v by piecewise constants on $\mathcal{T}^h$ and let h go to zero.)

9.x.15 Recall σ_z and δ^z from Sect. 8.1 and Sect. 8.2, respectively. Prove that

$$\int_\Omega \sigma_z^{n+\lambda} |\nabla \delta^z(x)|^2 \, dx + h_z^{-2} \int_\Omega \sigma_z^{n+\lambda} |\delta^z(x)|^2 \, dx \le C h_z^{\lambda-2}.$$

9.x.16 Recall σ_z from Sect. 8.1. Prove that

$$\int_\Omega \sigma_z^{-n-\lambda} \, dx \le C h_z^{-\lambda}.$$

Chapter 10

Variational Crimes

Consider the Dirichlet problem

$$-\Delta u = f \quad \text{in } \Omega \subseteq \mathbb{R}^2$$

(10.0.1)

$$u = 0 \quad \text{on } \partial\Omega.$$

We have already considered this problem when Ω is a convex polygonal domain with V_h being the set of piecewise polynomials that vanish on $\partial\Omega$. The error estimate is based on Ceá 's Theorem (cf. (2.8.2)), which uses the fact that

(10.0.2) $$V_h \subseteq V = \mathring{H}^1(\Omega).$$

In this chapter we consider two cases where (10.0.2) is violated. In the first case we consider (10.0.1) on a domain Ω with smooth, curved boundary. We consider two approaches to approximating such problems. Suppose we have a triangulation $\mathcal{T}^h$ where the "triangles" at the boundary have one curved side. Let V_h be a Lagrange finite element space associated with $\mathcal{T}^h$ satisfying the Dirichlet boundary condition at points on $\partial\Omega$. In general, we cannot expect the homogeneous Dirichlet boundary conditions to be satisfied exactly by members of V_h. We are therefore in the situation where $V_h \subseteq H^1(\Omega)$ but $V_h \not\subseteq V = \mathring{H}^1(\Omega)$. The other technique to be considered is the use of isoparametric finite elements described in Sect. 4.7.

The other case in which (10.0.2) is violated arises because of the use of "nonconforming" finite elements. This is illustrated with elements that are not C^0. Thus, (10.0.2) fails since the finite element functions are not sufficiently smooth; this can happen on a polygonal domain where the boundary conditions are satisfied exactly. In addition, the definition of the variational form must be altered.

The theory developed here will also find application in later chapters. For example, in Chapter 12 we consider "mixed" finite element approximations. We develop the basic convergence theory for mixed methods from the point of view of a variational crime.

10.1 Departure from the Framework

Let H denote a Hilbert space and let $V \subset H$ be a subspace. We first derive an abstract error estimate for variational problems in which $V_h \not\subset V$. We assume that $a(\cdot, \cdot)$ is a bilinear form defined on H but we do *not* assume it to be symmetric in the following lemma.

(10.1.1) Lemma. *Let V and V_h be subspaces of H. Assume that $a(\cdot, \cdot)$ is a continuous bilinear form on H which is coercive on V_h, with respective continuity and coercivity constants C and γ. Let $u \in V$ solve*

$$a(u, v) = F(v) \quad \forall\, v \in V,$$

where $F \in H'$. Let $u_h \in V_h$ solve

$$a(u_h, v) = F(v) \quad \forall\, v \in V_h.$$

Then

(10.1.2)
$$
\begin{aligned}
\|u - u_h\|_H &\le \left(1 + \frac{C}{\gamma}\right) \inf_{v \in V_h} \|u - v\|_H \\
&\quad + \frac{1}{\gamma} \sup_{w \in V_h \setminus \{0\}} \frac{|a(u - u_h, w)|}{\|w\|_H}.
\end{aligned}
$$

Proof. For any $v \in V_h$,

$$
\begin{aligned}
\|u - u_h\|_H &\le \|u - v\|_H + \|v - u_h\|_H && \text{(triangle inequality)} \\
&\le \|u - v\|_H + \frac{1}{\gamma} \sup_{w \in V_h \setminus \{0\}} \frac{|a(v - u_h, w)|}{\|w\|_H} && \text{(coercivity)} \\
&= \|u - v\|_H + \frac{1}{\gamma} \sup_{w \in V_h \setminus \{0\}} \frac{|a(v - u, w) + a(u - u_h, w)|}{\|w\|_H} \\
&\le \|u - v\|_H + \frac{1}{\gamma} \sup_{w \in V_h \setminus \{0\}} \frac{|a(v - u, w)|}{\|w\|_H} \\
&\quad + \frac{1}{\gamma} \sup_{w \in V_h \setminus \{0\}} \frac{|a(u - u_h, w)|}{\|w\|_H} && \text{(triangle inequality)} \\
&\le \|u - v\|_H + \frac{C}{\gamma}\|v - u\|_H && \text{(continuity)} \\
&\quad + \frac{1}{\gamma} \sup_{w \in V_h \setminus \{0\}} \frac{|a(u - u_h, w)|}{\|w\|_H} \\
&= \left(1 + \frac{C}{\gamma}\right)\|u - v\|_H + \frac{1}{\gamma} \sup_{w \in V_h \setminus \{0\}} \frac{|a(u - u_h, w)|}{\|w\|_H}.
\end{aligned}
$$

(10.1.3)

$\square$

The second term on the right-hand side of (10.1.2) would be zero if $V_h \subseteq V$. Therefore, it measures the effect of $V_h \not\subseteq V$.

Also note that, by continuity,

$$(10.1.4) \qquad \frac{|a(u - u_h, w)|}{\|w\|_H} \leq C \, \|u - u_h\|_H$$

so that

$$(10.1.5) \qquad \|u - u_h\|_H \geq \frac{1}{C} \sup_{w \in V_h \setminus \{0\}} \frac{|a(u - u_h, w)|}{\|w\|_H}.$$

Combining (10.1.5) and (10.1.2) gives

$$
(10.1.6)
\begin{aligned}
\max &\left\{ \frac{1}{C} \sup_{w \in V_h \setminus \{0\}} \frac{|a(u - u_h, w)|}{\|v\|_H} , \; \inf_{v \in V_h} \|u - v\|_H \right\} \\
&\leq \|u - u_h\|_H \\
&\leq \left(1 + \frac{C}{\gamma}\right) \inf_{v \in V_h} \|u - v\|_H + \frac{1}{\gamma} \sup_{w \in V_h \setminus \{0\}} \frac{|a(u - u_h, w)|}{\|w\|_H}.
\end{aligned}
$$

Inequality (10.1.6) indicates that the term $\inf_{v \in V_h} \|u - v\|_H$, together with $\sup_{w \in V_h \setminus \{0\}} |a(u - u_h, w)| / \|w\|_H$ truly reflect the size of the discretization error $\|u - u_h\|_H$.

When the variational form $a(\cdot, \cdot)$ is symmetric positive-definite, we can improve Lemma 10.1.1 by using the natural norm

$$\|v\|_a := \sqrt{a(v, v)}$$

as follows.

(10.1.7) Lemma. *Let V and V_h be subspaces of H and $\dim V_h < \infty$. Assume that $a(\cdot, \cdot)$ is a symmetric positive-definite bilinear form on H. Let $u \in V$ and $u_h \in V_h$ be as in Lemma 10.1.1. Then*

$$(10.1.8) \qquad \|u - u_h\|_a \leq \inf_{v \in V_h} \|u - v\|_a + \sup_{w \in V_h \setminus \{0\}} \frac{|a(u - u_h, w)|}{\|w\|_a}.$$

Proof. Exercise 10.x.5.   □

In Sect. 10.2 we will apply the abstract error estimate (10.1.2) to finite element approximations of (10.0.1) with interpolated boundary conditions. In Sect. 10.4 we will do the same for isoparametric finite element approximations of (10.0.1). In Sect. 10.5 we show that finite element approximations of the Stokes equations can be studied via (10.1.8).

The second case where (10.0.2) is violated arises from the use of non-conforming finite elements, where $V_h \not\subseteq H^1(\Omega)$. Since elements of V_h do not have global weak derivatives, we must use $a_h(\cdot, \cdot)$, a modification of $a(\cdot, \cdot)$, in the discretized problem. We have the following abstract error estimate.

(10.1.9) Lemma. *Assume dim $V_h < \infty$. Let $a_h(\cdot, \cdot)$ be a symmetric positive-definite bilinear form on $V + V_h$ which reduces to $a(\cdot, \cdot)$ on V. Let $u \in V$ solve*

$$a(u, v) = F(v) \quad \forall\, v \in V,$$

where $F \in V' \cap V_h'$. Let $u_h \in V_h$ solve

$$a_h(u_h, v) = F(v) \quad \forall\, v \in V_h.$$

Then

$$(10.1.10) \qquad \|u - u_h\|_h \leq \inf_{v \in V_h} \|u - v\|_h + \sup_{w \in V_h \setminus \{0\}} \frac{|a_h(u - u_h, w)|}{\|w\|_h},$$

where $\|\cdot\|_h = \sqrt{a_h(\cdot, \cdot)}$.

Proof. Let $\tilde{u}_h \in V_h$ satisfy

$$(10.1.11) \qquad a_h(\tilde{u}_h, v) = a_h(u, v) \quad \forall\, v \in V_h,$$

which implies that

$$(10.1.12) \qquad \|u - \tilde{u}_h\|_h = \inf_{v \in V_h} \|u - v\|_h.$$

Then

$$
\begin{aligned}
(10.1.13) \qquad \|u - u_h\|_h &\leq \|u - \tilde{u}_h\|_h + \|\tilde{u}_h - u_h\|_h \\
&= \|u - \tilde{u}_h\|_h + \sup_{w \in V_h \setminus \{0\}} \frac{|a_h(\tilde{u}_h - u_h, w)|}{\|w\|_h}.
\end{aligned}
$$

The estimate (10.1.10) follows from (10.1.11)-(10.1.13). $\qquad\square$

Again, the second term on the right-hand side of (10.1.10) is zero if $V_h \subseteq V$. This term, therefore, measures the effect of $V_h \not\subseteq V$. Also, an inequality analogous to (10.1.6) holds here. In Sect. 10.3, we will apply the abstract error estimate (10.1.10) to the nonconforming, piecewise linear (see Fig. 3.2) finite element approximation of (10.0.1).

10.2 Finite Elements with Interpolated Boundary Conditions

Let $\Omega \subseteq \mathbb{R}^2$ be a bounded domain with smooth boundary, and $\mathcal{T}^h$ be a triangulation of Ω, where each triangle at the boundary has at most one curved side (cf. Fig. 10.1).

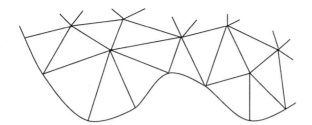

Fig. 10.1. Smooth domain triangulated with curved triangles

We assume that there exists $\rho > 0$ such that for each triangle $T \in \mathcal{T}^h$ we can find two concentric circular discs D_1 and D_2 such that

$$(10.2.1) \qquad D_1 \subseteq T \subseteq D_2 \quad \text{and} \quad \frac{\text{diam } D_2}{\text{diam } D_1} \leq \rho.$$

It follows from a homogeneity argument (see the proof of Lemma 4.5.3) that

$$(10.2.2) \qquad \|\phi\|_{W^{k-1}_\infty(D_2)} \leq C_{k,\rho}(\text{diam } D_2)^{1-k}\|\phi\|_{H^1(D_1)}$$

for any polynomial ϕ of degree $\leq k - 1$.

In order to describe the finite element space V_h, we need to use the nodes of the Lobatto quadrature formula. Therefore, we first briefly describe the basic facts concerning Lobatto quadrature. Let the polynomial $L_k(\xi)$ of degree k be defined by

$$(10.2.3) \qquad L_k(x) = \left(\frac{d}{dx}\right)^{k-2} (x(1-x))^{k-1}.$$

$L_k(\xi)$ has k distinct roots $0 = \xi_0 < \xi_1 < \dots < \xi_{k-1} = 1$ (cf. exercise 10.x.1).

For each j, $0 \leq j \leq k-1$, let P_j be the Lagrange interpolating polynomial of degree $k-1$ (see Remark 3.1.7) such that $P_j(\xi_i) = \delta_{ij}$ (the Kronecker delta), and let

$$(10.2.4) \qquad \omega_j = \int_0^1 P_j(x)\, dx.$$

(10.2.5) Lemma. *For any polynomial P of degree less than $2k - 2$, we have*

$$(10.2.6) \qquad \int_0^1 P(x)\, dx = \sum_{j=0}^{k-1} \omega_j P(\xi_j).$$

Proof. For any polynomial f of degree less than k, we have by the definition of ω_j that

$$(10.2.7) \qquad \int_0^1 f(x)\,dx = \int_0^1 \sum_{j=0}^{k-1} f(\xi_j) P_j(x)\,dx$$

$$= \sum_{j=0}^{k-1} \omega_j f(\xi_j).$$

Let P_I be the Lagrange interpolant of P of degree $\leq k-1$ such that $P_I(\xi_j) = P(\xi_j)$ for $0 \leq j \leq k-1$. Then (10.2.7) implies

$$\int_0^1 P(x)\,dx - \sum_{j=0}^{k-1} \omega_j P(\xi_j) = \int_0^1 (P - P_I)(x)\,dx.$$

Since $(P - P_I)$ vanishes at ξ_j, $0 \leq j \leq k-1$, we can write $(P - P_I)(x) = L_k(x)q(x)$, where $\deg q < k-2$. It follows from (10.2.3) and integration by parts $k-2$ times that

$$\int_0^1 L_k(x)q(x)\,dx = \int_0^1 \left(\left(\frac{d}{dx}\right)^{k-2} (x(1-x))^{k-1} \right) q(x)\,dx$$

$$(10.2.8) \qquad\qquad = (-1)^{k-2} \int_0^1 (x(1-x))^{k-1} \left(\frac{d}{dx}\right)^{k-2} q(x)\,dx$$

$$= 0,$$

because all boundary terms vanish and $\deg q < k-2$. $\square$

(10.2.9) Corollary. *Given k, there exists a positive constant C_k such that for all $h > 0$*

$$\left| \int_0^h f(x)\,dx - h \sum_{j=0}^{k-1} \omega_j f(h\xi_j) \right| \leq C_k \, h^{2k-1} \, \|f^{(2k-2)}\|_{L^\infty(0,k)}$$

for any C^{2k-2} function f on $[0,h]$.

Proof. Exercise 10.x.2. $\square$

We are now ready to define V_h. For each boundary edge

$$e = \{ x(s) \; : \; s \in [s_e, s_e + h_e], \; s = \text{ arc length} \},$$

let the boundary nodes be $x(s_e + h_e \xi_j)$, $j = 0, \ldots, k-1$. In Fig. 10.2, the case $k = 3$ is depicted. Note that for $h = \max_{T \in \mathcal{T}^h}(\operatorname{diam} T)$ small enough, the boundary nodes can be used as part of the nodal variables that determine $\mathcal{P}_{k-1}$ (see exercise 10.x.3 and (Scott 1975)). This will be assumed from now on.

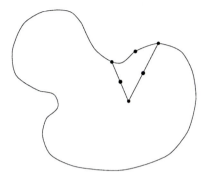

Fig. 10.2. Nodal variables for curved-triangle Lagrange quadratics

The finite element space V_h is defined by

(10.2.10)
$$V_h = \{v \in C^0(\overline{\Omega}) : v|_T \in \mathcal{P}_{k-1} \text{ and}$$
$$v \text{ vanishes at the boundary nodes}\}.$$

Let $a(v,w) = \int_\Omega \nabla v \cdot \nabla w \, dx$. Then the variational form of (10.0.1) is to find $u \in V = \mathring{H}^1(\Omega)$ such that

(10.2.11)
$$a(u,v) = F(v) \quad \forall v \in V = \mathring{H}^1(\Omega),$$

where $F(v) = \int_\Omega f v \, dx$ (see Chapter 5). The solution $u_h \in V_h$ of the discretized problem satisfies

(10.2.12)
$$a(u_h, v) = F(v) \quad \forall v \in V_h.$$

In order to use the abstract estimate (10.1.2) we must estimate

$$\sup \left\{ |a(u - u_h, w)| / \|w\|_{H^1(\Omega)} \; : \; w \in V_h \setminus \{0\} \right\}$$

and verify that $a(\cdot, \cdot)$ is coercive on V_h. By Green's Theorem, we have

(10.2.13)
$$a(u - u_h, w) = a(u, w) - (f, w)$$
$$= (-\Delta u, w) + \int_{\partial\Omega} \frac{\partial u}{\partial \nu} w \, ds - (f, w)$$
$$= \int_{\partial\Omega} \frac{\partial u}{\partial \nu} w \, ds.$$

We first do a local estimate.

(10.2.14) Lemma. *Let T be a triangle satisfying (10.2.1) with a curved edge e. Assume that $u \in W_\infty^{2k-1}(T)$, and $w \in \mathcal{P}_{k-1}$ vanishes at the Lobatto nodes along e. Then there exists a constant $C_{k,\rho}$ such that*

(10.2.15)
$$\left| \int_e \frac{\partial u}{\partial \nu} w \, ds \right| \le C_{k,\rho} h_e^{2k-1} (\operatorname{diam} D_2)^{1-k} \|u\|_{W_\infty^{2k-1}(T)} \|w\|_{H^1(T)}$$

where $h_e = $ length of e.

Proof. Let s denote arc length. In terms of the parameterization $x(s)$, $0 \leq s \leq h_e$,

$$(10.2.16) \qquad \int_e \frac{\partial u}{\partial \nu} w \, ds = \int_0^{h_e} \frac{\partial u}{\partial \nu}(x(s)) w(x(s)) \, ds.$$

Then

$$\left| \int_0^{h_e} \frac{\partial u}{\partial \nu}(x(s)) w(x(s)) \, ds \right|$$

$$\leq C_k \, h_e^{2k-1} \left\| \frac{\partial u}{\partial \nu} \right\|_{W_\infty^{2k-2}(T)} \|w\|_{W_\infty^{k-1}(T)} \qquad \text{(Corollary 10.2.9)}$$

$$\leq C_{k,\rho} \, h_e^{2k-1} (\operatorname{diam} D_2)^{1-k} \|u\|_{W_\infty^{2k-1}(T)} \|w\|_{H^1(T)}. \text{(using 10.2.2)}$$

Using (10.2.16) completes the lemma. $\qquad\qquad \square$

From the local estimate (10.2.15) we deduce the following global estimate.

(10.2.17) Lemma. *Assume that* $u \in W_\infty^{2k-1}(\Omega)$. *For small* h *and fixed* k, *there exists* $C_\rho > 0$ *such that*

$$\sup_{w \in V_h \backslash \{0\}} \frac{|a(u - u_h, w)|}{\|w\|_{H^1(\Omega)}} \leq C_\rho \, h^{k - \frac{1}{2}} \|u\|_{W_\infty^{2k-1}(\Omega)}.$$

Proof. Since $\partial \Omega$ is smooth, for h small enough, we have

$$(10.2.18) \qquad\qquad h_e < 2 \operatorname{diam} T < 2 \operatorname{diam} D_2$$

in Lemma 10.2.14. Therefore,

$$(10.2.19) \qquad \left| \int_e \frac{\partial u}{\partial \nu} w \, ds \right| \leq C_\rho \, h_e^k \|u\|_{W_\infty^{2k-1}(\Omega)} \|w\|_{H^1(T)}.$$

By summing over all boundary edges, it follows from (10.2.19) that

$$\left| \int_\Omega \frac{\partial u}{\partial \nu} w \, ds \right| \leq \sum_e \left| \int_e \frac{\partial u}{\partial \nu} w \, ds \right|$$

$$\leq C_\rho \, h^{k - \frac{1}{2}} \|u\|_{W_\infty^{2k-1}(\Omega)} \left(\sum_e h_e^{\frac{1}{2}} \|w\|_{H^1(T)} \right)$$

$$\leq C_\rho \, h^{k - \frac{1}{2}} \|u\|_{W_\infty^{2k-1}(\Omega)} \left(\sum_e h_e \right)^{\frac{1}{2}} \|w\|_{H^1(\Omega)}.$$

Since $\sum_e h_e =$ length of $\partial \Omega$, the lemma follows from (10.2.13). $\qquad \square$

We next turn to the question of coercivity of $a(\cdot, \cdot)$ on V_h. We need the following lemma, where Ω is assumed to be connected.

(10.2.20) Lemma. *There exists a positive constant β such that for all $v \in H^1(\Omega)$, we have*

$$(10.2.21) \qquad \beta \|v\|_{H^1(\Omega)} \le |v|_{H^1(\Omega)} + \left| \int_{\partial\Omega} v \, ds \right|.$$

Proof. Assume that the lemma is false. Then there exists a sequence $v_j \in H^1(\Omega)$ such that

$$(10.2.22) \qquad \|v_j\|_{H^1(\Omega)} = 1$$

and

$$(10.2.23) \qquad |v_j|_{H^1(\Omega)} + \left| \int_{\partial\Omega} v_j \, ds \right| < \frac{1}{j}.$$

From Friedrichs' inequality (4.3.15) and (10.2.23) we have

$$(10.2.24) \qquad \|v_j - \bar{v}_j\|_{L^2(\Omega)} \le C \, |v_j|_{H^1(\Omega)} \le \frac{C}{j}$$

where $\bar{v}_j = |\Omega|^{-1} \int_\Omega v_j \, dx$. Then (10.2.23) and (10.2.24) imply

$$(10.2.25) \qquad
\begin{aligned}
\lim_{j\to\infty} \|v_j - \bar{v}_j\|^2_{H^1(\Omega)} &= \lim_{j\to\infty} \left(|v_j - \bar{v}_j|^2_{H^1(\Omega)} + \|v_j - \bar{v}_j\|^2_{L^2(\Omega)} \right) \\
&= \lim_{j\to\infty} \left(|v_j|^2_{H^1(\Omega)} + \|v_j - \bar{v}_j\|^2_{L^2(\Omega)} \right) \\
&= 0.
\end{aligned}$$

It follows from (10.2.25) and the Trace Theorem 1.6.6 that

$$(10.2.26) \qquad \lim_{j\to\infty} \int_{\partial\Omega} (v_j - \bar{v}_j) \, ds = 0.$$

Hence, we have from (10.2.23) and (10.2.26) that

$$(10.2.27) \qquad \lim_{j\to\infty} \int_{\partial\Omega} \bar{v}_j \, ds = \lim_{j\to\infty} \left(\int_{\partial\Omega} (\bar{v}_j - v_j) \, ds + \int_{\partial\Omega} v_j \, ds \right) = 0.$$

Therefore, $\lim_{j\to\infty} \bar{v}_j = 0$ and we conclude from (10.2.25) that

$$(10.2.28) \qquad \lim_{j\to\infty} \|v_j\|_{H^1(\Omega)} = 0.$$

But (10.2.28) contradicts (10.2.22). $\qquad\square$

We are now ready to prove the coercivity of $a(\cdot, \cdot)$ on V_h.

(10.2.29) Lemma. *There exists a positive constant γ such that for h small enough, we have*

$$(10.2.30) \qquad a(v,v) \geq \gamma \|v\|_{H^1(\Omega)}^2 \quad \forall v \in V_h.$$

Proof. Without loss of generality we may assume that Ω is connected. From Lemma 10.2.17 we have

$$(10.2.31) \qquad \left| \int_{\partial\Omega} \frac{\partial u}{\partial \nu} v \, ds \right| \leq C_\rho \, h^{k-\frac{1}{2}} \|u\|_{W_\infty^{2k-1}(\Omega)} \|v\|_{H^1(\Omega)}$$

for any $u \in W_\infty^{2k-1}(\Omega)$.

Let $u_* \in C^\infty(\overline{\Omega})$ satisfy

$$(10.2.32) \qquad \frac{\partial u_*}{\partial \nu} \equiv 1 \quad \text{on} \quad \partial\Omega,$$

then (10.2.31) becomes

$$(10.2.33) \qquad \left| \int_{\partial\Omega} v \, ds \right| \leq C_\rho \, h^{k-\frac{1}{2}} \|v\|_{H^1(\Omega)}.$$

Combining (10.2.21) and (10.2.33) we have

$$(10.2.34) \qquad \beta \|v\|_{H^1(\Omega)} \leq |v|_{H^1(\Omega)} + C_\rho \, h^{k-\frac{1}{2}} \|v\|_{H^1(\Omega)},$$

and hence

$$(10.2.35) \qquad \left(\beta - C_\rho \, h^{k-\frac{1}{2}} \right) \|v\|_{H^1(\Omega)} \leq |v|_{H^1(\Omega)} = \sqrt{a(v,v)}.$$

The coercivity estimate (10.2.30) therefore holds if we let $\gamma = \beta^2/4$, provided that $C_\rho \, h^{k-\frac{1}{2}} < \beta/2$. □

In view of Lemmas 10.1.1, 10.2.17, 10.2.29, and standard interpolation error estimates (cf. Theorem 4.4.20), we have proved the following theorem.

(10.2.36) Theorem. *Assume that $u \in W_\infty^{2k-1}(\Omega)$ and (10.2.1) holds for a $\rho > 0$ independent of h. When (10.0.1) is discretized using V_h, the Lagrange finite element space of continuous piecewise polynomials with degree $\leq k-1$ which vanish at the Lobatto boundary nodes, we have the following error estimate:*

$$(10.2.37) \qquad \|u - u_h\|_{H^1(\Omega)} \leq C_\rho \, h^{k-1} \|u\|_{W_\infty^{2k-1}(\Omega)},$$

for h sufficiently small.

It is not necessary to assume that $u \in W_\infty^{2k-1}(\Omega)$ in order to conclude that $\|u - u_h\|_{H^1(\Omega)} \leq C_\rho \, h^{k-1}$. In particular, it can be shown (exercise 10.x.4 and (Scott 1975)) that this holds for $u \in H^k(\Omega)$, the best possible norm.

10.3 Nonconforming Finite Elements

In this section we look at a variational crime as in Lemma 10.1.9. Recall that the variational formulation of (10.0.1) is to find $u \in V = \mathring{H}^1(\Omega)$ such that

$$(10.3.1) \qquad a(u, v) = F(v) \quad \forall v \in V = \mathring{H}^1(\Omega),$$

where $a(u, v) = \int_\Omega \nabla u \cdot \nabla v \, dx$ and $F(v) = \int_\Omega fv \, dx$. We assume that Ω is a convex polygonal domain, and $f \in L^2(\Omega)$. Therefore, u belongs to $H^2(\Omega)$ by elliptic regularity.

Let $\mathcal{T}^h$ be a triangulation of Ω and let $h := \max_{T \in \mathcal{T}^h} \operatorname{diam} T$. The nonconforming, piecewise linear finite element space (see Fig. 3.2) is defined to be

$$(10.3.2) \qquad \begin{aligned} V_h := \{v : v|_T \text{ is linear for all } T \in \mathcal{T}^h, \\ v \text{ is continuous at the midpoints of the edges and} \\ v = 0 \text{ at the midpoints on } \partial\Omega\}. \end{aligned}$$

Note that since functions in V_h are no longer continuous, they are no longer in $\mathring{H}^1(\Omega)$. We must therefore use a modified variational form $a_h(\cdot, \cdot)$ in the discretized problem.

We define the following bilinear form on $V_h + V$

$$(10.3.3) \qquad a_h(v, w) = \sum_{T \in \mathcal{T}^h} \int_T \nabla v \cdot \nabla w \, dx$$

and its associated norm

$$(10.3.4) \qquad \|v\|_h := \sqrt{a_h(v, v)}.$$

The form $a_h(\cdot, \cdot)$ is coercive on $V = \mathring{H}^1(\Omega)$ because $a_h(\cdot, \cdot) \equiv a(\cdot, \cdot)$ on V. It is positive-definite on V_h because $a_h(v, v) = 0$ implies v is piecewise constant, and the zero boundary condition together with continuity at midpoints imply $v \equiv 0$.

The discretized problem is to find $u_h \in V_h$ such that

$$(10.3.5) \qquad a_h(u_h, v) = F(v) \quad \forall v \in V_h.$$

We want to estimate $\|u - u_h\|_h$, using the abstract error estimate (10.1.10). Since $u \in H^2(\Omega) \cap \mathring{H}^1(\Omega)$, Theorem 4.4.20 yields

$$(10.3.6) \qquad \inf_{v \in V_h} \|u - v\|_h \leq C h \, |u|_{H^2(\Omega)},$$

where C is a positive constant that depends only on the parameter ρ in (4.4.16). For the second term on the right-hand side of (10.1.10), we have

$$a_h(u - u_h, w) = \sum_{T \in \mathcal{T}^h} \int_T \nabla u \cdot \nabla w \, dx - \int_\Omega f w \, dx \quad \text{(by 10.3.3 and 10.3.5)}$$

$$= \sum_{T \in \mathcal{T}^h} \left[\int_{\partial T} \frac{\partial u}{\partial \nu} w \, ds - \int_T \Delta u \, w \, dx \right] - \int_T f w \, dx$$

$$= \sum_{T \in \mathcal{T}^h} \int_{\partial T} \frac{\partial u}{\partial \nu} w \, ds \qquad \text{(by 10.0.1)}$$

$$= \sum_{e \in \mathcal{E}^h} \int_e \frac{\partial u}{\partial \nu_e} [w] \, ds,$$

where $\mathcal{E}^h$ is the set of the edges in the triangulation, ν_e is a unit normal to the edge e and $[w]$ is the jump of the function w across e. (The function w is taken to be 0 outside of Ω.)

The next step is to estimate $\left| \int_e \frac{\partial u}{\partial \nu_e} [w] \, ds \right|$, which we do in the following lemma.

(10.3.7) Lemma. *Let G be the union of two neighboring triangles such that* diam $G = 1$. *Let $\zeta \in H^1(G)$, let $V = \{z : z|_{T_i} \text{ is linear and } z \text{ is continuous at the midpoint } m\}$ (cf. Fig. 10.3). Then there exists $C < \infty$ such that*

$$(10.3.8) \qquad \left| \int_{\overline{p_1 p_2}} \zeta \left(z|_{T_2} - z|_{T_1} \right) ds \right| \leq C |\zeta|_{H^1(G)} \left(|z|_{H^1(T_1)} + |z|_{H^1(T_2)} \right)$$

for all $z \in V$.

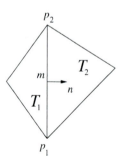

Fig. 10.3. Notation for Lemma 10.3.7

Proof. Let $[z] := z|_{T_2} - z|_{T_1}$, and let e denote $\overline{p_1 p_2}$. Note that $\int_e [z] \, ds = 0$ since $[z]$ is linear and vanishes at the midpoint of e. Given any constants c_1 and c_2, we have

$$\left| \int_e \zeta [z] \, ds \right| = \left| \int_e (\zeta - c_1)[z] \, ds \right|$$

$$= \left| \int_e (\zeta - c_1)[z - c_2] \, ds \right| \qquad ([c] = 0 \text{ for any constant})$$

$$\leq \|\zeta - c_1\|_{L^2(e)} \, \|[z - c_2]\|_{L^2(e)} \qquad \text{(Schwarz' inequality 2.1.5)}$$

$$\leq C \|\zeta - c_1\|_{H^1(G)} \left(\|z - c_2\|_{H^1(T_1)} + \|z - c_2\|_{H^1(T_2)} \right)$$

by the Trace Theorem 1.6.6. Taking the infimum over $c_1, c_2 \in \mathbb{R}$, the estimate (10.3.8) follows from the Bramble-Hilbert Lemma 4.3.8.   □

(10.3.9) Lemma. *Let* $\zeta \in H^1(T)$ *and* $V = \{z : z$ *is linear on* T *and* $z = 0$ *at* $m\}$ *(cf. Fig. 10.4). Then there exists* $C < \infty$ *such that*

(10.3.10)
$$\left| \int_{\overline{p_1 \, p_2}} \zeta \, z \, ds \right| \leq C \, |\zeta|_{H^1(T)} \, |z|_{H^1(T)}.$$

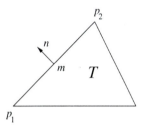

Fig. 10.4. Notation for Lemma 10.3.9

Proof. Exercise 10.x.7.   □

By applying Lemmas 10.3.7 and 10.3.9 with $\zeta = \frac{\partial u}{\partial \nu_e}$ and using a homogeneity argument (cf. exercise 10.x.8) we have

(10.3.11) $$|a_h(u - u_h, w)| \leq C \, h \, |u|_{H^2(\Omega)} \|w\|_h.$$

Combining (10.1.10), (10.3.6) and (10.3.11), we have therefore established the following theorem.

(10.3.12) Theorem. *Let* Ω *be a convex polygonal domain, let* $f \in L^2(\Omega)$ *and let* u_h *be the solution of the nonconforming, piecewise linear discretization (10.3.5) of (10.0.1). Then the following discretization error estimate holds:*

(10.3.13) $$\|u - u_h\|_h \leq C \, h \, |u|_{H^2(\Omega)},$$

where the positive constant C *only depends on the parameter* ρ *in (4.4.16).*

It is reasonable to ask why one would want to use nonconforming finite elements. There are situations where nonconforming methods are clearly desirable. One example is the incompressible fluid flow problem. If one uses the vector conforming, piecewise linear Lagrange finite element, the only function in V_h which satisfies the divergence-free condition is the zero function on generic meshes. In order to obtain a good approximate solution by a conforming method, one must therefore use higher-order polynomials. On the other hand, the vector nonconforming piecewise linear finite element space can be used to solve such a problem (cf. (12.4.12)).

Another example is the biharmonic equation (cf. Sect. 5.9). Since this is a fourth order problem, the conforming finite elements are C^1 elements, for example the Argyris finite element, which involves fifth degree polynomials. On the other hand, it can be solved using the nonconforming quadratic Morley finite element, which is depicted in Fig. 10.5. The properties of this finite element and its application to the biharmonic equation can be found in exercises 10.x.9 through 10.x.14 (see also (Shi 1990)). Note also that the Morley element can be used to construct preconditioners for the system resulted from the discretization of the biharmonic equation by the Argyris element (cf. (Brenner 1996)).

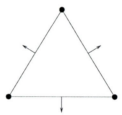

Fig. 10.5. Morley finite element

10.4 Isoparametric Finite Elements

The use of isoparametric finite elements (Sect. 4.7) also entails a variational crime; one can envisage either that (10.0.2) is violated or that there is a modified bilinear form. However, this approach is simpler to estimate than the previous one. When one interpolates the boundary conditions using polynomials, it can happen that $v \in V_h$ differs from zero significantly on $\partial\Omega$. The use of Lobatto interpolation points minimizes the effect of this by guaranteeing that the error oscillates in such a way as to cancel any deleterious effects. On the other hand, isoparametric finite elements are no longer polynomials in the coordinates of Ω, being transformed via a piecewise polynomial mapping from some domain Ω_h. Thus, they are able to match Dirichlet boundary conditions more closely in a pointwise sense.

We begin with the formulation for the model problem (10.0.1). Recall that we have a polyhedral approximation, Ω_h, to Ω, and a mapping F^h such that $F^h(\Omega_h)$ closely approximates Ω. Suppose that properties 1–3 of

Sect. 4.7 hold for the isoparametric mapping. In particular, we will assume there is an auxiliary mapping $F : \Omega_h \to \Omega$ and that $F_i^h = \mathcal{I}^h F_i$ for each component of the mapping. Further, we assume that conditions 1 and 3 in Sect. 4.7 hold as well for F. The construction of such an F is not trivial, but it is done in (Lenoir 1986). Define

$$a_h(v, w) = \int_{F^h(\Omega_h)} \nabla v(x) \cdot \nabla w(x) \, dx.$$

For the purposes of analysis, we will need the mapping $\Phi^h : \Omega \to F^h(\Omega_h)$ defined by $\Phi^h(x) = F^h(F^{-1}(x))$. Note that we can write

$$a_h(v, w) = \int_{\Omega} J_{\Phi^h}(x)^{-t} \nabla \hat{v}(x) \cdot J_{\Phi^h}(x)^{-t} \nabla \hat{w}(x) \det J_{\Phi^h}(x) \, dx$$

where for any function v defined on $F^h(\Omega_h)$ we set

(10.4.1) $$\hat{v}(x) := v\left(\Phi^h(x)\right)$$

and J_{Φ^h} denotes the Jacobian of Φ^h. The key point is that

(10.4.2) $$J_{\Phi^h} - I = \mathcal{O}(h^{k-1}).$$

Let V_h be defined as in (4.7.2), where $\tilde{\Omega} = \Omega_h$ and $\tilde{F} = F^h$, and define $u_h \in V_h$ by

(10.4.3) $$a_h(u_h, v) = \int_{F^h(\Omega_h)} f(x)v(x) \, dx.$$

First we derive an estimate analogous to (10.1.10). We define $\widehat{V}_h = \{\hat{v} : v \in V_h\}$, and let $\hat{v}_h$ denote the projection of u onto $\widehat{V}_h$:

(10.4.4) $$a(\hat{v}_h, \hat{w}) = a(u, \hat{w}) \quad \forall \hat{w} \in \widehat{V}_h.$$

Note that $\widehat{V}_h \subset V$, that is, the Dirichlet boundary conditions are satisfied exactly for $\hat{w} \in \widehat{V}_h$. Then for any $w \in V_h$

$$a(\hat{v}_h - \hat{u}_h, \hat{w}) = a(u - \hat{u}_h, \hat{w}) \qquad \text{(by 10.4.4)}$$

(10.4.5) $$= (f, \hat{w}) - a(\hat{u}_h, \hat{w})$$

$$= (f, \hat{w}) - \int_{F^h(\Omega_h)} fw \, dx + a_h(u_h, w) - a(\hat{u}_h, \hat{w}). \text{(by 10.4.3)}$$

We thus find

$$
\begin{aligned}
\|u - \hat{u}_h\|_a &\leq \|u - \hat{v}_h\|_a + \|\hat{v}_h - \hat{u}_h\|_a && \text{(triangle inequality)} \\
&= \inf_{v \in V_h} \|u - \hat{v}\|_a + \|\hat{v}_h - \hat{u}_h\|_a && \text{(see 10.1.12)}
\end{aligned}
$$

$$
(10.4.6) \qquad = \inf_{v \in V_h} \|u - \hat{v}\|_a + \sup_{w \in V_h \setminus \{0\}} \frac{|a(\hat{v}_h - \hat{u}_h, \hat{w})|}{\|\hat{w}\|_a} \qquad \text{(see 2.x.16)}
$$

$$
\leq \inf_{v \in V_h} \|u - \hat{v}\|_a + \sup_{w \in V_h \setminus \{0\}} \frac{|a_h(u_h, w) - a(\hat{u}_h, \hat{w})|}{\|\hat{w}\|_a}
$$

$$
+ \sup_{w \in V_h \setminus \{0\}} \frac{\left| (f, \hat{w}) - \int_{F^h(\Omega_h)} f w \, dx \right|}{\|\hat{w}\|_a}. \qquad \text{(by 10.4.5)}
$$

The second term above may be expanded as

$$
\begin{aligned}
a_h(u_h, w) - a(\hat{u}_h, \hat{w}) &= \int_\Omega J_{\Phi^h}^{-t} \nabla \hat{u}_h \cdot J_{\Phi^h}^{-t} \nabla \hat{w} \, \det J_{\Phi^h} \, dx \\
&\quad - \int_\Omega \nabla \hat{u}_h \cdot \nabla \hat{w} \, dx \\
&= \int_\Omega \left(J_{\Phi^h}^{-t} - I \right) \nabla \hat{u}_h \cdot J_{\Phi^h}^{-t} \nabla \hat{w} \, \det J_{\Phi^h} \, dx \\
&\quad + \int_\Omega \nabla \hat{u}_h \cdot J_{\Phi^h}^{-t} \nabla \hat{w} \, \det J_{\Phi^h} \, dx - \int_\Omega \nabla \hat{u}_h \cdot \nabla \hat{w} \, dx \\
&= \int_\Omega \left(J_{\Phi^h}^{-t} - I \right) \nabla \hat{u}_h \cdot J_{\Phi^h}^{-t} \nabla \hat{w} \, \det J_{\Phi^h} \, dx \\
&\quad + \int_\Omega \nabla \hat{u}_h \cdot \left((\det J_{\Phi^h}) J_{\Phi^h}^{-t} - I \right) \nabla \hat{w} \, dx.
\end{aligned}
$$

From (10.4.2) we conclude that

$$
(10.4.7) \qquad \sup_{w \in V_h \setminus \{0\}} \frac{|a_h(u_h, w) - a(\hat{u}_h, \hat{w})|}{\|\hat{w}\|_a} \leq C h^{k-1} \|\hat{u}_h\|_a.
$$

The term involving f can be estimated similarly:

$$
\begin{aligned}
\left| (f, \hat{w}) - \int_{F^h(\Omega_h)} f w \, dx \right| &= \left| \int_\Omega \left(f(x) - f(\Phi^h(x)) \right) \det J_{\Phi^h}(x) \, \hat{w}(x) \, dx \right| \\
&\leq \left| \int_\Omega \left(f(x) - f(\Phi^h(x)) \right) \det J_{\Phi^h}(x) \hat{w}(x) \, dx \right| \\
&\quad + \left| \int_\Omega f(x) \left(1 - \det J_{\Phi^h}(x) \right) \hat{w}(x) \, dx \right| \\
&\leq C \left(h^k \|f\|_{W_\infty^1(\Omega)} + h^{k-1} \|f\|_{L^2(\Omega)} \right) \|\hat{w}\|_{L^2(\Omega)}.
\end{aligned}
$$

Combining this with (10.4.6) and (10.4.7), we have the following.

(10.4.8) Theorem. *Let u be defined by (10.0.1) and let u_h be defined by (10.4.3) where V_h is defined in (4.7.2) (with $\tilde{\Omega} = \Omega_h$ and $\tilde{F} = F^h$). Then*

$$\|u - \hat{u}_h\|_{H^1(\Omega)} \le C\, h^{k-1} \left(\|u\|_{H^k(\Omega)} + \|f\|_{W^1_\infty(\Omega)} \right)$$

for h sufficiently small (see (10.4.1) for the definition of $\hat{u}_h$).

10.x Exercises

10.x.1 Prove that the polynomial L_k defined in (10.2.3) has distinct roots $0 = \xi_0 < \xi_1 < \ldots < \xi_{k-1} = 1$. (Hint: let $\xi_1 < \ldots < \xi_r$ be the roots in $(0,1)$ of odd multiplicity. Show that $L_k(x) \prod_{i=1}^r (x - \xi_i)$ does not change sign in $[0,1]$. Show that $r < k-2$ yields a contradiction using (10.2.8).)

10.x.2 Prove Corollary 10.2.9. (Hint: add and subtract $Q^{2k-2}f$ and apply the techniques of Sect. 4.4.)

10.x.3 Show that Lagrange elements on curved triangles are well defined for h sufficiently small. (Hint: use a dilation to map to a family of reference elements of size one with one curved edge. Observe that this edge is only $\mathcal{O}(h)$ from a fixed straight edge and use a perturbation argument.)

10.x.4 Show that Theorem 10.2.36 can be improved to say that

$$\|u - u_h\|_{H^1(\Omega)} \le C_\rho\, h^{k-1} \|u\|_{H^k(\Omega)}$$

for h sufficiently small. (Hint: approximate $\frac{\partial u}{\partial \nu}$ by a polynomial $P(s)$ and estimate separately the two terms

$$\int_{\partial\Omega} \left(\frac{\partial u}{\partial \nu} - P \right) w\, ds \quad \& \quad \int_{\partial\Omega} P w\, ds.$$

Use the fact that $w = \mathcal{O}(h^2)$ on $\partial\Omega$ if $\|w\|_{W^1_\infty(\Omega)} = 1$.)

10.x.5 Prove Lemma 10.1.7. (Hint: choose v to be the orthogonal projection of u onto V_h and follow (10.1.3).)

10.x.6 Can Theorem 10.4.8 be improved with respect to the norm on f in the error estimate? For example, could it say that

$$\|u - \hat{u}_h\|_{H^1(\Omega)} \le C_\rho\, h^{k-1} \left(\|u\|_{H^k(\Omega)} + \|f\|_{H^{k-2}(\Omega)} \right)$$

for $k \le 3 + n/2$? (Hint: consider a bound for $\|f - f \circ \Phi^h\|_{H^{-1}(\Omega)}$.)

10.x.7 Prove Lemma 10.3.9. (Hint: copy the proof of Lemma 10.3.7.)

10.x.8 Prove (10.3.11). (Hint: show that the constants in (10.3.8) and (10.3.10) can be replaced by Ch for non-degenerate triangles of size h.)

10.x.9 Show that the Morley finite element depicted in Fig. 10.5 is a finite element. Here K = triangle, $\mathcal{P}$ = {quadratic polynomials} and $\mathcal{N}$ = {evaluations at the vertices and evaluations of the normal derivatives at the midpoints}. (Hint: a quadratic that vanishes at triangle vertices attains an extremum at edge midpoints. Use this to show $\mathcal{N}v = \{0\}$ implies $\nabla v \equiv 0$ for $v \in \mathcal{P}$.)

10.x.10 Consider the biharmonic problem described in Sect. 5.9 and define a nonconforming bilinear form $a_h(u,v)$ by

$$\sum_{T \in \mathcal{T}^h} \int_T \Delta u \, \Delta v - (1-\nu)\left(2u_{xx}v_{yy} + 2u_{yy}v_{xx} - 4u_{xy}v_{xy}\right) dx dy$$

where $\mathcal{T}^h$ is a triangulation of Ω and $0 < \nu < 1$. Let V_h be the Morley finite element space associated with $\mathcal{T}^h$. Show that $a_h(\cdot,\cdot)$ is non-degenerate on $V \cup V_h$. (Hint: use (5.9.2) and see the discussion following (10.3.4).)

10.x.11 Let G be the union of two non-degenerate triangles having a common edge e, and diam $G = 1$. Let $V = \{z : z|_{T_i} \text{ is quadratic}, z$ is continuous at the end points of e, and ∇z is continuous at the midpoint of $e\}$. Prove that there exists $C < \infty$ such that

$$\inf_{P \in \mathcal{P}_1} \|z - P\|_{L^2(G)} \le C \sum_{i=1}^{2} |z|_{H^2(T_i)}$$

for all $z \in V$. (Hint: show that $\|z\|_{L^2(T)} \le C|z|_{H^2(T)}$ for all $z \in \mathcal{P}_2$ that vanish at the vertices of an edge of T and that have a vanishing normal derivative at the midpoint of that edge.)

10.x.12 Let $D^k f$ $(1 \le k \le 3)$ denote derivatives of f of order k which come from the integration by parts formula

$$\int_T \left(\Delta z \Delta \zeta - (1-\nu)(2z_{xx}\zeta_{yy} + 2z_{yy}\zeta_{xx} - 4z_{xy}\zeta_{xy})\right) dx dy$$
$$= \int_{\partial T} (D^1 z D^2 \zeta + z D^3 \zeta) \, ds + \int_T z \Delta^2 \zeta \, dx dy.$$

Let G be the union of two non-degenerate triangles having a common edge e, and diam $G = 1$. Let $V = \{z : z|_{T_i} \text{ is quadratic}, z$ is continuous at the end points of e, and ∇z is continuous at the midpoint of $e\}$. Prove that there exists $C < \infty$ such that

$$\left| \int_e D^1(z|_{T_1} - z|_{T_2}) D^2\zeta \, ds + \int_e (z|_{T_1} - z|_{T_2}) D^3\zeta \, ds \right|$$

$$\leq C|z|_G \left(|\zeta|_{H^3(G)} + \|\Delta^2\zeta\|_{L^2(G)} \right),$$

where $z \in V$, $\zeta \in H^3(G)$, $\Delta^2\zeta \in L^2(G)$ and

$$|z|_G^2 = \sum_{i=1}^2 \int_{T_i} \left((\Delta z)^2 - (1-\nu)(4z_{xx}z_{yy} - 4z_{xy}^2) \right) \, dx dy \, .$$

(Hint: use the integration by parts formula to prove that the boundary term is bounded by $|z|_G|\zeta|_G + \|z\|_{L^2(G)}\|\Delta^2\zeta\|_{L^2(G)}$ and then apply the Bramble-Hilbert Lemma 4.3.8 and exercise 10.x.11, utilizing the fact that the boundary term does not change if we add a quadratic to ζ and any smooth function to z.)

10.x.13 Let $a_h(\cdot, \cdot)$ and V_h be as in exercise 10.x.10, and let $u_h \in V_h$ be such that

$$a_h(u_h, v) = \int_\Omega fv \, dx \quad \forall v \in V_h.$$

Assuming that the data f and solution u of (5.9.4) satisfy $f \in L^2(\Omega)$ and $u \in H^3(\Omega)$, show that

$$\|u - u_h\|_h \leq C h \left(|u|_{H^3(\Omega)} + h\|f\|_{L^2(\Omega)} \right).$$

where the associated energy norm is defined by $\|u\|_h := \sqrt{a_h(u, u)}$. (Hint: apply exercise 10.x.12.)

10.x.14 Consider the nonconforming method for the biharmonic problem described in exercise 10.x.13 based on the Morley element. Assuming only that the solution u of (5.9.4) is in $H^3(\Omega)$, can you show that

$$\|u - u_h\|_h \leq C h |u|_{H^3(\Omega)}$$

without assuming that $f \in L^2(\Omega)$?

10.x.15 The use of numerical quadrature to approximate the right-hand-side, as in exercise 0.x.11, is a very simple form of variational crime. Suppose the variational problem is as in Sect. 5.4, namely,

$$a(u_h, v) = (f, v) \quad \forall v \in V_h$$

where V_h consists of piecewise polynomials of degree $\leq r$, and suppose that $\tilde{u}_h \in V_h$ is defined by

$$a(\tilde{u}_h, v) = Q(fv) \quad \forall v \in V_h$$

where the quadrature approximation Q satisfies

$$\left| \int_\Omega v(x)w(x)\,dx - Q(vw) \right| \le Ch^k \sum_{T \in \mathcal{T}^h} \|v\|_{H^k(T)} \|w\|_{H^k(T)}$$

for some k. Prove that

$$\|u_h - \tilde{u}_h\|_{H^1(\Omega)} \le Ch^{k-r+1} \|f\|_{H^k(\Omega)}.$$

(Hint: see exercises 0.x.13 and 0.x.14, and use inverse estimates.)

10.x.16 In addition to the assumption on the quadrature rule in exercise 10.x.15, assume that $Q(w) = \sum_{T \in \mathcal{T}^h} Q_T(w)$ where

$$|Q_T(w)| \le C \operatorname{meas}(T) \|w\|_{L^\infty(T)} \quad \forall T \in \mathcal{T}^h.$$

Improve the estimate in exercise 10.x.15 to read

$$\|u_h - \tilde{u}_h\|_{H^1(\Omega)} \le Ch^{k-r+1} \|f\|_{H^{k-r+1}(\Omega)}$$

under the assumption that $k - r + 1 > n/2$. (Hint: subtract an interpolant from f.)

10.x.17 The use of numerical quadrature to approximate the variational form in the case of variable coefficients (see Sect. 5.7) leads to a variational crime of the sort covered by Lemma 10.1.9. Let $a(\cdot, \cdot)$ be as in (5.6.3) and define

$$a_h(u, v) := Q\left(\sum_{i,j=1}^n a_{ij}(x) \frac{\partial u}{\partial x_i}(x) \frac{\partial v}{\partial x_j}(x) \right)$$

where Q satisfies the condition in exercise 10.x.15. Assuming V_h consists of piecewise polynomials of degree $\le r$, that $a(\cdot, \cdot)$ is coercive on V, and that $k > 2r - 2$, show that $a_h(\cdot, \cdot)$ is coercive on V_h. (Hint: use inverse estimates to show that

$$|a(v, v) - a_h(v, v)| \le Ch^{k-2r+2} \|v\|^2_{H^1(\Omega)} \quad \forall v \in V_h.$$

Note that we do not need to assume that the weights in the quadrature rule are positive.)

10.x.18 Under the assumptions of exercise 10.x.17, let $\tilde{u}_h \in V_h$ be defined by

$$a_h(\tilde{u}_h, v) = (f, v) \quad \forall v \in V_h.$$

Prove that

$$\|u - \tilde{u}_h\|_{H^1(\Omega)} \le Ch^r \|u\|_{H^{k+1}(\Omega)} \left(1 + \max_{ij} \|a_{ij}\|_{W^k_\infty(\Omega)} \right).$$

(Hint: see exercises 0.x.13 and 0.x.14, apply Lemma 10.1.9 and use inverse estimates.)

10.x.19 Under the assumptions of exercises 10.x.16 and 10.x.17, improve the norms on u and a_{ij} in exercise 10.x.18. (Hint: subtract interpolants from u and a_{ij}.)

10.x.20 Consider the variational approximation of Poisson's equation as described in Sect. 5.4 using piecewise linear functions on a simplicial mesh $\mathcal{T}^h$ but with a quadrature approximation for the right-hand-side as in exercise 10.x.15. In particular, let $Q(w) = \sum_{T \in \mathcal{T}^h} Q_T(w)$ with

$$Q_T(w) := \text{meas}(T) \sum_{i=1}^{n} w(z_i^T) \quad \forall T \in \mathcal{T}^h$$

where $\{z_i^T : i = 1, \ldots, n\}$ are the vertices of T. (This is the n-dimensional version of the trapezoidal rule.) Show that the "difference stencil" corresponding to a regular mesh on $\Omega = [0, 1] \times [0, 1]$ consisting of $45°$ right triangles (cf. Sect. 0.5) is the standard 5-point difference stencil in the interior of Ω. Describe the stencil at the boundary for both Neumann and Dirichlet conditions. Compute the difference stencil for a regular mesh in three dimensions.

10.x.21 Prove that the n-dimensional trapezoidal rule defined in exercise 10.x.20 satisfies the condition in exercise 10.x.16 as well as

$$\left| \int_T v(x)w(x)\,dx - Q_T(vw) \right| \le Ch^2 \|v\|_{H^2(T)} \|w\|_{H^2(T)},$$

provided $n \le 3$ where $h := \text{diam}(T)$. (Hint: show that Q_T is exact for linear functions and apply Sobolev's inequality and approximation results from Chapter 4.)

10.x.22 A two-dimensional generalization of the difference method (0.5.3) (also see exercise 0.x.11) for Poisson's equation can be generated using piecewise linears together with the trapezoidal rule on the following mesh. Let $0 = x_0 < x_1 < \cdots < x_m = 1$ and $0 = y_0 < y_1 < \cdots < y_n = 1$ be partitions of $[0, 1]$, and define a triangular mesh based on the vertices

$$\{(x_i, y_j) : 0 \le i \le m,\ 0 \le j \le n\}$$

with, say, all edges either parallel to one of the axes or running diagonally from (x_i, y_j) to (x_{i+1}, y_{j+1}). Show that the difference method is essentially second-order accurate in the maximum norm, that is,

$$\max_{i,j} |u(x_i, y_j) - U_{ij}| \le Ch^2 |\log h|$$

under suitable hypotheses on u and f. (Hint: use exercises 10.x.18, 10.x.20 and 10.x.21 and the fact that $\|v\|_{L^\infty(\Omega)} \le C |\log h| \|v\|_{H^1(\Omega)}$ for piecewise linear functions, v. Also see Sect. 8.5 and (Scott 1976).)

10.x.23 Consider the quadrature rule in exercise 6.x.11. For which piecewise polynomial degrees will this give optimal order approximations for problems with variable coefficients? State and prove an appropriate convergence theorem. (Hint: use exercise 10.x.18.)

10.x.24 State and prove a convergence theorem combining the effect of quadrature for variable coefficients and the right-hand side, i.e., where $\tilde{u}_h \in V_h$ is defined by

$$a_h(\tilde{u}_h, v) = Q(fv) \quad \forall v \in V_h.$$

10.x.25 Let V_h be the nonconforming $\mathcal{P}_1$ finite element space associated with the triangulation $\mathcal{T}^h$ defined in (10.3.2) and $\tilde{V}_h \subset H_0^1(\Omega)$ be the $\mathcal{P}_2$ Lagrange finite element space associated with $\mathcal{T}^h$. Let $E_h : V_h \longrightarrow \tilde{V}_h$ be defined by
  (i) $(E_h v)(m) = v(m)$ at a midpoint m,
  (ii) $(E_h v)(p) = $ average value of v at the midpoints adjacent to the vertex p,
and $F_h : \tilde{V}_h \longrightarrow V_h$ be defined by

$$(F_h \tilde{v})(m) = \tilde{v}(m) \quad \text{at a midpoint } m.$$

Show that
  (a) $F_h E_h v = v$ for all $v \in V_h$.
  (b) $\|v - E_h v\|_{L^2(\Omega)} \le Ch\|v\|_h$ for all $v \in V_h$,
  (c) $\|\tilde{v} - F_h \tilde{v}\|_{L^2(\Omega)} \le Ch|v|_{H^1(\Omega)}$ for all $\tilde{v} \in \tilde{V}_h$.

10.x.26 Use the results from exercise 10.x.25 to prove the following Poincaré inequality for the nonconforming $\mathcal{P}_1$ finite element:

$$\|v\|_{L^2(\Omega)} \le C\|v\|_h \quad \forall v \in V_h.$$

(Hint: Use the Poincaré inequality (cf. Proposition 5.3.5) and inverse estimates (cf. Theorem 4.5.11 and Remark 4.5.20).)

10.x.27 The $\mathcal{P}_2$ Lagrange finite element is a *conforming relative* of the non-conforming $\mathcal{P}_1$ finite element in the sense that the shape functions and nodal variables of the latter are also shape functions and nodal variables of the former. Find a conforming relative for the rotated $\mathcal{Q}_1$ element (cf. exercise 3.x.15) and construct E_h and F_h with similar properties. Derive a Poincaré inequality for the rotated $\mathcal{Q}_1$ element.

10.x.28 Find a conforming relative (cf. exercise 10.x.27) for the Morley finite element.

Chapter 11

Applications to Planar Elasticity

In most physical applications, quantities of interest are governed by a system of partial differential equations, not just a single equation. So far, we have only considered single equations (for a scalar quantity), although much of the theory relates directly to systems. We consider one such system coming from solid mechanics in this chapter.

We apply the theory developed in Chapters 1 through 4 and Chapter 10 to boundary value problems in linear planar elasticity. Much of the work is in establishing the coercivity of the variational formulation. Once this is done, applications of the basic theory are immediate for the general case.

A new phenomenon arises because we have a *system* of partial differential equations. We discuss the phenomenon of "locking" which arises when the elastic material becomes nearly incompressible, and show that it can be overcome, if appropriate finite elements are used.

11.1 The Boundary Value Problems

We begin with some notation. We adopt the convention that an undertilde denotes vector-valued operators, functions, and their associated spaces. Double undertildes are used for matrix-valued functions and operators.

Let p, $\underset{\sim}{v} = (v_1, v_2)^t$ and $\underset{\approx}{\tau} = (\tau_{ij})_{1 \leq i,j \leq 2}$ be functions of two variables. We define

$$\operatorname*{grad}_{\sim} p = \begin{pmatrix} \partial p / \partial x_1 \\ \partial p / \partial x_2 \end{pmatrix}, \quad \operatorname*{curl}_{\sim} p = \begin{pmatrix} \partial p / \partial x_2 \\ -\partial p / \partial x_1 \end{pmatrix},$$

$$\operatorname{div} \underset{\sim}{v} = \partial v_1 / \partial x_1 + \partial v_2 / \partial x_2, \quad \operatorname{rot} \underset{\sim}{v} = -\partial v_1 / \partial x_2 + \partial v_2 / \partial x_1,$$

$$\operatorname*{grad}_{\approx} \underset{\sim}{v} = \begin{pmatrix} \partial v_1 / \partial x_1 & \partial v_1 / \partial x_2 \\ \partial v_2 / \partial x_1 & \partial v_2 / \partial x_2 \end{pmatrix}, \quad \operatorname*{curl}_{\approx} \underset{\sim}{v} = \begin{pmatrix} \partial v_1 / \partial x_2 & -\partial v_1 / \partial x_1 \\ \partial v_2 / \partial x_2 & -\partial v_2 / \partial x_1 \end{pmatrix},$$

$$\underset{\approx}{\epsilon}(\underset{\sim}{v}) = \frac{1}{2} \left(\operatorname*{grad}_{\approx} \underset{\sim}{v} + \left(\operatorname*{grad}_{\approx} \underset{\sim}{v} \right)^t \right),$$

and

$$\text{div}\underset{\sim}{\tau} = \begin{pmatrix} \partial\tau_{11}/\partial x_1 + \partial\tau_{12}/\partial x_2 \\ \partial\tau_{21}/\partial x_1 + \partial\tau_{22}/\partial x_2 \end{pmatrix}.$$

We also define

$$\underset{\approx}{\delta} = \begin{pmatrix} 1 & 0 \\ 0 & 1 \end{pmatrix}, \quad \underset{\approx}{\chi} = \begin{pmatrix} 0 & -1 \\ 1 & 0 \end{pmatrix},$$

and the following inner product between matrices

$$\underset{\approx}{\sigma} : \underset{\approx}{\tau} = \sum_{i=1}^{2}\sum_{j=1}^{2}\sigma_{ij}\tau_{ij}.$$

The trace of a matrix is defined by

$$\text{tr}(\underset{\approx}{\tau}) = \underset{\approx}{\tau} : \underset{\approx}{\delta} = \tau_{11} + \tau_{22},$$

and we have the following relation (cf. exercise 11.x.1)

(11.1.1) $$\underset{\approx}{\epsilon}(\underset{\sim}{v}) = \text{grad}\,\underset{\sim}{v} - \frac{1}{2}(\text{rot}\,\underset{\sim}{v})\,\underset{\approx}{\chi}.$$

We consider an isotropic elastic material in the configuration space $\Omega \subseteq \mathbb{R}^2$. Let $\underset{\sim}{u}(x)$ be the displacement and $\underset{\sim}{f}(x)$ be the body force. Then in the static theory of linear elasticity, the equation satisfied by $\underset{\sim}{u}$ is

(11.1.2) $$-\text{div}\,\underset{\approx}{\sigma}(\underset{\sim}{u}) = \underset{\sim}{f} \quad \text{in } \Omega,$$

where the stress tensor $\underset{\approx}{\sigma}(\underset{\sim}{u})$ is defined by

(11.1.3) $$\underset{\approx}{\sigma}(\underset{\sim}{u}) = 2\mu\underset{\approx}{\epsilon}(\underset{\sim}{u}) + \lambda\,\text{tr}\left(\underset{\approx}{\epsilon}(\underset{\sim}{u})\right)\underset{\approx}{\delta}.$$

The positive constants μ and λ are called the Lamé constants. We assume that $(\mu, \lambda) \in [\mu_1, \mu_2] \times (0, \infty)$ where $0 < \mu_1 < \mu_2$.

Let Γ_1 and Γ_2 be two open subsets of $\partial\Omega$ such that $\partial\Omega = \overline{\Gamma}_1 \cup \overline{\Gamma}_2$ and $\Gamma_1 \cap \Gamma_2 = \emptyset$. We impose the displacement boundary condition on Γ_1

(11.1.4) $$\underset{\sim}{u}|_{\Gamma_1} = \underset{\sim}{g},$$

and the traction boundary condition on Γ_2 (where $\underset{\sim}{\nu}$ is the unit outer normal)

(11.1.5) $$\left(\underset{\approx}{\sigma}(\underset{\sim}{u})\underset{\sim}{\nu}\right)\Big|_{\Gamma_2} = \underset{\sim}{t}.$$

If $\Gamma_1 = \emptyset$ (resp. $\Gamma_2 = \emptyset$), the boundary value problem is called a pure traction (resp. displacement) problem.

Note that $\underset{\approx}{\epsilon}$ has a nontrivial kernel. Let

(11.1.6) $$\underset{\sim}{\text{RM}} := \left\{\underset{\sim}{v} : \underset{\sim}{v} = \underset{\sim}{c} + b\,(x_2, -x_1)^t, \; \underset{\sim}{c} \in \mathbb{R}^2, \; b \in \mathbb{R}\right\}$$

be the space of infinitesimal rigid motions. Then it is easily verified (cf. exercise 11.x.2) that

$$(11.1.7) \qquad \underset{\approx}{\epsilon}(\underset{\sim}{v}) = \underset{\approx}{0} \quad \forall \underset{\sim}{v} \in \underset{\sim}{\mathrm{RM}}.$$

Therefore, $\underset{\sim}{\mathrm{RM}}$ is in the kernel of the homogeneous pure traction problem, and $\underset{\sim}{f}$, $\underset{\sim}{t}$ must satisfy certain constraints for the pure traction problem to be solvable.

Lebesgue and Sobolev spaces and associated norms (and inner products, where appropriate) are defined analogously. For example, the space $\underset{\approx}{L^2}(\Omega)$ has the inner product

$$(\underset{\approx}{\sigma}, \underset{\approx}{\tau})_{L^2(\Omega)} := \int_\Omega \underset{\approx}{\sigma} : \underset{\approx}{\tau} \, dx.$$

The space $\underset{\sim}{H^1}(\Omega)$ has the inner product

$$(\underset{\sim}{u}, \underset{\sim}{v})_{H^1(\Omega)} := (\underset{\approx}{\mathrm{grad}}\,\underset{\sim}{u}, \underset{\approx}{\mathrm{grad}}\,\underset{\sim}{v})_{L^2(\Omega)} + (\underset{\sim}{u}, \underset{\sim}{v})_{L^2(\Omega)}.$$

We leave the definition of other inner products (including the one most recently used) and norms as an exercise.

11.2 Weak Formulation and Korn's Inequality

Assume that $\underset{\sim}{u} \in \underset{\sim}{H^2}(\Omega)$ satisfies (11.1.2). Then given any $\underset{\sim}{v} \in \underset{\sim}{H^1}(\Omega)$ and $\underset{\sim}{v}|_{\Gamma_1} = 0$, it follows from integration by parts (assuming it can be done) that

$$\int_\Omega \underset{\sim}{f} \cdot \underset{\sim}{v} \, dx$$

$$= -\int_\Omega \mathrm{div}\underset{\approx}{\sigma}(\underset{\sim}{u}) \cdot \underset{\sim}{v} \, dx$$

$$= \int_\Omega \left\{ 2\mu\underset{\approx}{\epsilon}(\underset{\sim}{u}) + \lambda \, \mathrm{tr}\left(\underset{\approx}{\epsilon}(\underset{\sim}{u})\right) \underset{\approx}{\delta} \right\} : \underset{\approx}{\mathrm{grad}}\,\underset{\sim}{v} \, dx - \int_{\Gamma_2} \underset{\approx}{\sigma}(\underset{\sim}{u})\underset{\sim}{\nu} \cdot \underset{\sim}{v} \, ds.$$

Therefore, we have the following weak formulation of (11.1.2) through (11.1.5).

Find $\underset{\sim}{u} \in \underset{\sim}{H^1}(\Omega)$ such that $\underset{\sim}{u}|_{\Gamma_1} = \underset{\sim}{g}$ and

$$(11.2.1) \qquad a(\underset{\sim}{u}, \underset{\sim}{v}) = \int_\Omega \underset{\sim}{f} \cdot \underset{\sim}{v} \, dx + \int_{\Gamma_2} \underset{\sim}{t} \cdot \underset{\sim}{v} \, ds$$

for all $\underset{\sim}{v} \in \underset{\sim}{V}$, where

$$a(\underset{\sim}{u}, \underset{\sim}{v}) := \int_{\Omega} \left(2\mu \underset{\approx}{\epsilon}(\underset{\sim}{u}) : \underset{\approx}{\epsilon}(\underset{\sim}{v}) + \lambda \operatorname{div} \underset{\sim}{u} \operatorname{div} \underset{\sim}{v} \right) dx \,,$$
(11.2.2)
$$V := \{\underset{\sim}{v} \in \underset{\sim}{H}^1(\Omega) : \underset{\sim}{v}|_{\Gamma_1} = \underset{\sim}{0}\} \,.$$

The first question we must ask is: does (11.2.1) have a unique solution? In other words, we need to establish the coercivity of the bounded bilinear form $a(\cdot, \cdot)$ on $\underset{\sim}{V}$. We begin with a lemma regarding an underdetermined boundary value problem (note: there is no uniqueness) that will be used frequently in this chapter and in Chapter 12. Its proof uses Sobolev-space techniques beyond what we have developed in Chapter 1, but we include a sketch of its proof for the sake of completeness. For the rest of this section we assume that either Ω is a polygon or $\partial\Omega$ is smooth.

(11.2.3) Lemma. *There exists a positive constant C such that for all $p \in L^2(\Omega)$ there is a $\underset{\sim}{v} \in \underset{\sim}{H}^1(\Omega)$ satisfying*

$$\operatorname{div} \underset{\sim}{v} = p$$

and

$$\|\underset{\sim}{v}\|_{\underset{\sim}{H}^1(\Omega)} \le C \, \|p\|_{L^2(\Omega)} \,.$$

If, furthermore, p satisfies $\int_{\Omega} p \, dx = 0$, then we may assume that $\underset{\sim}{v} \in \overset{\circ}{\underset{\sim}{H}}{}^1(\Omega)$.

Proof. This lemma holds for Ω with smooth boundary and for polygonal Ω. Here we give the proof for the smooth boundary case and refer the reader to (Girault & Raviart 1986, Arnold, Scott & Vogelius 1988) for the polygonal case. There exists a unique $w \in H^2(\Omega)$ satisfying

$$-\Delta w = p \quad \text{in } \Omega$$
$$w = 0 \quad \text{on } \partial\Omega \,.$$

By elliptic regularity (see Sect. 5.5) we have

$$\|w\|_{H^2(\Omega)} \le C_{\Omega} \, \|p\|_{L^2(\Omega)} \,.$$

Then $\underset{\sim}{v} = -\operatorname{grad} w$ satisfies the conditions of the Lemma.

Now suppose $\int_{\Omega} p \, dx = 0$. Then there exists a unique $w \in H^2(\Omega)$ satisfying

$$-\Delta w = p \quad \text{in } \Omega$$
(11.2.4)
$$\frac{\partial w}{\partial \nu} = 0 \quad \text{on } \partial\Omega,$$

and

$$\int_{\Omega} w \, dx = 0$$

(see Sect. 5.2). By elliptic regularity we have

$$\|w\|_{H^2(\Omega)} \le C_{\Omega} \, \|p\|_{L^2(\Omega)} \,.$$

Let $\underset{\sim}{v}_1 = -\text{grad}\, w$. Then $\underset{\sim}{v}_1 \in \underset{\sim}{H}^1(\Omega)$,

(11.2.5) $\text{div}\, \underset{\sim}{v}_1 = p$

and

(11.2.6) $\|\underset{\sim}{v}_1\|_{\underset{\sim}{H}^1(\Omega)} \leq C_\Omega \|p\|_{L^2(\Omega)}$.

Recall that $\underset{\sim}{\nu}$ is the unit outer normal vector to $\partial\Omega$. Observe that

(11.2.7) $\underset{\sim}{v}_1|_{\partial\Omega} \cdot \underset{\sim}{\nu} = -\text{grad}\, w|_{\partial\Omega} \cdot \underset{\sim}{\nu} = 0$

by (11.2.4). Let $\underset{\sim}{\tau}$ be the positively oriented unit tangent vector. Then the trace theorem (cf. Adams 1975) implies that there exists $\psi \in H^2(\Omega)$ such that

$$\psi|_{\partial\Omega} = 0$$

$$\frac{\partial \psi}{\partial \nu}\Big|_{\partial\Omega} = \underset{\sim}{v}_1|_{\partial\Omega} \cdot \underset{\sim}{\tau}$$

and

(11.2.8) $\|\psi\|_{H^2(\Omega)} \leq C_\tau \|\underset{\sim}{v}_1\|_{\underset{\sim}{H}^1(\Omega)}$

for some positive constant C_τ. Let $\underset{\sim}{v}_2 = \text{curl}\, \psi$. Then we have

(11.2.9) $\underset{\sim}{v}_2|_{\partial\Omega} \cdot \underset{\sim}{\nu} = \text{grad}\, \psi|_{\partial\Omega} \cdot \underset{\sim}{\tau} = 0,$

and

(11.2.10) $\underset{\sim}{v}_2|_{\partial\Omega} \cdot \underset{\sim}{\tau} = -\text{grad}\, \psi|_{\partial\Omega} \cdot \underset{\sim}{\nu} = -\underset{\sim}{v}_1|_{\partial\Omega} \cdot \underset{\sim}{\tau}$.

Combining (11.2.7), (11.2.9) and (11.2.10) we conclude that

(11.2.11) $\underset{\sim}{v}_2|_{\partial\Omega} = -\underset{\sim}{v}_1|_{\partial\Omega}$.

Let $\underset{\sim}{v} = \underset{\sim}{v}_1 + \underset{\sim}{v}_2$. Then $\text{div}\, \underset{\sim}{v} = \text{div}\, \underset{\sim}{v}_1 = p$ by (11.2.5), $\underset{\sim}{v}|_{\partial\Omega} = \underset{\sim}{0}$ by (11.2.11), and

$$
\begin{aligned}
\|\underset{\sim}{v}\|_{\underset{\sim}{H}^1(\Omega)} &\leq \|\underset{\sim}{v}_1\|_{\underset{\sim}{H}^1(\Omega)} + \|\underset{\sim}{v}_2\|_{\underset{\sim}{H}^1(\Omega)} && \text{(triangle inequality)} \\
&\leq \|\underset{\sim}{v}_1\|_{\underset{\sim}{H}^1(\Omega)} + C\|\psi\|_{H^2(\Omega)} && (\underset{\sim}{v}_2 = \text{curl}\, \psi) \\
&\leq C \|\underset{\sim}{v}_1\|_{\underset{\sim}{H}^1(\Omega)} && \text{(by 11.2.8)} \\
&\leq C \|p\|_{L^2(\Omega)} . && \text{(by 11.2.6)}
\end{aligned}
$$

$\square$

Let $\widehat{\underset{\sim}{H}}^k(\Omega)$ be defined by

$$\widehat{\underset{\sim}{H}}^k(\Omega) := \{\underset{\sim}{v} \in \underset{\sim}{H}^k(\Omega) : \int_\Omega \underset{\sim}{v}\, dx = \underset{\sim}{0}, \int_\Omega \text{rot}\, \underset{\sim}{v}\, dx = 0\}.$$

These are closed subspaces of $\underset{\sim}{H}^k(\Omega)$ for $k \geq 1$.

(11.2.12) Theorem. (Second Korn Inequality) *There exists a positive constant C such that*

$$(11.2.13) \qquad \|\underset{\approx}{\epsilon}(\underset{\sim}{v})\|_{L^2(\Omega)} \geq C \|\underset{\sim}{v}\|_{H^1(\Omega)} \quad \forall \underset{\sim}{v} \in \overset{\circ}{\widehat{H}}{}^1(\Omega).$$

Proof. Let $\underset{\sim}{v} \in \widehat{H}^1(\Omega)$. Since $\int_\Omega \operatorname{rot} \underset{\sim}{v}\, dx = 0$, Lemma 11.2.3 gives the existence of $\underset{\sim}{w} \in \overset{\circ}{H}{}^1(\Omega)$ such that

$$(11.2.14) \qquad \begin{aligned} \operatorname{div} \underset{\sim}{w} &= \operatorname{rot} \underset{\sim}{v} \\ \|\underset{\sim}{w}\|_{H^1(\Omega)} &\leq C_1 \|\underset{\sim}{v}\|_{H^1(\Omega)} \end{aligned}$$

for some positive constant C_1. Then

$$\int_\Omega \underset{\approx}{\epsilon}(\underset{\sim}{v}) : \left(\operatorname*{grad}_{\approx} \underset{\sim}{v} - \operatorname*{curl}_{\approx} \underset{\sim}{w}\right) dx$$

$$= \int_\Omega \left(\operatorname*{grad}_{\approx} \underset{\sim}{v} - \frac{1}{2}(\operatorname{rot} \underset{\sim}{v})\underset{\approx}{\chi}\right) : \left(\operatorname*{grad}_{\approx} \underset{\sim}{v} - \operatorname*{curl}_{\approx} \underset{\sim}{w}\right) dx \qquad \text{(by 11.1.1)}$$

$$= \|\operatorname*{grad}_{\approx} \underset{\sim}{v}\|_{L^2(\Omega)}^2 - \int_\Omega \operatorname*{grad}_{\approx} \underset{\sim}{v} : \operatorname*{curl}_{\approx} \underset{\sim}{w}\, dx$$

$$\qquad - \frac{1}{2} \int_\Omega (\operatorname{rot} \underset{\sim}{v}) \left(\underset{\approx}{\chi} : \operatorname*{grad}_{\approx} \underset{\sim}{v} - \underset{\approx}{\chi} : \operatorname*{curl}_{\approx} \underset{\sim}{w}\right) dx$$

$$= \|\operatorname*{grad}_{\approx} \underset{\sim}{v}\|_{L^2(\Omega)}^2 - \frac{1}{2} \int_\Omega (\operatorname{rot} \underset{\sim}{v}) \left(\operatorname{rot} \underset{\sim}{v} - \operatorname{div} \underset{\sim}{w}\right) dx$$

$$\qquad\qquad\qquad\qquad\qquad\qquad\qquad \text{(by 11.x.3, 11.x.4, 11.x.5)}$$

$$= \|\operatorname*{grad}_{\approx} \underset{\sim}{v}\|_{L^2(\Omega)}^2 . \qquad\qquad\qquad\qquad\qquad \text{(by 11.2.14)}$$

Therefore, Schwarz' inequality (2.1.5) and (11.2.14) imply

$$(11.2.15) \qquad \begin{aligned} \|\operatorname*{grad}_{\approx} \underset{\sim}{v}\|_{L^2(\Omega)}^2 &\leq \|\underset{\approx}{\epsilon}(\underset{\sim}{v})\|_{L^2(\Omega)} \|\operatorname*{grad}_{\approx} \underset{\sim}{v} - \operatorname*{curl}_{\approx} \underset{\sim}{w}\|_{L^2(\Omega)} \\ &\leq C \|\underset{\approx}{\epsilon}(\underset{\sim}{v})\|_{L^2(\Omega)} \|\underset{\sim}{v}\|_{H^1(\Omega)} . \end{aligned}$$

The theorem now follows from (11.2.15) and Friedrichs' inequality (4.3.15) because $\int_\Omega \underset{\sim}{v}\, dx = 0$. $\qquad\qquad\qquad\qquad\qquad\qquad\qquad\qquad\square$

(11.2.16) Theorem. (Korn's Inequality) *There exists a positive constant α such that*

$$(11.2.17) \qquad \|\underset{\approx}{\epsilon}(\underset{\sim}{v})\|_{L^2(\Omega)} + \|\underset{\sim}{v}\|_{L^2(\Omega)} \geq \alpha \|\underset{\sim}{v}\|_{H^1(\Omega)} \quad \forall \underset{\sim}{v} \in H^1(\Omega).$$

Proof. Observe first that $\underset{\sim}{H}^1(\Omega) = \widehat{\underset{\sim}{H}}{}^1(\Omega) \oplus \underset{\sim}{\text{RM}}$ (cf. exercise 11.x.6), where "$\oplus$" is understood only in the algebraic sense (they are not orthogonal in $\underset{\sim}{H}^1(\Omega)$). Therefore, given any $\underset{\sim}{v} \in H^1(\Omega)$, there exists a unique pair $(\underset{\sim}{z}, \underset{\sim}{w}) \in \widehat{\underset{\sim}{H}}{}^1(\Omega) \times \underset{\sim}{\text{RM}}$ such that

$$\underset{\sim}{v} = \underset{\sim}{z} + \underset{\sim}{w}.$$

In particular, $\underset{\sim}{w}$ is of the form given in (11.1.6) with

(11.2.18)

$$b := \frac{-1}{2 \operatorname{meas}(\Omega)} \int_{\Omega} \operatorname{rot} \underset{\sim}{v}\, dx\,,$$

$$\underset{\sim}{c} := \frac{1}{\operatorname{meas}(\Omega)} \int_{\Omega} \left(\underset{\sim}{v}(x) - b\,(x_2, -x_1)^t \right)\, dx\,.$$

Therefore, $\|\underset{\sim}{w}\|_{H^1(\Omega)} \le C\|\underset{\sim}{v}\|_{H^1(\Omega)}$, where C depends only on $\operatorname{meas}(\Omega)$. By the triangle inequality, there exists a positive constant C_1 (cf. also exercise 11.x.7), such that

(11.2.19)

$$C_1 \left(\|\underset{\sim}{z}\|_{H^1(\Omega)} + \|\underset{\sim}{w}\|_{H^1(\Omega)} \right) \le \|\underset{\sim}{v}\|_{H^1(\Omega)}\,.$$

We establish the theorem by contradiction. If we assume that (11.2.17) does not hold for any positive constant C, then there exists a sequence $\{\underset{\sim}{v}_n\} \subseteq H^1(\Omega)$ such that

(11.2.20)

$$\|\underset{\sim}{v}_n\|_{H^1(\Omega)} = 1$$

and

(11.2.21)

$$\|\underset{\approx}{\epsilon}(\underset{\sim}{v}_n)\|_{L^2(\Omega)} + \|\underset{\sim}{v}_n\|_{L^2(\Omega)} < \frac{1}{n}\,.$$

For each n, let $\underset{\sim}{v}_n = \underset{\sim}{z}_n + \underset{\sim}{w}_n$, where $\underset{\sim}{z}_n \in \hat{H}^1(\Omega)$ and $\underset{\sim}{w}_n \in \operatorname{\underset{\sim}{RM}}$. Then

$$\|\underset{\approx}{\epsilon}(\underset{\sim}{z}_n)\|_{L^2(\Omega)} = \|\underset{\approx}{\epsilon}(\underset{\sim}{v}_n)\|_{L^2(\Omega)} \qquad \text{(by 11.1.7)}$$

$$< \frac{1}{n}\,. \qquad \text{(by 11.2.21)}$$

The second Korn inequality then implies that $\underset{\sim}{z}_n \to \underset{\sim}{0}$ in $H^1(\Omega)$.

It follows from (11.2.19) and (11.2.20) that $\{\underset{\sim}{w}_n\}$ is a bounded sequence in $H^1(\Omega)$. But since $\operatorname{\underset{\sim}{RM}}$ is three-dimensional, $\{\underset{\sim}{w}_n\}$ has a convergent subsequence $\{\underset{\sim}{w}_{n_j}\}$ in $H^1(\Omega)$. Then the subsequence $\{\underset{\sim}{v}_{n_j} = \underset{\sim}{z}_{n_j} + \underset{\sim}{w}_{n_j}\}$ converges in $H^1(\Omega)$ to some $\underset{\sim}{v} \in \operatorname{\underset{\sim}{RM}}$. We conclude from (11.2.20) and (11.2.21) that

$$\|\underset{\sim}{v}\|_{H^1(\Omega)} = 1 \quad \text{and} \quad \|\underset{\sim}{v}\|_{L^2(\Omega)} = 0,$$

which is a contradiction. □

(11.2.22) Corollary. *Let* $\underset{\sim}{V}$ *be defined by* (11.2.2) *where* $\operatorname{meas}(\Gamma_1) > 0$. *There exists a positive constant* C *such that*

(11.2.23)

$$\|\underset{\approx}{\epsilon}(\underset{\sim}{v})\|_{L^2(\Omega)} \ge C\,\|\underset{\sim}{v}\|_{H^1(\Omega)} \quad \forall\, \underset{\sim}{v} \in \underset{\sim}{V}.$$

Proof. The same proof by contradiction yields the existence of some $\underset{\sim}{v} \in$ $\underset{\sim}{RM} \cap \underset{\sim}{V}$ such that

(11.2.24) $\|\underset{\sim}{v}\|_{H^1(\Omega)} = 1.$

But the only $\underset{\sim}{v} \in \underset{\sim}{RM}$ satisfying $\underset{\sim}{v}|_{\Gamma_1} = \underset{\sim}{0}$ is $\underset{\sim}{v} \equiv \underset{\sim}{0}$ because meas $(\Gamma_1) > 0$ (cf. exercise 11.x.9). This contradicts (11.2.24).   □

In the case where $\Gamma_2 = \emptyset$ we immediately find the following.

(11.2.25) Corollary. (First Korn Inequality) *There exists a positive constant C such that*

(11.2.26) $\|\underset{\approx}{\epsilon}(\underset{\sim}{v})\|_{L^2(\Omega)} \geq C\|\underset{\sim}{v}\|_{H^1(\Omega)} \quad \forall \underset{\sim}{v} \in \overset{\circ}{\underset{\sim}{H}}{}^1(\Omega).$

(11.2.27) *Remark.* Korn's inequality actually holds for $\Omega \subseteq \mathbb{R}^n$ with Lipschitz boundary. For a proof of the general case we refer the reader to (Duvaut & Lions 1972, Nitsche 1981).

Observe now that the coercivity of $a(\cdot, \cdot)$ on $\underset{\sim}{V}$ (when meas($\Gamma_1) > 0$) follows immediately from Corollary 11.2.22, and the unique solvability of (11.2.1) follows.

(11.2.28) Theorem. *Assume that* $\underset{\sim}{f} \in \underset{\sim}{H}^{-1}(\Omega)$, $\underset{\sim}{g} = \underset{\sim}{w}|_{\Gamma_1}$ *where* $\underset{\sim}{w} \in \underset{\sim}{H}^1(\Omega)$, $\underset{\sim}{t} \in \underset{\sim}{L}^2(\Gamma_2)$, *and* meas($\Gamma_1) > 0$. *Then the variational problem* (11.2.1) *has a unique solution.*

Proof. Let $\underset{\sim}{u}^* = \underset{\sim}{u} - \underset{\sim}{w}$. Then (11.2.1) is equivalent to the problem of finding $\underset{\sim}{u}^* \in \underset{\sim}{V}$ such that for all $\underset{\sim}{v} \in \underset{\sim}{V}$,

(11.2.29)
$$a(\underset{\sim}{u}^*, \underset{\sim}{v}) = \int_\Omega \underset{\sim}{f} \cdot \underset{\sim}{v}\, dx + \int_{\Gamma_2} \underset{\sim}{t} \cdot \underset{\sim}{v}\, ds$$
$$- \int_\Omega \{2\mu\underset{\approx}{\epsilon}(\underset{\sim}{w}) : \underset{\approx}{\epsilon}(\underset{\sim}{v}) + \lambda \operatorname{div} \underset{\sim}{w} \operatorname{div} \underset{\sim}{v}\}\, dx$$
$$=: F(\underset{\sim}{v}),$$

where $F \in \underset{\sim}{V}'$ by the conditions on $\underset{\sim}{f}$, $\underset{\sim}{t}$ and $\underset{\sim}{w}$. Therefore, by Theorem 2.5.6, (11.2.29) has a unique solution, which implies that (11.2.1) also has a unique solution.   □

Note that $\underset{\sim}{V}$ becomes $\underset{\sim}{H}^1(\Omega)$ in the case of the pure traction problem. Since the homogeneous pure traction problem has a nontrivial kernel $\underset{\sim}{RM}$, $\underset{\sim}{f}$ and $\underset{\sim}{t}$ must satisfy some compatibility conditions before problem (11.2.1) can be solved.

(11.2.30) Theorem. *Assume* $\underset{\sim}{f} \in L^2(\Omega)$ *and* $\underset{\sim}{t} \in L^2(\Gamma)$. *Then the variational problem*

(11.2.31)

$$\text{find } \underset{\sim}{u} \in H^1(\Omega) \text{ such that}$$
$$a(\underset{\sim}{u}, \underset{\sim}{v}) = \int_\Omega \underset{\sim}{f} \cdot \underset{\sim}{v} \, dx + \int_\Gamma \underset{\sim}{t} \cdot \underset{\sim}{v} \, ds \quad \forall \underset{\sim}{v} \in H^1(\Omega)$$

is solvable if and only if the following compatibility condition holds:

(11.2.32)

$$\int_\Omega \underset{\sim}{f} \cdot \underset{\sim}{v} \, dx + \int_\Gamma \underset{\sim}{t} \cdot \underset{\sim}{v} \, ds = 0 \quad \forall \underset{\sim}{v} \in \underset{\sim}{RM}.$$

When (11.2.31) *is solvable, there exists a unique solution in* $\widehat{H}^1(\Omega)$.

Proof. (*Necessary:*) If (11.2.31) is solvable, then

$$\int_\Omega \underset{\sim}{f} \cdot \underset{\sim}{v} \, dx + \int_\Gamma \underset{\sim}{t} \cdot \underset{\sim}{v} \, ds = a(\underset{\sim}{u}, \underset{\sim}{v}) \quad \forall \underset{\sim}{v} \in \underset{\sim}{RM}$$
$$= 0. \qquad\qquad \text{(by 11.1.7)}$$

(*Sufficient:*) Assume that (11.2.32) holds. By the second Korn inequality (11.2.13) and Theorem 2.5.6, there exists a unique $\underset{\sim}{u}^* \in \widehat{H}^1(\Omega)$ such that

$$a(\underset{\sim}{u}^*, \underset{\sim}{v}) = \int_\Omega \underset{\sim}{f} \cdot \underset{\sim}{v} \, dx + \int_\Gamma \underset{\sim}{t} \cdot \underset{\sim}{v} \, ds \quad \forall \underset{\sim}{v} \in \widehat{H}^1(\Omega).$$

But (11.1.7) and the compatibility condition (11.2.32) imply that

$$a(\underset{\sim}{u}^*, \underset{\sim}{v}) = \int_\Omega \underset{\sim}{f} \cdot \underset{\sim}{v} \, dx + \int_\Gamma \underset{\sim}{t} \cdot \underset{\sim}{v} \, ds \quad \forall \underset{\sim}{v} \in \underset{\sim}{RM}.$$

Since $\underset{\sim}{H}^1(\Omega) = \widehat{H}^1(\Omega) \oplus \underset{\sim}{RM}$, $\underset{\sim}{u}^*$ is a solution of (11.2.31). $\qquad\square$

Under certain conditions on the boundary of Ω, the fact that $\underset{\sim}{f} \in L^2(\Omega)$, $\underset{\sim}{g} = \underset{\sim}{w}_1|_{\Gamma_1}$ where $\underset{\sim}{w}_1 \in H^2(\Omega)$, and $\underset{\sim}{t} = \underset{\sim}{w}_2|_{\Gamma_2}$ where $\underset{\sim}{w}_2 \in H^1(\Omega)$ imply that the solution $\underset{\sim}{u}$ of (11.2.1) belongs to $H^2(\Omega)$, and then the techniques of Proposition 5.1.9 show that $\underset{\sim}{u}$ actually satisfies the boundary value problem (11.1.2) through (11.1.5). For example, this is true when $\partial\Omega$ is smooth and $\overline{\Gamma}_1 \cap \overline{\Gamma}_2 = \emptyset$ (cf. (Valent 1988)), or if Ω is a convex polygon and either Γ_1 or Γ_2 is empty (cf. (Grisvard 1986 and 1989)). Moreover, in these cases (cf. (Vogelius 1983) and (Brenner & Sung 1992)), there exists a positive constant C independent of $(\mu, \lambda) \in [\mu_1, \mu_2] \times (0, \infty)$ such that

(11.2.33)

$$\|\underset{\sim}{u}\|_{H^2(\Omega)} + \lambda \|\text{div } \underset{\sim}{u}\|_{H^1(\Omega)}$$
$$\leq C \left(\|\underset{\sim}{f}\|_{L^2(\Omega)} + \|\underset{\sim}{w}_1\|_{H^2(\Omega)} + \|\underset{\sim}{w}_2\|_{H^1(\Omega)} \right).$$

11.3 Finite Element Approximation and Locking

For simplicity, we assume that Ω is a convex polygonal domain, and either Γ_1 or Γ_2 is empty. For the pure displacement problem $(\Gamma_2 = \emptyset)$ we only consider the case of homogeneous boundary conditions.

Let $\mathcal{T}^h$ be a non-degenerate family of triangulations of Ω. For the pure displacement problem $(\Gamma_2 = \emptyset)$, we use the finite element space

$$(11.3.1) \qquad \mathring{\underset{\sim}{V}}_h := \{\underset{\sim}{v} \in \mathring{\underset{\sim}{H}}^1(\Omega) : \underset{\sim}{v}|_T \text{ is linear } \forall T \in \mathcal{T}^h\},$$

and for the pure traction problem $(\Gamma_1 = \emptyset)$ we use (cf. exercise 11.x.13)

$$(11.3.2) \qquad \widehat{\underset{\sim}{V}}_h := \{\underset{\sim}{v} \in \widehat{\underset{\sim}{H}}^1(\Omega) : \underset{\sim}{v}|_T \text{ is linear } \forall T \in \mathcal{T}^h\}.$$

By the theory developed in Chapters 2 and 4 we obtain the following theorems.

(11.3.3) Theorem. *Let* $\underset{\sim}{u} \in \underset{\sim}{H}^2(\Omega) \cap \mathring{\underset{\sim}{H}}^1(\Omega)$ *satisfy the pure displacement problem and* $\underset{\sim}{u}_h \in \mathring{\underset{\sim}{V}}_h$ *satisfy*

$$a(\underset{\sim}{u}_h, \underset{\sim}{v}) = \int_\Omega \underset{\sim}{f} \cdot \underset{\sim}{v}\, dx \quad \forall \underset{\sim}{v} \in \mathring{\underset{\sim}{V}}_h.$$

Then there exists a positive constant $C_{(\mu,\lambda)}$ *such that*

$$(11.3.4) \qquad \|\underset{\sim}{u} - \underset{\sim}{u}_h\|_{H^1(\Omega)} \le C_{(\mu,\lambda)}\, h\, \|\underset{\sim}{u}\|_{H^2(\Omega)}.$$

(11.3.5) Theorem. *Let* $\underset{\sim}{u} \in \widehat{\underset{\sim}{H}}^2(\Omega)$ *satisfy the pure traction problem. Let* $\underset{\sim}{u}_h \in \widehat{\underset{\sim}{V}}_h$ *satisfy*

$$a(\underset{\sim}{u}_h, \underset{\sim}{v}) = \int_\Omega \underset{\sim}{f} \cdot \underset{\sim}{v}\, dx + \int_\Gamma \underset{\sim}{t} \cdot \underset{\sim}{v}\, ds \quad \forall \underset{\sim}{v} \in \widehat{\underset{\sim}{V}}_h.$$

Then there exists a positive constant $C_{(\mu,\lambda)}$ *such that*

$$\|\underset{\sim}{u} - \underset{\sim}{u}_h\|_{H^1(\Omega)} \le C_{(\mu,\lambda)}\, h\, \|\underset{\sim}{u}\|_{H^2(\Omega)}.$$

For a convergence theorem in the general case $\emptyset \ne \Gamma_1 \ne \partial\Omega$, see exercise 11.x.25.

For fixed μ and λ, Theorems 11.3.3 and 11.3.5 give satisfactory convergent finite element approximations to the elasticity problem. But the performance of these finite element methods deteriorates as λ approaches ∞. This is known as the phenomenon of *locking*, which we will explain in the rest of this section.

Let $\Omega = (0,1) \times (0,1)$. We consider the pure displacement boundary value problem for $\mu = 1$:

$$\text{div} \left\{ 2\epsilon(\underset{\sim}{u}^\lambda) + \lambda \text{tr}\, (\epsilon(\underset{\sim}{u}^\lambda))\underset{\approx}{\delta} \right\} = \underset{\sim}{f} \quad \text{in } \Omega$$

(11.3.6)

$$\underset{\sim}{u}^\lambda|_{\partial\Omega} = \underset{\sim}{0}.$$

Note that for given $\underset{\sim}{f}$, as $\lambda \to \infty$, (11.2.33) implies that $\|\text{div}\, \underset{\sim}{u}^\lambda\|_{H^1(\Omega)} \to 0$. In other words, we are dealing with an elastic material that becomes nearly incompressible. To emphasize the dependence on λ we will denote the stress tensor (11.1.3) by $\underset{\approx}{\sigma}_\lambda(\underset{\sim}{v})$ and the variational form (11.2.2) by $a_\lambda(\underset{\sim}{v}, \underset{\sim}{w})$, that is,

$$\underset{\approx}{\sigma}_\lambda(\underset{\sim}{v}) = 2\underset{\approx}{\epsilon}(\underset{\sim}{v}) + \lambda \, \text{tr}\, (\underset{\approx}{\epsilon}(\underset{\sim}{v}))\underset{\approx}{\delta}$$

$$a_\lambda(\underset{\sim}{v}, \underset{\sim}{w}) = \int_\Omega \left\{ 2\underset{\approx}{\epsilon}(\underset{\sim}{v}) : \underset{\approx}{\epsilon}(\underset{\sim}{w}) + \lambda \, \text{div}\underset{\sim}{v}\, \text{div}\, \underset{\sim}{w} \right\} dx \,.$$

Let $\mathcal{T}^h$ be a regular triangulation of Ω (cf. Fig. 11.1), and $\underset{\sim}{V}_h$ be defined as in (11.3.1). For each $\underset{\sim}{u} \in H^2(\Omega) \cap H_0^1(\Omega)$, we define $\underset{\sim}{u}_h^\lambda \in \underset{\sim}{V}_h$ to be the unique solution of

$$a_\lambda(\underset{\sim}{u}_h^\lambda, \underset{\sim}{v}) = \int_\Omega \left[-\text{div}\underset{\approx}{\sigma}_\lambda(\underset{\sim}{u}) \right] \cdot \underset{\sim}{v}\, dx \quad \forall \underset{\sim}{v} \in \underset{\sim}{V}_h.$$

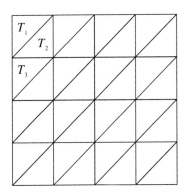

Fig. 11.1. a regular triangulation of the unit square

Let $L_{\lambda,h}$ be defined by

$$L_{\lambda,h} := \sup \left\{ \frac{|\underset{\sim}{u} - \underset{\sim}{u}_h^\lambda|_{H^1(\Omega)}}{\|\text{div}\underset{\approx}{\sigma}_\lambda(\underset{\sim}{u})\|_{L^2(\Omega)}} : \underset{\sim}{0} \neq \underset{\sim}{u} \in H^2(\Omega) \cap \overset{\circ}{H}{}^1(\Omega) \right\}.$$

We want to show that there exists a positive constant C independent of h such that

(11.3.7)
$$\liminf_{\lambda\to\infty} L_{\lambda,h} \geq C.$$

The meaning of (11.3.7) is: no matter how small h is, if λ is large enough, then we can find $\underset{\sim}{u} \in H^2(\Omega) \cap \overset{\circ}{H}{}^1(\Omega)$ such that the relative error

$|\underset{\sim}{u} - u_h|_{H^1(\Omega)}/\|\text{div}\underset{\approx}{\sigma}_\lambda(\underset{\sim}{u})\|_{L^2(\Omega)}$ is bounded below by a constant independent of h. In other words, the performance of the finite element method will deteriorate for large λ.

To prove (11.3.7), we begin with the observation that

$$(11.3.8) \qquad \left\{\underset{\sim}{v} \in \overset{\circ}{V}_h : \text{div}\,\underset{\sim}{v} = 0\right\} = \{\underset{\sim}{0}\}$$

(cf. exercise 11.x.14). Therefore, the map $\underset{\sim}{v} \to \text{div}\,\underset{\sim}{v}$ is a one-to-one map from the finite-dimensional space $\overset{\circ}{V}_h$ into $L^2(\Omega)$, and there exists a positive constant $C_1(h)$ such that

$$(11.3.9) \qquad \|\underset{\sim}{v}\|_{H^1(\Omega)} \le C_1(h)\,\|\text{div}\,\underset{\sim}{v}\|_{L^2(\Omega)} \quad \forall \underset{\sim}{v} \in \overset{\circ}{V}_h.$$

Let ψ be a C^∞ function on $\overline{\Omega}$ such that $\text{curl}\,\psi = \underset{\sim}{0}$ on the boundary of Ω and $\|\underset{\approx}{\epsilon}(\text{curl}\,\psi)\|_{L^2(\Omega)} = 1$. Let $\underset{\sim}{u} := \text{curl}\,\psi$. Then $\underset{\sim}{u} \in H^2(\Omega) \cap \overset{\circ}{H}^1(\Omega)$, and we have

$$(11.3.10) \qquad \text{div}\,\underset{\sim}{u} = 0,$$

$$(11.3.11) \qquad \|\underset{\approx}{\epsilon}(\underset{\sim}{u})\|_{L^2(\Omega)} = 1,$$

$$(11.3.12) \qquad \underset{\approx}{\sigma}_\lambda(\underset{\sim}{u}) = 2\underset{\approx}{\epsilon}(\underset{\sim}{u}).$$

It follows from (11.3.10), (11.3.11) and the integration by parts at the beginning of Sect. 11.2 that

$$(11.3.13) \qquad -\int_\Omega \text{div}\,\underset{\approx}{\epsilon}(\underset{\sim}{u}) \cdot \underset{\sim}{u}\,dx = \int_\Omega \underset{\approx}{\epsilon}(\underset{\sim}{u}) : \underset{\approx}{\epsilon}(\underset{\sim}{u})\,dx = 1.$$

Hence we deduce by (11.3.12) and (11.3.13) that

$$(11.3.14) \qquad \lim_{\lambda \to \infty} \text{div}\,\underset{\approx}{\sigma}_\lambda(\underset{\sim}{u}) = 2\text{div}\,\underset{\approx}{\epsilon}(\underset{\sim}{u}) \ne \underset{\sim}{0}.$$

By Corollary 2.5.10,

$$(11.3.15) \qquad \sqrt{a_\lambda(\underset{\sim}{u} - \underset{\sim}{u}_h^\lambda, \underset{\sim}{u} - \underset{\sim}{u}_h^\lambda)} = \min_{\underset{\sim}{v} \in \overset{\circ}{V}_h} \sqrt{a_\lambda(\underset{\sim}{u} - \underset{\sim}{v}, \underset{\sim}{u} - \underset{\sim}{v})}$$

$$\le \sqrt{a_\lambda(\underset{\sim}{u}, \underset{\sim}{u})}.$$

From (11.3.10) and (11.3.11) we obtain

$$(11.3.16) \qquad a_\lambda(\underset{\sim}{u}, \underset{\sim}{u}) = 2.$$

Therefore, for λ sufficiently large we have

$$(11.3.17) \qquad \sqrt{a_\lambda(\underset{\sim}{u} - \underset{\sim}{u}_h^\lambda, \underset{\sim}{u} - \underset{\sim}{u}_h^\lambda)} \le 2.$$

It follows from (11.3.10) and (11.3.17) that

$$\sqrt{\lambda}\,\|\mathrm{div}\,\underset{\sim}{u}_h^\lambda\|_{L^2(\Omega)} = \sqrt{\lambda}\,\|\,\mathrm{div}\,(\underset{\sim}{u} - \underset{\sim}{u}_h^\lambda)\|_{L^2(\Omega)}$$

$$\leq \sqrt{a_\lambda(\underset{\sim}{u} - \underset{\sim}{u}_h^\lambda, \underset{\sim}{u} - \underset{\sim}{u}_h^\lambda)}$$

$$\leq 2$$

for sufficiently large λ, which implies that

$$\lim_{\lambda \to \infty} \|\,\mathrm{div}\,\underset{\sim}{u}_h^\lambda\|_{L^2(\Omega)} = 0.$$

By (11.3.9) we therefore have

(11.3.18)
$$\lim_{\lambda \to \infty} \|\underset{\sim}{u}_h^\lambda\|_{\underset{\sim}{H}^1(\Omega)} = 0.$$

Finally, we obtain (cf. exercise 11.x.16)

(11.3.19)
$$\liminf_{\lambda \to \infty} L_{\lambda,h} \geq \liminf_{\lambda \to \infty} \frac{|\underset{\sim}{u} - \underset{\sim}{u}_h^\lambda|_{\underset{\sim}{H}^1(\Omega)}}{\|\mathrm{div}\,\underset{\approx}{\sigma}_\lambda(\underset{\sim}{u})\|_{\underset{\sim}{L}^2(\Omega)}}$$

$$= \frac{|\underset{\sim}{u}|_{\underset{\sim}{H}^1(\Omega)}}{\|\mathrm{div}\,\underset{\approx}{\sigma}(\underset{\sim}{u})\|_{\underset{\sim}{L}^2(\Omega)}} > 0.$$

This concludes our discussion of locking for this particular example. For more information on locking we refer the reader to (Babuška & Suri 1992).

11.4 A Robust Method for the Pure Displacement Problem

Roughly speaking, the reason behind the locking phenomenon described in the previous section is the following. The finite element space $\overset{\circ}{V}_h$ is a poor choice for approximating nearly incompressible material because the set $\{\underset{\sim}{v} \in \overset{\circ}{V}_h : \mathrm{div}\,\underset{\sim}{v} = 0\} = \{\underset{\sim}{0}\}$. Since $\mathrm{div}\,\underset{\sim}{u} = 0$, for $a_\lambda(\underset{\sim}{u} - \underset{\sim}{u}_h^\lambda, \underset{\sim}{u} - \underset{\sim}{u}_h^\lambda)$ to be small, $\|\mathrm{div}\,\underset{\sim}{u}_h^\lambda\|_{L^2(\Omega)}$ must be small. But then $\|\underset{\sim}{u}_h^\lambda\|_{\underset{\sim}{H}^1(\Omega)}$ is small and hence $\underset{\sim}{u}_h^\lambda$ cannot approximate $\underset{\sim}{u}$ very well in the H^1-norm.

We will now show that locking can be overcome if we use a nonconforming piecewise linear finite element space. The key to its success lies in its extra freedom which allows good approximation of functions with divergence-zero constraints. We refer the readers to the papers (Arnold, Brezzi & Douglas 1984), (Arnold, Douglas & Brezzi 1984) and (Stenberg 1988) for more sophiscated robust finite element methods.

For simplicity we will consider (11.1.2) with homogeneous pure displacement boundary condition on a convex polygonal domain Ω. (See (Falk

1991) for the treatment of the pure traction boundary condition.) Observe that the pure displacement problem can be written as

$$-\mu\,\Delta\underset{\sim}{u} - (\mu+\lambda)\,\text{grad}\,(\text{div}\,\underset{\sim}{u}) = \underset{\sim}{f} \quad \text{in } \Omega$$

(11.4.1)

$$\underset{\sim}{u} = \underset{\sim}{0} \quad \text{on } \partial\Omega,$$

where $\underset{\sim}{f} \in L^2(\Omega)$. It has the following weak formulation:

(11.4.2)   find $\underset{\sim}{u} \in \overset{\circ}{H}{}^1(\Omega)$ such that $a^s(\underset{\sim}{u},\underset{\sim}{v}) = \int_\Omega \underset{\sim}{f}\cdot\underset{\sim}{v}\,dx \quad \forall \underset{\sim}{v} \in \overset{\circ}{H}{}^1(\Omega),$

where the bounded bilinear form $a^s(\cdot,\cdot)$ on $\overset{\circ}{H}{}^1(\Omega)$ is defined by

(11.4.3) $a^s(\underset{\sim}{v}_1,\underset{\sim}{v}_2) := \mu\int_\Omega \text{grad}\,\underset{\sim}{u} : \text{grad}\,\underset{\sim}{v}\,dx + (\mu+\lambda)\int_\Omega (\text{div}\,\underset{\sim}{u})(\text{div}\,\underset{\sim}{v})\,dx.$

Note that $a(\cdot,\cdot)$ and $a^s(\cdot,\cdot)$ differ only in the corresponding natural boundary conditions. Thus, they are completely equivalent for the pure displacement problem.

The coercivity of $a^s(\cdot,\cdot)$ follows from Poincaré's inequality (5.3.5). Therefore, (11.4.2) has a unique solution which actually belongs to $H^2(\Omega)\cap \overset{\circ}{H}{}^1(\Omega)$ (cf. the discussion at the end of Sect. 11.2). Moreover, from (11.2.33) we have the elliptic regularity estimate

(11.4.4) $\|\underset{\sim}{u}\|_{H^2(\Omega)} + \lambda\,\|\text{div}\,\underset{\sim}{u}\|_{H^1(\Omega)} \le C\,\|\underset{\sim}{f}\|_{L^2(\Omega)},$

where the positive constant C is independent of $(\mu,\lambda) \in [\mu_1,\mu_2] \times (0,\infty)$.

Let $\mathcal{T}^h$ be a non-degenerate family of triangulations of Ω, and let $\underset{\sim}{V}{}_h^* := V_h \times V_h$ where V_h is defined in (11.3.2) based on the element depicted in Fig. 3.2, that is,

(11.4.5) $\underset{\sim}{V}{}_h^* = \Big\{ \underset{\sim}{v} : \underset{\sim}{v} \in \underset{\sim}{L}{}^2(\Omega),\ \underset{\sim}{v}|_T$ is linear for all $T \in \mathcal{T}^h$, $\underset{\sim}{v}$ is continuous at the midpoints of interelement boundaries and $\underset{\sim}{v} = \underset{\sim}{0}$ at the midpoints of edges along $\partial\Omega\Big\}.$

Since $\underset{\sim}{V}{}_h^* \not\subseteq \underset{\sim}{H}{}^1(\Omega)$ (i.e., $\underset{\sim}{V}{}_h^*$ is nonconforming), any differential operator on $\underset{\sim}{V}{}_h^*$ must be defined piecewise. For $\underset{\sim}{v} \in \underset{\sim}{V}{}_h^*$ we define grad_h and div_h by

(11.4.6) $(\text{grad}_h\underset{\sim}{v})|_T = \text{grad}\,(\underset{\sim}{v}|_T) \quad \text{and} \quad (\text{div}_h\underset{\sim}{v})|_T = \text{div}\,(\underset{\sim}{v}|_T)$

for all $T \in \mathcal{T}^h$. The discretized problem is the following.

(11.4.7)   Find $\underset{\sim}{u}_h \in \underset{\sim}{V}{}_h^*$ such that $a_h^s(\underset{\sim}{u}_h,\underset{\sim}{v}) = \int_\Omega \underset{\sim}{f}\cdot\underset{\sim}{v}\,dx \quad \forall \underset{\sim}{v} \in \underset{\sim}{V}{}_h^*,$

where the symmetric positive definite bilinear form $a_h^s(\cdot, \cdot)$ on $\underset{\sim}{V}_h^* + \overset{\circ}{\underset{\sim}{H}}{}^1(\Omega)$ is defined by

(11.4.8)
$$a_h^s(\underset{\sim}{v}, \underset{\sim}{w}) := \mu \int_\Omega \underset{\approx}{\mathrm{grad}}_h \underset{\sim}{v} : \underset{\approx}{\mathrm{grad}}_h \underset{\sim}{w} \, dx$$
$$+ (\mu + \lambda) \int_\Omega (\mathrm{div}_h \underset{\sim}{v})(\mathrm{div}_h \underset{\sim}{w}) \, dx.$$

Note this definition of the nonconforming bilinear form is identical to one given using sums of integrals over each $T \in \mathcal{T}^h$ (cf. (11.3.3)). Equation (11.4.7) has a unique solution because $a_h^s(\cdot, \cdot)$ is positive definite, which can be proved in a way similar to what was done in Chapter 10.

We define the nonconforming energy norm $\| \cdot \|_h$ on $\underset{\sim}{V}_h^* + \overset{\circ}{\underset{\sim}{H}}{}^1(\Omega)$ by

(11.4.9)
$$\|\underset{\sim}{v}\|_h = a_h^s(\underset{\sim}{v}, \underset{\sim}{v})^{1/2}.$$

It is clear that

(11.4.10)
$$\|\underset{\approx}{\mathrm{grad}}_h \underset{\sim}{v}\|_{\underset{\approx}{L}^2(\Omega)} \le \mu_1^{-1/2} \|\underset{\sim}{v}\|_h.$$

Our goal is to show that the finite element method (11.4.7) is robust in the sense that the error estimates are uniform with respect to $(\mu, \lambda) \in [\mu_1, \mu_2] \times (0, \infty)$. For this we need two ingredients. The first is the property of the divergence operator stated in Lemma 11.2.3, which holds in the case of polygonal domains by the results in (Girault & Raviart 1986, Arnold, Scott & Vogelius 1988). The other ingredient is an interpolation operator $\Pi_h : \underset{\sim}{H}^2(\Omega) \cap \overset{\circ}{\underset{\sim}{H}}{}^1(\Omega) \longrightarrow \underset{\sim}{V}_h^*$ with the property that

(11.4.11)
$$\mathrm{div}\, \underset{\sim}{\phi} = 0 \Longrightarrow \mathrm{div}_h(\Pi_h \underset{\sim}{\phi}) = 0.$$

We can define Π_h by

(11.4.12)
$$(\Pi_h \underset{\sim}{\phi})(m_e) := \frac{1}{|e|} \int_e \underset{\sim}{\phi} \, ds,$$

where m_e is the midpoint of edge e. Then

(11.4.13)
$$\mathrm{div}(\Pi_h \underset{\sim}{\phi})|_T = \frac{1}{|T|} \int_T \mathrm{div}\, \underset{\sim}{\phi} \, dx \quad \forall\, T \in \mathcal{T}^h,$$

and there exists a positive constant C independent of h such that

(11.4.14) $$\|\underset{\sim}{\phi} - \Pi_h \underset{\sim}{\phi}\|_{\underset{\approx}{L}^2(\Omega)} + h\|\underset{\approx}{\mathrm{grad}}_h(\underset{\sim}{\phi} - \Pi_h \underset{\sim}{\phi})\|_{\underset{\approx}{L}^2(\Omega)} \le C h^2 |\underset{\sim}{\phi}|_{\underset{\sim}{H}^2(\Omega)}$$

(cf. exercises 11.x.18 and 11.x.19 and Sect. 4.8).

(11.4.15) Theorem. *There exists a positive constant C independent of h and $(\mu, \lambda) \in [\mu_1, \mu_2] \times (0, \infty)$ such that*

(11.4.16)
$$\|\underset{\sim}{u} - \underset{\sim}{u}_h\|_h \le C\, h\, \|\underset{\sim}{f}\|_{\underset{\sim}{L}^2(\Omega)}.$$

Proof. In this proof, C represents a generic constant independent of h and $(\mu, \lambda) \in [\mu_1, \mu_2] \times (0, \infty)$. From (11.1.10) we have

$$(11.4.17) \quad \|\underset{\sim}{u} - \underset{\sim}{u}_h\|_h \leq \inf_{\underset{\sim}{v} \in \underset{\sim}{V}_h} \|\underset{\sim}{u} - \underset{\sim}{v}\|_h + \sup_{\underset{\sim}{v} \in \underset{\sim}{V}_h \setminus \{0\}} \frac{|a_h^s(\underset{\sim}{u}, \underset{\sim}{v}) - \int_\Omega \underset{\sim}{f} \cdot \underset{\sim}{v}\, dx|}{\|\underset{\sim}{v}\|_h}.$$

By the same techniques employed in the proof of Theorem 11.3.12 (i.e., using the Bramble-Hilbert Lemma and homogeneity arguments) we have

$$(11.4.18) \quad \left| \int_\Omega \operatorname*{grad}_{\approx} \underset{\sim}{u} : \operatorname*{grad}_{\approx}{}_h \underset{\sim}{v}\, dx + \int_\Omega \Delta \underset{\sim}{u} \cdot \underset{\sim}{v}\, dx \right| \\ \leq C\, h\, |\underset{\sim}{u}|_{H^2(\Omega)} \|\operatorname*{grad}_{\approx}{}_h \underset{\sim}{v}\|_{L^2(\Omega)},$$

and

$$(11.4.19) \quad \left| \int_\Omega \operatorname{div} \underset{\sim}{u}\, \operatorname{div}_h \underset{\sim}{v}\, dx + \int_\Omega \operatorname*{grad}_{\sim} (\operatorname{div} \underset{\sim}{u}) \cdot \underset{\sim}{v}\, dx \right| \\ \leq C\, h\, |\operatorname{div} \underset{\sim}{u}|_{H^1(\Omega)} \|\operatorname*{grad}_{\approx}{}_h \underset{\sim}{v}\|_{L^2(\Omega)}.$$

Combining (11.4.1), (11.4.8), (11.4.18), (11.4.19), (11.4.4) and (11.4.10) we have

$$(11.4.20) \quad \left| a_h^s(\underset{\sim}{u}, \underset{\sim}{v}) - \int_\Omega \underset{\sim}{f} \cdot \underset{\sim}{v}\, dx \right| \\ \leq C\, h\, \|\operatorname*{grad}_{\approx}{}_h \underset{\sim}{v}\|_{L^2(\Omega)} \left(\mu\, |\underset{\sim}{u}|_{H^2(\Omega)} + (\mu + \lambda)\, |\operatorname{div} \underset{\sim}{u}|_{H^1(\Omega)} \right) \\ \leq C\, h\, \|\underset{\sim}{v}\|_h \|\underset{\sim}{f}\|_{L^2(\Omega)}.$$

By an analog of Lemma 11.2.3 (cf. (Brenner & Sung 1992)), there exists $\underset{\sim}{u}_1 \in H^2(\Omega) \cap \overset{\circ}{H}{}^1(\Omega)$ such that

$$(11.4.21) \quad \operatorname{div} \underset{\sim}{u}_1 = \operatorname{div} \underset{\sim}{u}$$

and

$$(11.4.22) \quad \|\underset{\sim}{u}_1\|_{H^2(\Omega)} \leq C\, \|\operatorname{div} \underset{\sim}{u}\|_{H^1(\Omega)}.$$

Combining (11.4.22) and (11.4.4) we have

$$(11.4.23) \quad \|\underset{\sim}{u}_1\|_{H^2(\Omega)} \leq \frac{C}{1 + \lambda} \|\underset{\sim}{f}\|_{L^2(\Omega)}.$$

Note that (11.4.11) and (11.4.21) imply that

$$(11.4.24) \quad \operatorname{div}_h \Pi_h \underset{\sim}{u}_1 = \operatorname{div}_h \Pi_h \underset{\sim}{u}.$$

Then

$$\inf_{\underset{\sim}{v}\in V_h} \|\underset{\sim}{u} - \underset{\sim}{v}\|_h$$

$$\leq \|\underset{\sim}{u} - \Pi_h\underset{\sim}{u}\|_h$$

$$= \left(\mu\|\underset{\approx}{\text{grad}}_h(\underset{\sim}{u} - \Pi_h\underset{\sim}{u})\|^2_{L^2(\Omega)} + (\mu+\lambda)\|\text{div}_h(\underset{\sim}{u} - \Pi_h\underset{\sim}{u})\|^2_{L^2(\Omega)}\right)^{1/2}$$

(11.4.25)   (by 11.4.9)

$$= \Big(\mu\|\underset{\approx}{\text{grad}}_h(\underset{\sim}{u} - \Pi_h\underset{\sim}{u})\|^2_{L^2(\Omega)}$$

$$+ (\mu+\lambda)\|\text{div}_h(\underset{\sim}{u}_1 - \Pi_h\underset{\sim}{u}_1)\|^2_{L^2(\Omega)}\Big)^{1/2}\quad \text{(by 11.4.21 \& 11.4.24)}$$

$$\leq C\,h\,\|\underset{\sim}{f}\|_{L^2(\Omega)}\,.\qquad\qquad \text{(by 11.4.14, 11.4.23 \& 11.4.4)}$$

The theorem now follows by combining (11.4.17), (11.4.20) and (11.4.25).

□

11.x Exercises

11.x.1 Verify (11.1.1).

11.x.2 Show that $\{\underset{\sim}{v} \in \underset{\sim}{H}^1(\Omega) : \underset{\approx}{\epsilon}(\underset{\sim}{v}) = \underset{\approx}{0}\} = \underset{\sim}{\text{RM}}$.

11.x.3 Show that $\text{rot}\,\underset{\sim}{v} = \underset{\approx}{\chi} : \underset{\approx}{\text{grad}}\,\underset{\sim}{v}$.

11.x.4 Show that $\text{div}\,\underset{\sim}{v} = \underset{\approx}{\chi} : \underset{\approx}{\text{curl}}\,\underset{\sim}{v}$.

11.x.5 Show that $\int_\Omega \underset{\approx}{\text{grad}}\,\underset{\sim}{v} : \underset{\approx}{\text{curl}}\,\underset{\sim}{w}\,dx = 0$ for $\underset{\sim}{v} \in \underset{\sim}{H}^1(\Omega)$ and $\underset{\sim}{w} \in \underset{\sim}{\mathring{H}}^1(\Omega)$.

11.x.6 Show that $\underset{\sim}{H}^1(\Omega) = \underset{\sim}{\hat{H}}^1(\Omega) \oplus \underset{\sim}{\text{RM}}$. (Hint: use (11.2.18).)

11.x.7 Let A, B be closed subspaces of a Banach space V such that $V = A \oplus B$. Show that there exists a positive constant C such that given any $v \in V$, we have $v = v_1 + v_2$, where $v_1 \in A$, $v_2 \in B$ and $C(\|v_1\| + \|v_2\|) \leq \|v\|$. (Hint: use the Open Mapping Theorem (Rudin 1987).)

11.x.8 Show that the second Korn inequality follows from the Korn inequality.

11.x.9 Show that if $\underset{\sim}{v} \in \underset{\sim}{\text{RM}}$ and $\underset{\sim}{v} \not\equiv \underset{\sim}{0}$, then $\underset{\sim}{v} = \underset{\sim}{0}$ at at most one point.

11.x.10 Establish the first Korn inequality (11.2.26) directly through integration by parts and Poincaré's inequality.

11.x.11 Prove Theorem 11.2.12 on $\underset{\sim}{H}^1_\perp(\Omega) := \{\underset{\sim}{v} \in \underset{\sim}{H}^1(\Omega) : (\underset{\sim}{v}, \underset{\sim}{w})_{L^2(\Omega)} = 0 \quad \forall\,\underset{\sim}{w} \in \underset{\sim}{\text{RM}}\}$.

11.x.12 Does (11.3.7) contradict (11.3.4)? If not, what is the implication?

11.x.13 Show that given any $\underset{\sim}{v} \in \hat{H}^2(\Omega)$, there exists a $\underset{\sim}{v}_h \in \hat{V}_h$ such that

$$\|\underset{\sim}{v} - \underset{\sim}{v}_h\|_{H^1(\Omega)} \leq C\,h\,\|\underset{\sim}{v}\|_{H^2(\Omega)},$$

where the constant C is independent of h. (Hint: use (11.2.18) to correct the standard interpolant.)

11.x.14 Verify (11.3.8). (Hint: referring to Fig. 11.1, show that $\underset{\sim}{v} = \underset{\sim}{0}$ on T_1 and div $\underset{\sim}{v} = 0$ on T_2 and T_3 implies that $\underset{\sim}{v} = \underset{\sim}{0}$ on T_2 and T_3. Then repeat this argument.)

11.x.15 Verify (11.3.12) and (11.3.16).

11.x.16 Carry out the estimate of $\liminf_{\lambda\to\infty} L_{\lambda,h}$ in (11.3.19). (Hint: use (11.3.11), (11.3.14), and (11.3.18).)

11.x.17 Show that equation (11.1.2) is equivalent to (11.4.1).

11.x.18 Establish (11.4.13). Show that it implies (11.4.11).

11.x.19 Prove the interpolation error estimate (11.4.14). (Hint: use the Bramble-Hilbert Lemma and a homogeneity argument. If necessary, consult (Crouzeix & Raviart 1973).)

11.x.20 Work out the details of the proofs of (11.4.18) and (11.4.19).

11.x.21 Work out the details of (11.4.20).

11.x.22 Use a duality argument to show that there exists a positive constant C independent of h and $(\mu, \lambda) \in [\mu_1, \mu_2] \times (0, \infty)$ such that

$$\|\underset{\sim}{u} - \underset{\sim}{u}_h\|_{L^2(\Omega)} \leq C\,h^2\,\|\underset{\sim}{f}\|_{L^2(\Omega)}.$$

(Hint: consult (Brenner & Sung 1992).)

11.x.23 Prove the following discrete version of Lemma 11.2.3. Let Ω be a convex polygonal domain. Given any $p \in L^2(\Omega)$ such that $p|_T$ is a constant for all $T \in \mathcal{T}^h$ and $\int_\Omega p\,dx = 0$, there exists a positive constant C and $\underset{\sim}{v} \in V_h^*$ such that (i) $\text{div}_h \underset{\sim}{v} = p$ and (ii) $\|\text{grad}_h \underset{\sim}{v}\|_{L^2(\Omega)} \leq C\,\|p\|_{L^2(\Omega)}$. (Hint: use Lemma 11.2.3, the interpolation operator Π_h and (11.4.13).)

11.x.24 Let V_h^* be defined by (11.4.5). The subspace $W_h := \{\underset{\sim}{v} \in V_h^* : \text{div}_h \underset{\sim}{v} = 0\}$ of V_h^* is useful for incompressible fluid flow problems. Denote by ψ_e the piecewise linear function that takes the value 1 at the midpoint of the edge e and 0 at all other midpoints. The first kind of basis functions of W_h are associated with internal edges. Let e be an internal edge of $\mathcal{T}^h$ and t_e be a unit vector tangential to e, then $\underset{\sim}{\phi}_e := \psi_e\,\underset{\sim}{t}_e$. The second kind of basis functions are associated with internal vertices. Let p be an internal vertex and let $e_1, e_2, \ldots, e_l$ be the edges in $\mathcal{T}^h$ that have p as an endpoint, then $\underset{\sim}{\phi}_p := \sum_{i=1}^{l} \frac{1}{|e_i|} \psi_{e_i} \underset{\sim}{n}_{e_i}$, where $\underset{\sim}{n}_{e_i}$ is a unit vector normal to e_i pointing in the counterclockwise direction with respect to p (cf. Fig. 11.2). Show that a basis for $\underset{\sim}{W}_h$ is given by the union of the two sets

$$\{\underset{\sim}{\phi}_e : e \text{ is an internal edge of } \mathcal{T}^h\}$$

and

$$\{\underset{\sim}{\phi}_p : p \text{ is an internal vertex of } \mathcal{T}^h\}.$$

(Hint: if necessary, consult (Thomasset 1981).)

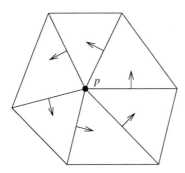

Fig. 11.2. basis function $\underset{\sim}{\phi}_p$

11.x.25 Formulate and prove a convergence theorem for the piecewise linear approximation of the variational problem (11.2.1) in the general case $\emptyset \neq \Gamma_1 \neq \partial\Omega$. (Hint: see Sect. 5.4.)

Chapter 12

Mixed Methods

The name "mixed method" is applied to a variety of finite element methods which have more than one approximation space. Typically one or more of the spaces play the role of Lagrange multipliers which enforce constraints. The name and many of the original concepts for such methods originated in solid mechanics where it was desirable to have a more accurate approximation of certain derivatives of the displacement. However, for the Stokes equations which govern viscous fluid flow, the natural Galerkin approximation is a mixed method.

One characteristic of mixed methods is that not all choices of finite element spaces will lead to convergent approximations. Standard approximability alone is insufficient to guarantee success. In fact, we will study mixed methods in the context of a variational crime.

We will focus on mixed methods in which there are two bilinear forms and two approximation spaces. There are two key conditions (cf. (Babuška 1971), (Brezzi 1974)) that lead to the success of a mixed method. Both are in some sense coercivity conditions for the bilinear forms. One of these will look like a standard coercivity condition, while the other, often called the *inf-sup* condition, takes a new form.

12.1 Examples of Mixed Variational Formulations

The Stokes equations for steady flow of a (very) viscous fluid are

(12.1.1)
$$-\Delta \underset{\sim}{u} + \operatorname*{grad} p = \underset{\sim}{f}$$
$$\operatorname{div} \underset{\sim}{u} = 0$$

in $\Omega \subset \mathbb{R}^n$, where $\underset{\sim}{u}$ denotes the fluid velocity and p denotes the pressure. Here we use the "under-tilde" notation introduced in the previous chapter for vectors, vector operators and matrices. However, we also use the obvious extensions to three dimensions in some cases. To make the dimensionality clear in all cases, we use the notation $H^s(\Omega)^n$ instead of $\underset{\sim}{H}^s(\Omega)$, $L^2(\Omega)^n$ instead of $\underset{\sim}{L}^2(\Omega)$, etc.

Equations (12.1.1) represent the limiting case of zero Reynolds' number for the Navier-Stokes equations (13.4.1) to be discussed subsequently. It is unusual to have a nonzero forcing term, $\underset{\sim}{f}$, but the linearity of these equations allow us to convert the more likely case of inhomogeneous boundary data to a homogeneous one. Thus, we assume, for example, that we have Dirichlet boundary conditions, $\underset{\sim}{u} = 0$, on $\partial\Omega$.

Integrating by parts, we can derive a variational identity for suitable $\underset{\sim}{v}$ and q:

$$(12.1.2) \qquad a(\underset{\sim}{u}, \underset{\sim}{v}) + b(\underset{\sim}{v}, p) = \int_\Omega \underset{\sim}{f} \cdot \underset{\sim}{v}\, dx \, ,$$

$$(12.1.3) \qquad b(\underset{\sim}{u}, q) = 0 \, ,$$

where the forms $a(\cdot, \cdot)$ and $b(\cdot, \cdot)$ are defined as

$$(12.1.4) \qquad a(\underset{\sim}{u}, \underset{\sim}{v}) := \int_\Omega \sum_{i=1}^n \operatorname{grad} u_i \cdot \operatorname{grad} v_i \, dx$$

$$(12.1.5) \qquad b(\underset{\sim}{v}, q) := - \int_\Omega (\operatorname{div} \underset{\sim}{v})\, q\, dx.$$

Let $\underset{\sim}{V} = \mathring{H}^1(\Omega)^n$ and $\Pi = \{q \in L^2(\Omega) : \int_\Omega q\, dx = 0\}$. It is well known (Girault & Raviart 1979 & 1986, Temam 1984) that the following has a unique solution.

Given $\underset{\sim}{F} \in \underset{\sim}{V}'$, find functions $\underset{\sim}{u} \in \underset{\sim}{V}$ and $p \in \Pi$ such that

$$(12.1.6) \qquad \begin{aligned} a(\underset{\sim}{u}, \underset{\sim}{v}) + b(\underset{\sim}{v}, p) &= F(\underset{\sim}{v}) \quad \forall \underset{\sim}{v} \in \underset{\sim}{V} \, , \\ b(\underset{\sim}{u}, q) &= 0 \quad \forall q \in \Pi \, . \end{aligned}$$

The well-posedness of this problem follows in part from the coercivity of $a(\cdot, \cdot)$ on $\mathring{H}^1(\Omega)^n$ (see exercise 12.x.1).

A model for fluid flow in a porous medium occupying a domain Ω takes the form (5.6.6), namely,

$$(12.1.7) \qquad -\sum_{i,j=1}^n \frac{\partial}{\partial x_i}\left(a_{ij}(x)\frac{\partial p}{\partial x_j}(x)\right) = f(x) \text{ in } \Omega,$$

where p is the pressure (again we take an inhomogeneous right-hand-side for simplicity). Darcy's Law postulates that the fluid velocity $\underset{\sim}{u}$ is related to the gradient of p by

$$\sum_{j=1}^n a_{ij}(x)\frac{\partial p}{\partial x_j}(x) = u_i(x) \quad \forall i = 1, \ldots, n.$$

The coefficients a_{ij}, which we assume form a symmetric, positive-definite system (almost everywhere), are related to the porosity of the medium. Of course, numerous other physical models also take the form (12.1.7).

A variational formulation for (12.1.7) of the form (12.1.6) can be derived by letting $\underset{\approx}{A}(x)$ denote the (almost everywhere defined) inverse of the coefficient matrix (a_{ij}) and by writing $\text{grad}\, p = \underset{\approx}{A}\underset{\sim}{u}$. Define

$$(12.1.8) \qquad a(\underset{\sim}{u}, \underset{\sim}{v}) := \int_\Omega \sum_{i,j=1}^n A_{ij} u_i v_j \, dx$$

with $b(\cdot, \cdot)$ as before. Then the solution to (12.1.7) solves

$$(12.1.9) \qquad \begin{aligned} a(\underset{\sim}{u}, \underset{\sim}{v}) + b(\underset{\sim}{v}, p) &= 0 \quad \forall \underset{\sim}{v} \in V, \\ b(\underset{\sim}{u}, q) &= F(q) \quad \forall q \in \Pi \end{aligned}$$

where $\Pi = L^2(\Omega)$ and

$$V := \left\{ \underset{\sim}{v} \in L^2(\Omega)^n \; : \; \text{div}\, \underset{\sim}{v} \in L^2(\Omega) \right\}.$$

The latter space, called $H(\text{div})$, has a natural norm given by

$$(12.1.10) \qquad \left\|\underset{\sim}{v}\right\|_{H(\text{div})}^2 = \left\|\underset{\sim}{v}\right\|_{L^2(\Omega)^n}^2 + \left\|\text{div}\,\underset{\sim}{v}\right\|_{L^2(\Omega)}^2 ,$$

(see exercise 12.x.2). Unlike the previous problem, the bilinear form $a(\cdot, \cdot)$ is not coercive on all of $\underset{\sim}{V}$. However, it is coercive on the critical subspace of divergence-zero functions. The role of this subspace will be made clear in the abstract setting to which we now turn our attention.

12.2 Abstract Mixed Formulation

We now abstract the key features of the above two problems. We have two Hilbert spaces V and Π and two bilinear forms $a(\cdot, \cdot) : V \times V \to \mathbb{R}$ and $b(\cdot, \cdot) : V \times \Pi \to \mathbb{R}$. It is natural to assume that these forms are continuous:

$$(12.2.1) \qquad \begin{aligned} a(u, v) &\leq C \|u\|_V \|v\|_V \quad \forall u, v \in V \\ b(v, p) &\leq C \|v\|_V \|p\|_\Pi \quad \forall v \in V, p \in \Pi, \end{aligned}$$

but for the moment we postpone discussion of the appropriate notion of coercivity for either form.

There is no unique relationship between V and Π, but there is an operator between them that provides a link. We will assume that $\mathcal{D} : V \to \Pi$ is continuous,

$$(12.2.2) \qquad \|\mathcal{D}v\|_\Pi \leq C \|v\|_V,$$

and, moreover, we make the simplifying assumption that

$$(12.2.3) \qquad b(v, p) = (\mathcal{D}v, p)_\Pi.$$

We consider the variational problem to find $u \in V$ and $p \in \Pi$ such that

(12.2.4)
$$a(u,v) + b(v,p) = F(v) \quad \forall v \in V$$
$$b(u,q) = G(q) \quad \forall q \in \Pi$$

where $F \in V'$ and $G \in \Pi'$.

We note that alternative formulations exist (Ciarlet & Lions 1991) in which there is an operator $\mathcal{D}' : \Pi \to V$ and $b(v,p) = (v, \mathcal{D}'p)_V$.

Define a closed subspace of V via

(12.2.5)
$$Z = \{v \in V \ : \ b(v,q) = 0 \quad \forall q \in \Pi\}.$$

Suppose for the moment that $G = 0$ in (12.2.4). Then the way (12.2.4) determines u is equivalent to the following: find $u \in Z$ such that

(12.2.6)
$$a(u,v) = F(v) \quad \forall v \in Z.$$

This is well-posed provided $a(\cdot,\cdot)$ is coercive, namely

(12.2.7)
$$\alpha\|v\|_V^2 \le a(v,v)$$

for all $v \in Z$, in view of the Lax-Milgram Theorem 2.7.7.

The coercivity condition (12.2.7) holds for both problems considered in the previous section. For the Stokes formulation based on the form $a(\cdot,\cdot)$ in (12.1.4), it holds for all $v \in V$, although for other formulations (see exercise 12.x.3) it will hold only on the subset Z of divergence zero functions. For the scalar elliptic problem it is essential that we are allowed to restrict to $z \in Z$ where in this case

$$Z = \left\{\underset{\sim}{v} \in H(\mathrm{div}) \ : \ \mathrm{div}\,\underset{\sim}{v} = 0\right\}.$$

For $\underset{\sim}{v} \in Z$, $\|\underset{\sim}{v}\|_{H(\mathrm{div})} = \|\underset{\sim}{v}\|_{L^2(\Omega)^n}$ and coercivity follows from the fact that the coefficients a_{ij} are bounded.

If G in (12.2.4) is not zero, we can reduce to the case $G = 0$ as follows. Suppose $u_0 \in V$ is any solution to $b(u_0, q) = G(q) \quad \forall q \in \Pi$. For the examples considered previously, this amounts to solving $\mathrm{div}\,\underset{\sim}{u}_0 = g$ which can be done in a variety of ways (cf. Lemma 11.2.3 and (Arnold, Scott & Vogelius 1988)). Then the solution u to (12.2.4) is of the form $u = u_1 + u_0$ where $u_1 \in Z$ satisfies

$$a(u_1, v) = F(v) - a(u_0, v) \quad \forall v \in Z$$

which is again of the form (12.2.6). Thus, for the remainder of the section we will focus on the problem

(12.2.8)
$$a(u,v) + b(v,p) = F(v) \quad \forall v \in V$$
$$b(u,q) = 0 \quad \forall q \in \Pi$$

where $F \in V'$. We return to the general case in Sect. 12.5.

With u well defined by (12.2.6), we then determine $p \in \Pi$ such that

$$(12.2.9) \qquad b(v,p) = -a(u,v) + F(v) \quad \forall v \in V.$$

The well-posedness of this problem follows from a new kind of coercivity. To motivate the new coercivity condition, let us re-examine the use of conditions such as (12.2.7). Such a condition implies

$$\alpha\|v\|_V \le \sup_{w \in V} \frac{a(w,v)}{\|w\|_V} \quad \forall v \in V$$

(take $w = v$), and the latter condition is sufficient for the types of estimates where we have employed a coercivity condition. Applying this idea to $b(\cdot,\cdot)$, we consider the condition (cf. (Babuška 1971), (Brezzi 1974))

$$(12.2.10) \qquad \beta\|p\|_\Pi \le \sup_{w \in V} \frac{b(w,p)}{\|w\|_V} \quad \forall p \in \Pi.$$

Problem (12.2.9) is of the form

$$(12.2.11) \qquad b(v,p) = \widetilde{F}(v) \quad \forall v \in V$$

where $\widetilde{F}(v) = 0$ for all $v \in Z$. Condition (12.2.10) (with $\beta > 0$) implies uniqueness of a solution. Existence of a solution also follows from (12.2.10), but this requires a bit more explanation.

(12.2.12) Lemma. *Suppose that (12.2.1) and (12.2.10) hold with $\beta > 0$. Then (12.2.11) has a unique solution.*

Proof. Let $Z^\perp$ denote the orthogonal complement (Sect. 2.2) of Z in V (recall V is a Hilbert space). Since the behavior of $b(\cdot,\cdot)$ is trivial on Z, we may as well restrict our attention to $v \in Z^\perp$ in (12.2.11). Recall (Proposition 2.4.4) that we can consider $Z^\perp$ as a Hilbert space with the inner-product $(\cdot,\cdot)_V$ inherited from V. Given $p \in \Pi$, the linear form $v \to b(v,p)$ is continuous on $Z^\perp$, so the Riesz Representation Theorem 2.4.2 guarantees the existence of $Tp \in Z^\perp$ such that

$$(12.2.13) \qquad (Tp,v)_V = b(v,p) \quad \forall v \in Z^\perp.$$

Moreover, part of Theorem 2.4.2 assures that T is linear, and (2.4.4) implies

$$\|Tp\|_V = \sup_{v \in Z^\perp} \frac{b(v,p)}{\|v\|_V} \le C\|p\|_\Pi,$$

where the inequality is the assumption (12.2.1). Let R denote the image of T in $Z^\perp$. If we can show that $R = Z^\perp$, then another application of the Riesz Representation Theorem completes the proof, since it implies we can always represent $\widetilde{F}$ as

$$\widetilde{F}(v) = (u,v)_V \quad \forall v \in Z^\perp$$

for some $u \in Z^{\perp}$. We simply pick p such that $Tp = u$.

To show that R is all of $Z^{\perp}$, we begin by showing it to be closed. Suppose that $p_j \in \Pi$ is a sequence with the property that $Tp_j \to w$ in $Z^{\perp}$. Then $\{Tp_j\}$ is Cauchy in $Z^{\perp}$, and

$$\beta \|p_j - p_k\|_{\Pi} \leq \sup_{w \in Z^{\perp}} \frac{b(w, p_j - p_k)}{\|w\|_V} \qquad \text{(by 12.2.10)}$$

$$= \sup_{w \in Z^{\perp}} \frac{(w, Tp_j - Tp_k)_V}{\|w\|_V} \qquad \text{(by 12.2.13)}$$

$$= \|Tp_j - Tp_k\|_V \qquad \text{(exercise 2.x.16)}$$

so that $\{p_j\}$ is Cauchy in Π. Let $q = \lim_{j \to \infty} p_j$. By the continuity of T, $Tq = w$, so R is closed. If $R \neq Z^{\perp}$, pick $0 \neq v \in R^{\perp}$. Then $b(v, q) = (v, Tq)_V = 0$ for all $q \in \Pi$. But this implies that $v \in Z$, a contradiction. So $R = Z^{\perp}$ and the proof is complete. $\qquad\qquad \square$

Condition (12.2.10) holds for the problems introduced in the previous section, and the proof is similar in both cases. It follows from the fact (Lemma 11.2.3) that one can solve the underdetermined system div $\underset{\sim}{w} = p$ for $\underset{\sim}{w} \in H^1(\Omega)^n$ with

$$\left\| \underset{\sim}{w} \right\|_{H^1(\Omega)^n} \leq (1/\beta) \|p\|_{L^2(\Omega)}$$

by taking $\beta = 1/C$ where C is the constant in Lemma 11.2.3. If further $\int_{\Omega} p \, dx = 0$, then we may take $\underset{\sim}{w} \in \overset{\circ}{H}{}^1(\Omega)$, that is, we may assume $\underset{\sim}{w}$ vanishes on the boundary (cf. (Arnold, Scott & Vogelius 1988)). Thus, in either of the cases studied in the previous section, there is a $\underset{\sim}{w} \in V$ such that

$$\beta \|p\|_{L^2(\Omega)} = \beta \frac{b(\underset{\sim}{w}, p)}{\|p\|_{L^2(\Omega)}} \leq \frac{b(\underset{\sim}{w}, p)}{\left\|\underset{\sim}{w}\right\|_{H^1(\Omega)^n}} \leq \sqrt{n} \frac{b(\underset{\sim}{w}, p)}{\left\|\underset{\sim}{w}\right\|_{H(\text{div})}},$$

and this verifies condition (12.2.10) for both cases.

12.3 Discrete Mixed Formulation

Now let $V_h \subset V$ and $\Pi_h \subset \Pi$ and consider the variational problem to find $u_h \in V_h$ and $p_h \in \Pi_h$ such that

(12.3.1)
$$a(u_h, v) + b(v, p_h) = F(v) \quad \forall v \in V_h,$$
$$b(u_h, q) = 0 \quad \forall q \in \Pi_h.$$

(The case of an inhomogeneous right-hand side in the second equation is considered in Sect. 12.5.) Similarly, define

(12.3.2) $\qquad Z_h = \{v \in V_h \; : \; b(v, q) = 0 \quad \forall q \in \Pi_h\}.$

Then (12.3.1) is equivalent to the following. Find $u_h \in Z_h$ such that

$$(12.3.3) \qquad a(u_h, v) = F(v) \quad \forall v \in Z_h,$$

and then determine $p_h \in \Pi_h$ such that

$$(12.3.4) \qquad b(v, p_h) = -a(u_h, v) + F(v) \quad \forall v \in V_h.$$

If $Z_h \subset Z$, we can apply Céa's Theorem 2.8.1 to obtain

$$(12.3.5) \qquad \|u - u_h\|_V \leq \frac{C}{\alpha} \inf_{v \in Z_h} \|u - v\|_V.$$

If $Z_h \not\subset Z$, then we have a variational crime and must apply the theory developed in Sect. 10.1. We find from Lemma 10.1.1 that

$$\|u - u_h\|_V \leq \left(1 + \frac{C}{\alpha}\right) \inf_{v \in Z_h} \|u - v\|_V + \frac{1}{\alpha} \sup_{w \in Z_h \setminus \{0\}} \frac{|a(u - u_h, w)|}{\|w\|_V}$$

provided (12.2.7) holds as well on Z_h. We now identify the latter term. For $w \in Z_h$

$$(12.3.6) \qquad \begin{aligned} a(u - u_h, w) &= a(u, w) - F(w) &&\text{(by 12.3.3)} \\ &= -b(w, p) &&\text{(by 12.2.8 since } w \text{ is in } V) \\ &= -b(w, p - q) \quad \forall q \in \Pi_h. &&(w \text{ is in } Z_h) \end{aligned}$$

But inequality (12.2.1) implies

$$|b(w, p - q)| \leq C\|w\|_V \|p - q\|_\Pi.$$

Since q was arbitrary, we find

$$|a(u - u_h, w)| \leq C\|w\|_V \inf_{q \in \Pi_h} \|p - q\|_\Pi.$$

Thus, we have proved the following result (cf. (Brezzi 1974, Remark 2.1)).

(12.3.7) Theorem. *Let $V_h \subset V$ and $\Pi_h \subset \Pi$, and define Z and Z_h by (12.2.5) and (12.3.2), respectively. Suppose that (12.2.7) holds for all $z \in Z \cup Z_h$. Let u and p be determined by (12.2.8), and let u_h be determined equivalently by (12.3.1) or (12.3.3). Then*

$$\|u - u_h\|_V \leq \left(1 + \frac{C}{\alpha}\right) \inf_{v \in Z_h} \|u - v\|_V + \frac{C}{\alpha} \inf_{q \in \Pi_h} \|p - q\|_\Pi$$

where C is given in (12.2.1).

The main point of this theorem is that the error $u - u_h$ depends only on approximability of the spaces Z_h and Π_h and the coercivity condition (12.2.7). Bounds regarding p_h require more, in fact p_h may not even be stably determined. We note that the approximation properties of Z_h may

not be very good, but nevertheless, u_h is at least stably determined if
(12.2.7) holds.

12.4 Convergence Results for Velocity Approximation

As the first application of the above theory, we consider families of spaces for
approximating the Stokes equations (12.1.1) in two dimensions. Condition
(12.2.7) holds for the form (12.1.4) for all $\underset{\sim}{z} \in \underset{\sim}{V}$, so we only need to prove
approximability.

Let V_h^k denote C^0 piecewise polynomials of degree k on a non-
degenerate triangulation of a polygonal domain $\Omega \subset \mathbb{R}^2$ of maximum triangle diameter h. Let

$$(12.4.1) \qquad \underset{\sim}{V}_h = \{\underset{\sim}{v} \in V_h^k \times V_h^k \ : \ \underset{\sim}{v} = 0 \text{ on } \partial\Omega\}$$

and let Π_h be any subset of $\Pi = \{q \in L^2(\Omega) \ : \ \int_\Omega q(x)\,dx = 0\}$ which
satisfies

$$(12.4.2) \qquad \inf_{q \in \Pi_h} \|p - q\|_{L^2(\Omega)} \leq Ch^s\|p\|_{H^s(\Omega)}, \quad \forall p \in \Pi \cap H^s(\Omega)$$

for all $0 \leq s \leq k$. One family of such spaces is the Taylor-Hood family of
pressure spaces (Brezzi & Falk 1991), as described in the following result.

(12.4.3) Lemma. *Condition* (12.4.2) *holds for*

$$(12.4.4) \qquad \Pi_h = \left\{q \in V_h^{k-1} \ : \ \int_\Omega q(x)\,dx = 0\right\}.$$

Proof. From Corollary 4.4.24 (also see Theorem 4.8.12), we have

$$\inf_{q \in V_h^{k-1}} \|p - q\|_{L^2(\Omega)} \leq Ch^s\|p\|_{H^s(\Omega)}, \ 0 \leq s \leq k.$$

The infimum is achieved by $P_{V_h^{k-1}}p$, the $L^2(\Omega)$ projection of p onto V_h^{k-1}.
Since constant functions are contained in V_h^{k-1}, we have $\int_\Omega P_{V_h^{k-1}}p(x)\,dx = \int_\Omega p(x)\,dx = 0$, so that $P_{V_h^{k-1}}p \in \Pi_h$. $\square$

(12.4.5) Theorem. *Let* $\underset{\sim}{V}_h$ *be as given in* (12.4.1) *and let* Π_h *satisfy* (12.4.2).
Let $\underset{\sim}{u}$ *and* $\underset{\sim}{u}_h$ *be as in Theorem* 12.3.7. *Suppose that* $k \geq 4$. *Then for* $0 \leq s \leq k$

$$\|\underset{\sim}{u} - \underset{\sim}{u}_h\|_{H^1(\Omega)^2} \leq Ch^s \left(\|\underset{\sim}{u}\|_{H^{s+1}(\Omega)^2} + \|p\|_{H^s(\Omega)}\right)$$

provided the solution $(\underset{\sim}{u}, p)$ *to* (12.1.6) *satisfies* $(\underset{\sim}{u}, p) \in H^{s+1}(\Omega)^2 \times H^s(\Omega)$.

Proof. In view of (12.4.2) and Theorem 12.3.7, we only need to show that

(12.4.6) $\inf\limits_{\substack{v \in Z_h}} \|u - v\|_{H^1(\Omega)^2} \le Ch^s\|u\|_{H^{s+1}(\Omega)^2}$.

Observe that $u \in Z$ implies that $u = \operatorname{curl}\psi$ for $\psi \in H^{s+2}(\Omega) \cap \mathring{H}^2(\Omega)$ (Arnold, Scott & Vogelius 1988), where curl was defined in Sect. 10.1. Using the Argyris element ((3.2.10) and following) on $\mathcal{T}^h$ (also see (Morgan & Scott 1975)), there is a C^1 piecewise polynomial, $\psi_h \in \mathring{H}^2(\Omega)$, of degree $s + 1$ such that

$$\|\psi - \psi_h\|_{H^2(\Omega)} \le Ch^s|\psi|_{H^{s+2}(\Omega)} .$$

Since $\operatorname{curl}\psi_h \in Z_h$ ($V_h \cap Z \subset Z_h$ always), we have completed the proof. Note that $|\psi|_{H^{s+2}(\Omega)^2} \le C|u|_{H^{s+1}(\Omega)^2}$ since $u = (\psi_x, -\psi_y)$.   □

(12.4.7) Remark. In the next section, we prove the result for $k = 2$. The case $k = 3$ is treated by (Brezzi & Falk 1991).

To approximate the scalar elliptic problem (12.1.7) by a mixed method, we have to contend with the fact that the corresponding form $a(\cdot, \cdot)$ is not coercive on all of V, as it was in the case of the Stokes problem. It is clearly coercive on the space

$$Z = \{v \in H(\operatorname{div}) \; : \; \operatorname{div} v = 0\}$$

so that (12.2.6) is well-posed. However, some care is required to assure that it is well-posed as well on Z_h, as given in (12.3.2). One simple solution is to insure that $Z_h \subset Z$ and we will present one way this can be done.

Returning to the general notation of the previous section, we note that Π_h is naturally paired with $\mathcal{D}V_h$ (in both of the examples studied so far, $\mathcal{D} = -\operatorname{div}$). If we take $\Pi_h = \mathcal{D}V_h$, then the definition (12.3.2) of Z_h guarantees $Z_h \subset Z$, and this, in turn, guarantees coercivity. For example, we could take $V_h = V_h^k \times V_h^k$ (cf. (12.4.1)), and the proof of (12.4.6) shows that

(12.4.8) $\inf\limits_{\substack{v \in Z_h}} \|u - v\|_{H^1(\Omega)^2} \le Ch^k\|u\|_{H^{k+1}(\Omega)^2}$,

since this holds with Z_h replaced by $V_h \cap \{v \in H^1(\Omega)^2 \; : \; \operatorname{div} v = 0\}$, and the latter is a subset of Z_h. In the next chapter we will study algorithms that allow one to compute using $\Pi_h = \mathcal{D}V_h$ without having explicit information about the structure of Π_h. As a corollary to (12.3.5) and (12.4.8), we have the following result.

(12.4.9) Theorem. *Consider the mixed-method for the scalar elliptic problem (12.1.7) for $n = 2$, where the form $a(\cdot, \cdot)$ is given in (12.1.8). Let (cf. (12.4.1))*

$$V_h = V_h^k \times V_h^k$$

and let $\Pi_h = \operatorname{div} V_h$. Suppose that $k \ge 4$. Then for $0 \le s \le k$

$$\left\|\underset{\approx}{a}\underset{\sim}{\mathrm{grad}}\, p - \underset{\sim h}{u}\right\|_{L^2(\Omega)^2} \le Ch^{s+1}\left\|\underset{\approx}{a}\underset{\sim}{\mathrm{grad}}\, p\right\|_{H^{s+1}(\Omega)^2}$$

provided the solution p to (12.1.7) satisfies $\underset{\approx}{a}\underset{\sim}{\mathrm{grad}}\, p \in H^{s+1}(\Omega)$.

Similarly, the following result is a consequence of (12.3.5) and (12.4.6).

(12.4.10) Theorem. Let V_h be as given in (12.4.1) and let $\Pi_h = \mathrm{div}\, V_h$. Let $\underset{\sim}{u}$ and $\underset{\sim h}{u}$ be as in Theorem 12.3.7. Suppose that $k \ge 4$. Then for $0 \le s \le k$

$$\left\|\underset{\sim}{u} - \underset{\sim h}{u}\right\|_{H^1(\Omega)^2} \le Ch^s\left\|\underset{\sim}{u}\right\|_{H^{s+1}(\Omega)^2}$$

provided the solution $\underset{\sim}{u}$ to (12.1.6) satisfies $\underset{\sim}{u} \in H^{s+1}(\Omega)^2$.

The above theory can also be developed for nonconforming finite element approximation of the Stokes equations. Define (cf. (11.4.6))

$$a_h(\underset{\sim}{u}, \underset{\sim}{v}) := \sum_{T \in \mathcal{T}^h} \int_T \underset{\approx}{\mathrm{grad}_h}\underset{\sim}{v} : \underset{\approx}{\mathrm{grad}_h}\underset{\sim}{w}\, dx$$

$$b_h(\underset{\sim}{v}, q) := -\sum_{T \in \mathcal{T}^h} \int_T \mathrm{div}_h\underset{\sim}{v}\, q\, dx$$

and let $\|\underset{\sim}{v}\|_h := \sqrt{a_h(\underset{\sim}{v}, \underset{\sim}{v})}$. Let V_h^* be the space defined in (11.4.5), that is, $V_h^* := V_h \times V_h$ where V_h is the space defined in (10.3.2). Let Π_h denote piecewise constant functions, q, on $\mathcal{T}^h$ satisfying $\int_\Omega q\, dx = 0$. Define $\underset{\sim h}{u} \in V_h^*$ and $p_h \in \Pi_h$ by

$$(12.4.11)\qquad \begin{aligned} a_h(\underset{\sim h}{u}, \underset{\sim}{v}) + b_h(\underset{\sim}{v}, p_h) &= \int_\Omega \underset{\sim}{f} \cdot \underset{\sim}{v}\, dx \quad \forall \underset{\sim}{v} \in V_h^*, \\ b_h(\underset{\sim h}{u}, q) &= 0 \quad \forall q \in \Pi_h. \end{aligned}$$

(12.4.12) Theorem. If $\underset{\sim}{u} \in H^2(\Omega)^2$ and $p \in H^1(\Omega)$, then

$$\left\|\underset{\sim}{u} - \underset{\sim h}{u}\right\|_h \le Ch\left(\left\|\underset{\sim}{u}\right\|_{H^2(\Omega)^2} + \|p\|_{H^1(\Omega)}\right).$$

Proof. Let $Z_h = \{\underset{\sim}{v} \in V_h^* : b_h(\underset{\sim}{v}, q) = 0 \ \forall q \in \Pi_h\}$. From (10.1.10) we have

$$\|\underset{\sim}{u} - \underset{\sim h}{u}\|_h \le \inf_{\underset{\sim}{v} \in Z_h} \|\underset{\sim}{u} - \underset{\sim}{v}\|_h + \sup_{\underset{\sim}{w} \in Z_h \backslash \{0\}} \frac{|a_h(\underset{\sim}{u} - \underset{\sim h}{u}, \underset{\sim}{w})|}{\|\underset{\sim}{w}\|_h}.$$

Note that $a_h(\cdot, \cdot)$ is coercive on $V = H^1(\Omega)^2$ since $a_h(\cdot, \cdot) = a(\cdot, \cdot)$ on V. It is non-degenerate on V_h^* since $a_h(\underset{\sim}{v}, \underset{\sim}{v}) = 0$ implies $\underset{\sim}{v}$ is piecewise constant, and the zero boundary condition together with continuity at midpoints imply $\underset{\sim}{v} \equiv \underset{\sim}{0}$. Then

$$a_h(\underset{\sim}{u} - \underset{\sim}{u}_h, \underset{\sim}{w}) = a_h(\underset{\sim}{u}, \underset{\sim}{w}) - \int_\Omega \underset{\sim}{f} \cdot \underset{\sim}{w}\, dx \qquad \text{(by 12.4.11)}$$

$$= a_h(\underset{\sim}{u}, \underset{\sim}{w}) - \int_\Omega (-\Delta\underset{\sim}{u} + \operatorname{grad} p) \cdot \underset{\sim}{w}\, dx \qquad \text{(by 12.1.1)}$$

$$= \sum_e \int_e \sum_{i=1,2} \frac{\partial u_i}{\partial \nu} [w_i] - p[\underset{\sim}{w} \cdot \underset{\sim}{\nu}]\, ds - b_h(\underset{\sim}{w}, p)$$

after integrating by parts, where for simplicity of notation we make the convention that $\underset{\sim}{w}$ is defined to be zero outside Ω, so that the "jump" $[\underset{\sim}{w}]$ is the same as $\underset{\sim}{w}$ on the boundary edges. From Lemmas 10.3.7 and 10.3.9, we find

$$\left| \sum_e \int_e \sum_{i=1,2} \frac{\partial u_i}{\partial \nu} [w_i] - p[\underset{\sim}{w} \cdot \underset{\sim}{\nu}]\, ds \right| \leq Ch \left(\|\underset{\sim}{u}\|_{H^2(\Omega)^2} + \|p\|_{H^1(\Omega)} \right) \|\underset{\sim}{w}\|_h.$$

Following the derivation of Theorem 12.3.7, we find that

$$|b_h(\underset{\sim}{w}, p)| \leq \sqrt{2} \|\underset{\sim}{w}\|_h \inf_{q \in \Pi_h} \|p - q\|_{L^2(\Omega)}$$

$$\leq Ch \|\underset{\sim}{w}\|_h \|p\|_{H^1(\Omega)}$$

for all $\underset{\sim}{w} \in Z_h$. Thus, we only need to show that

$$\inf_{\underset{\sim}{v} \in Z_h} \|\underset{\sim}{u} - \underset{\sim}{v}\|_h \leq Ch \|\underset{\sim}{u}\|_{H^2(\Omega)^2}.$$

Let M_h denote the Morley space depicted in Fig. 10.5. Then $\operatorname{curl} M_h \subset Z_h$ (exercise 12.x.11). Choosing $\underset{\sim}{v} = \operatorname{curl} \mathcal{I}^h \phi$ where $\underset{\sim}{u} = \operatorname{curl} \phi$ (as in the proof of Theorem 12.4.5) completes the proof.   □

12.5 The Discrete Inf-Sup Condition

In the previous section, we saw that error estimates could be derived for the (velocity) error, $u - u_h$, in terms of approximation properties of the spaces Z_h and Π_h. We now consider the well-posedness of the problem for p_h, (12.3.4). As a by-product, we will simplify the approximation problem for Z_h, reducing it to one for V_h. In Theorems 12.4.5, 12.4.9 and 12.4.12, the approximation problem for Z_h was easy to resolve, but in other instances it is far more complex. We show that a counterpart of (12.2.10) restricted to (V_h, Π_h) is crucial both to approximability for Z_h and to solvability for p_h, starting with the latter.

(12.5.1) Lemma. (Inf-Sup Condition) *In order for* (12.3.4) *to have a unique solution, it is necessary and sufficient that*

$$(12.5.2) \qquad\qquad 0 < \beta := \inf_{q \in \Pi_h} \sup_{v \in V_h} \frac{|b(v,q)|}{\|v\|_V \|q\|_\Pi}.$$

Proof. If $\beta = 0$, then the finite-dimensionality of Π_h would imply that there is a $q \in \Pi_h$ such that $b(v,q) = 0$ for all $v \in V_h$ (exercise 12.x.7). This proves the necessity of the inf-sup condition.

On the other hand, $\beta > 0$ implies uniqueness for (12.3.4) since (12.5.2) is equivalent to

$$(12.5.3) \qquad\qquad \beta \|q\|_\Pi \le \sup_{v \in V_h} \frac{|b(v,q)|}{\|v\|_V} \quad \forall q \in \Pi_h.$$

We now consider the solvability of (12.3.4).

The right-hand side of (12.3.4) vanishes for all $v \in Z_h$. Thus, (12.3.4) is equivalent to

$$(12.5.4) \qquad\qquad b(v, p_h) = -a(u_h, v) + F(v) \quad \forall v \in Z_h^\perp$$

where

$$(12.5.5) \qquad Z_h^\perp := \{v \in V_h \; : \; (v,z)_V = 0 \quad \forall z \in Z_h\}.$$

Once we see that $\dim Z_h^\perp = \dim \Pi_h$, then (12.5.4) represents a square system, and uniqueness implies existence. But we may view Z_h as the kernel of the mapping

$$T : V_h \to \mathbb{R}^{\dim \Pi_h} \text{ where } (Tv)_i := b(v, q_i)$$

and $\{q_i \; : \; i = 1, \dots, \dim \Pi_h\}$ is a basis for Π_h. If T were not onto, then there would be a nontrivial vector of coefficients, (c_i), such that

$$\sum_{i=1}^{\dim \Pi_h} c_i \, (Tv)_i = 0$$

for all $v \in V_h$. Define $0 \ne q \in \Pi_h$ by $q = \sum_{i=1}^{\dim \Pi_h} c_i q_i$. Then $b(v,q) = 0$ for all $v \in V_h$, contradicting $\beta > 0$. Thus, T must be onto, and consequently

$$\dim Z_h = \dim \text{ kernel } T = \dim V_h - \dim \text{ image } T = \dim V_h - \dim \Pi_h.$$

Of course, $\dim V_h = \dim Z_h + \dim Z_h^\perp$ since $V_h = Z_h \oplus Z_h^\perp$. Therefore, $\dim Z_h^\perp = \dim V_h - \dim Z_h = \dim \Pi_h$. □

Let us introduce a solution operator related to the problem (12.3.4). Define $M : Z_h^\perp \to \Pi_h$ by

$$(12.5.6) \qquad b(v, Mu) = (u, v)_V \quad \forall v \in V_h.$$

It is easy to see (exercise 12.x.8) that

$$\|M\|_{V \to \Pi} := \sup_{0 \neq u \in V_h} \frac{\|Mu\|_\Pi}{\|u\|_V} \leq \frac{1}{\beta}$$

where β is the constant in (12.5.2).

There is an operator $L : \Pi_h \to Z_h^\perp$ (see (12.5.5)) defined by

$$(12.5.7) \qquad b(Lp, q) = (p, q)_\Pi \quad \forall q \in \Pi_h$$

since (12.5.7) represents a square system (see the proof of Lemma 12.5.1), and uniqueness is guaranteed by the fact that $Z_h \cap Z_h^\perp = \{0\}$. L is adjoint to M in the sense that

$$(12.5.8) \qquad (p, Mu)_\Pi = b(Lp, Mu) = (u, Lp)_V \quad \forall u \in V_h, \; p \in \Pi_h.$$

The norms of M and L are related by

$$\begin{aligned}
\|L\|_{\Pi \to V} &:= \sup_{0 \neq p \in \Pi_h} \frac{\|Lp\|_V}{\|p\|_\Pi} \\
&= \sup_{0 \neq p \in \Pi_h} \sup_{0 \neq u \in V_h} \frac{(u, Lp)_V}{\|u\|_V \|p\|_\Pi} \quad \text{(exercise 2.x.16)} \\
&= \sup_{0 \neq p \in \Pi_h} \sup_{0 \neq u \in V_h} \frac{(Mu, p)_\Pi}{\|u\|_V \|p\|_\Pi} \quad \text{(by 12.5.8)} \\
&= \sup_{0 \neq u \in V_h} \frac{\|Mu\|_\Pi}{\|u\|_V} \quad \text{(exercise 2.x.16)} \\
&= \|M\|_{V \to \Pi} \, .
\end{aligned}$$

We can characterize Lp as the minimum-norm solution to the underdetermined problem to find $v \in V_h$ such that

$$b(v, q) = (p, q)_\Pi \quad \forall q \in \Pi_h.$$

More precisely, $Lp \in Z_h^\perp$ implies (see exercise 12.x.9) that

$$(12.5.9) \qquad \|Lp\|_V = \min_{v \in Z_h^p} \|v\|_V$$

where $Z_h^p := \{v \in V_h \; : \; b(v, q) = (p, q)_\Pi \; \forall q \in \Pi_h\}$. Therefore,

$$\begin{aligned}
\frac{1}{\|L\|_{\Pi \to V}} &= \inf_{0 \neq p \in \Pi_h} \frac{\|p\|_\Pi}{\|Lp\|_V} \\
&= \inf_{0 \neq p \in \Pi_h} \sup_{0 \neq v \in Z_h^p} \frac{\|p\|_\Pi}{\|v\|_V} \quad \text{(by 12.5.9)}
\end{aligned}$$

$$= \inf_{0 \neq p \in \Pi_h} \sup_{0 \neq v \in Z_h^p} \frac{(p, p)_\Pi}{\|p\|_\Pi \|v\|_V}$$

$$= \inf_{0 \neq p \in \Pi_h} \sup_{0 \neq v \in Z_h^p} \frac{b(v, p)}{\|p\|_\Pi \|v\|_V} \quad \text{(definition of } Z_h^p\text{)}$$

$$\leq \inf_{0 \neq p \in \Pi_h} \sup_{0 \neq v \in V_h} \frac{b(v, p)}{\|p\|_\Pi \|v\|_V} \quad (Z_h^p \subset V_h)$$

which implies that $\|L\|_{\Pi \to V} \geq 1/\beta$. We summarize the above results in the following.

(12.5.10) Lemma. *Condition* (12.5.2) *is equivalent to the existence of operators L and M, defined in* (12.5.7) *and* (12.5.6) *respectively, that satisfy* $\|L\|_{\Pi \to V} = \|M\|_{V \to \Pi} = 1/\beta$.

Proof. The existence of L and M together with evaluation of their norms has just been demonstrated. The converse is left to the reader in exercise 12.x.10.   □

Error estimates for $p - p_h$ are derived as follows. By subtracting the first equations in each of (12.2.8) and (12.3.1) we find the relation

$$(12.5.11) \qquad b(v, p - p_h) = -a(u - u_h, v) \quad \forall v \in V_h.$$

For any $q \in \Pi_h$, we find

$$\beta \|q - p_h\|_\Pi \leq \sup_{v \in V_h} \frac{|b(v, q - p_h)|}{\|v\|_V} \qquad \text{(by 12.5.3)}$$

$$= \sup_{v \in V_h} \frac{|b(v, p - p_h) + b(v, q - p)|}{\|v\|_V}$$

$$= \sup_{v \in V_h} \frac{|-a(u - u_h, v) + b(v, q - p)|}{\|v\|_V} \qquad \text{(by 12.5.11)}$$

$$\leq C\big(\|u - u_h\|_V + \|q - p\|_\Pi\big). \qquad \text{(by 12.2.1)}$$

The following result is then a consequence of the triangle inequality.

(12.5.12) Theorem. *Let $V_h \subset V$ and $\Pi_h \subset \Pi$. Let u and p be determined equivalently by* (12.2.8) *or* (12.2.6) *and* (12.2.9), *and let u_h be determined equivalently by* (12.3.1) *or* (12.3.3). *Suppose that $\beta > 0$ in* (12.5.2). *Then there is a unique solution, p_h, to* (12.3.4) *which satisfies*

$$\|p - p_h\|_\Pi \leq \frac{C}{\beta} \|u - u_h\|_V + \left(1 + \frac{C}{\beta}\right) \inf_{q \in \Pi_h} \|p - q\|_\Pi$$

where C is given in (12.2.1).

Combining Theorems 12.5.12 and 12.3.7, we obtain the following.

(12.5.13) Corollary. *Under the conditions of Theorems 12.5.12 and 12.3.7,*

$$\|p - p_h\|_\Pi \le c \inf_{v \in Z_h} \|u - v\|_V + (1 + c) \inf_{q \in \Pi_h} \|p - q\|_\Pi,$$

where $c := \frac{C}{\beta}\left(1 + \frac{C}{\alpha}\right)$.

We now show how (12.5.2) is involved in the approximability of Z_h. In doing so, we return to the full problem (12.2.4), and we consider the approximation of it by the following.

> For $V_h \subset V$ and $\Pi_h \subset \Pi$, and $F \in V'$ and $G \in \Pi'$,
> find $u_h \in V_h$ and $p_h \in \Pi_h$ such that

(12.5.14)
$$a(u_h, v) + b(v, p_h) = F(v) \quad \forall v \in V_h$$
$$b(u_h, q) = G(q) \quad \forall q \in \Pi_h.$$

Observe that u_h lies in the affine set

(12.5.15)
$$Z_h^G := \left\{ v \in V_h \ : \ b(v, q) = G(q) \quad \forall q \in \Pi_h \right\}.$$

In particular, $Z_h^0 = Z_h$. Similarly, we can define

(12.5.16)
$$Z^G := \left\{ v \in V \ : \ b(v, q) = G(q) \quad \forall q \in \Pi \right\}.$$

If (u, p) denotes the solution to (12.2.4), then $u \in Z^G$, and $Z^0 = Z$.

We now study the relation between approximation of u from Z_h^G and the full space V_h. For any $v \in V_h$, let $w \in V_h$ satisfy

$$b(w, q) = b(u - v, q) \quad \forall q \in \Pi_h.$$

Thus, $w = LP_{\Pi_h}\mathcal{D}(u - v)$ (see 12.5.7), and Lemma 12.5.10 and (12.2.2) imply

$$\|w\|_V \le \frac{1}{\beta}\|\mathcal{D}(u - v)\|_\Pi \le \frac{C}{\beta}\|u - v\|_V.$$

By the definition of w we have $v + w \in Z_h^G$, provided $u \in Z^G$. Moreover,

$$\|u - (v + w)\|_V \le \|u - v\|_V + \|w\|_V$$
$$\le \left(1 + \frac{C}{\beta}\right)\|u - v\|_V.$$

Thus, we have proved the following.

(12.5.17) Theorem. *Let $V_h \subset V$ and $\Pi_h \subset \Pi$, and define Z^G and Z_h^G by (12.5.16) and (12.5.15), respectively. Suppose that $\beta > 0$ in (12.5.2). Then for all $u \in Z^G$*

$$\inf_{z \in Z_h^G} \|u - z\|_V \le \left(1 + \frac{C}{\beta}\right) \inf_{v \in V_h} \|u - v\|_V.$$

(12.5.18) Corollary. *Let (u, p) denote the solution to (12.2.4), and let (u_h, p_h) denote the solution to (12.5.14). There is a constant c depending only on the constants C in (12.2.1), α in (12.2.7) and β in (12.5.2) such that*

$$\|u - u_h\|_V + \|p - p_h\|_\Pi \leq c\left(\inf_{v \in V_h} \|u - v\|_V + \inf_{q \in \Pi_h} \|p - q\|_\Pi \right).$$

Proof. Modifying the proof of Lemma 10.1.1 (cf. exercise 12.x.13), we find

(12.5.19)
$$\|u - u_h\|_V \leq \left(1 + \frac{C}{\alpha}\right) \inf_{v \in Z_h^G} \|u - v\|_V$$
$$+ \frac{1}{\alpha} \sup_{0 \neq w \in Z_h^0} \frac{|a(u - u_h, w)|}{\|w\|_V}.$$

From the proof of Theorem 12.3.7, we have

$$|a(u - u_h, w)| \leq C\|w\|_V \inf_{q \in \Pi_h} \|p - q\|_\Pi$$

for all $w \in Z_h^0$. Thus, the estimate for $u - u_h$ follows from Theorem 12.5.17. The estimate for $p - p_h$ follows from Theorem 12.5.12. $\qquad\square$

There is no universal way to verify (12.5.2), but there is one simple situation that we indicate here. Suppose that the following assumption holds.

(12.5.20)   There is an operator $T_h : V \to V_h$ with the properties that
$$\mathcal{D}T_h v = P_{\Pi_h} \mathcal{D}v \quad \text{and} \quad \|T_h v\|_V \leq B\|v\|_V \quad \forall v \in V,$$

for some constant $B < \infty$, where P_{Π_h} denotes the projection onto Π_h with respect to the inner product $(\cdot, \cdot)_\Pi$. Suppose further that $\mathcal{D}$ is boundedly invertible, that is,

(12.5.21)   for any $p \in \Pi$ there exists $v \in V$ such that
$$\mathcal{D}v = p \quad \text{and} \quad \|v\|_V \leq C\|p\|_\Pi.$$

Then for any $p \in \Pi_h$ we find

$$
\begin{aligned}
(p, p)_\Pi &= (p, \mathcal{D}v) & &\text{(from 12.5.21)}\\
&= (p, P_{\Pi_h} \mathcal{D}v) & &\text{(since } p \in \Pi_h\text{)}\\
&= (p, \mathcal{D}T_h v) & &\text{(from 12.5.20)}\\
&= b(T_h v, p). & &\text{(from 12.2.3)}
\end{aligned}
$$

Therefore,

$$\|p\|_{\Pi} = \frac{b(T_h v, p)}{\|p\|_{\Pi}}$$

$$\leq C \frac{b(T_h v, p)}{\|v\|_V} \qquad \text{(by 12.5.21)}$$

$$\leq C B \frac{b(T_h v, p)}{\|T_h v\|_V}. \qquad \text{(by 12.5.20)}$$

Thus, we have proved the following result.

(12.5.22) Lemma. *Suppose conditions* (12.5.20) *and* (12.5.21) *hold. Then* (12.5.2) *holds with* $\beta = \frac{1}{CB}$.

In the next section, we give an example of the construction of the required operator T_h. Such a construction is not obvious in the general case, but we now indicate that (12.5.2) implies such a T_h must always exist. Given $u \in V$, define $u_h = T_h u$ by solving

$$a(u_h, w) + b(v, p_h) = a(u, v) \quad \forall v \in V_h$$
$$b(u_h, q) = b(u, q) \quad \forall q \in \Pi_h,$$

which corresponds to the approximation of (12.2.4) with $F(v) := a(u, v)$ and $G(q) := b(u, q)$. The second equation is simply the statement that $\mathcal{D}u_h = P_{\Pi_h} \mathcal{D}u$ in view of (12.2.3). The solution to (12.2.4) with this data is of course the pair $(u, 0)$, so Corollary 12.5.18 implies that

$$\|u - u_h\|_V \leq c \|u\|_V$$

so that (12.5.20) holds with $B = 1 + c$ by the triangle inequality. Therefore, we have proved the following theorem.

(12.5.23) Theorem. *Suppose that* (12.5.21) *holds. Then conditions* (12.5.2) *and* (12.5.20) *are equivalent.*

12.6 Verification of the Inf-Sup Condition

We now derive (12.5.2) for the particular case of the Taylor-Hood spaces, that is, V_h given by (12.4.1) and Π_h as in (12.4.4), for approximating the Stokes equations. We limit our discussion to the case $k = 2$. We use a general technique that reduces the problem to local estimates.
    Let $p \in \Pi_h$. Since $\Pi_h \subset H^1(\Omega)$ in this case, we have

$$b(\underset{\sim}{v}, p) = \int_{\Omega} \underset{\sim}{v} \cdot \underset{\sim}{\text{grad}}\, p \, dx.$$

For each edge e of a triangle in $\mathcal{T}^h$, let τ_e denote a unit vector tangential to e, ν_e denote a unit normal and m_e denote the edge midpoint. For each interior edge, choose

$$\underline{v}(m_e) \cdot \underline{\tau}_e = \left(|T_{e,1}| + |T_{e,2}|\right)\left[\underline{\tau}_e \cdot \operatorname{grad} p(m_e)\right],$$

$$\underline{v}(m_e) \cdot \underline{\nu}_e = 0,$$

where $T_{e,1}$ and $T_{e,2}$ are the two triangles in $\mathcal{T}^h$ having e as a common edge. Note that $\underline{\tau}_e \cdot \operatorname{grad} p(m_e)$ is unambiguous, since the tangential component of $\operatorname{grad} p$ is continuous across the edge. We set $\underline{v} = \underline{0}$ at all vertices and all boundary edge midpoints. Let r_T denote the number of interior edges of $T \in \mathcal{T}^h$, that is, $r_T = 3$ if T is in the interior and $r_T = 2$ if one edge lies on $\partial\Omega$. We assume that no triangle in $\mathcal{T}^h$ has two edges on $\partial\Omega$. For a given $T \in \mathcal{T}^h$, let e^i $(1 \le i \le r_T)$ denote the interior edges. Then we have

$$\int_T \underline{v} \cdot \operatorname{grad} p \, dx = \frac{|T|}{3} \sum_{i=1}^{r_T} (\underline{v} \cdot \operatorname{grad} p)(m_{e^i}) \qquad \text{(from exercise 6.x.11)}$$

$$= \frac{|T|}{3} \sum_{i=1}^{r_T} \left(\sum_{j=1}^{2} |T_{e^i,j}| \, \big| \, \underline{\tau}_{e^i} \cdot \operatorname{grad} p(m_{e^i}) \, \big|^2 \right) \qquad \text{(definition of } \underline{v})$$

$$= \frac{|T|}{3} \sum_{i=1}^{r_T} \left(\sum_{j=1}^{2} |T_{e^i,j}| \, \big| \, \underline{\tau}_{e^i} \cdot \left(\operatorname{grad} p|_T\right) \big|^2 \right) \qquad \text{(grad } p \text{ is constant on } T)$$

$$\ge |T|^2 \, c_\rho \, \big| \left(\operatorname{grad} p|_T\right) \big|^2 \qquad \text{(} T \text{ is non-degenerate)}$$

$$= c_\rho |T| \int_T \big| \operatorname{grad} p \big|^2 \, dx.$$

Summing over $T \in \mathcal{T}^h$ yields

(12.6.1) $$b(\underline{v}, p) \ge c_\rho \sum_{T \in \mathcal{T}^h} |T| \, |p|^2_{H^1(T)}.$$

The definition of $\underline{v}$ also implies (cf. exercise 12.x.15)

(12.6.2) $$\|\underline{v}\|_{H^1(\Omega)^2} \le C \left[\sum_{T \in \mathcal{T}^h} |T| \, |p|^2_{H^1(T)} \right]^{1/2}.$$

Let $K > 0$ be an arbitrary constant (to be chosen later), and define

$$\Pi_h^K = \left\{ q \in \Pi_h : \|q\|_{L^2(\Omega)} \le K \left[\sum_{T \in \mathcal{T}^h} |T| \, |q|^2_{H^1(T)} \right]^{1/2} \right\}.$$

For $p \in \Pi_h^K$ and with $\underline{v}$ defined as above,

$$\frac{b(\underset{\sim}{v},p)}{\|\underset{\sim}{v}\|_{H^1(\Omega)^2}} \geq c_\rho \frac{\sum_{T \in \mathcal{T}^h} |T| \, |p|_{H^1(T)}^2}{\|\underset{\sim}{v}\|_{H^1(\Omega)^2}} \qquad \text{(by 12.6.1)}$$

$$\geq \frac{c_\rho}{K} \frac{\left[\sum_{T \in \mathcal{T}^h} |T| \, |p|_{H^1(T)}^2\right]^{1/2} \|p\|_{L^2(\Omega)}}{\|\underset{\sim}{v}\|_{H^1(\Omega)^2}} \qquad (p \in \Pi_h^K)$$

$$\geq \frac{c_\rho}{CK} \|p\|_{L^2(\Omega)} . \qquad \text{(by 12.6.2)}$$

This proves (12.5.3) for $p \in \Pi_h^K$ for a given K.

For $p \notin \Pi_h^K$, we proceed as follows. From Lemma 11.2.3 or (Girault & Raviart 1986, Arnold, Scott & Vogelius 1988) we may pick $\underset{\sim}{u} \in \mathring{H}^1(\Omega)^2$ such that

(12.6.3)
$$-\text{div}\,\underset{\sim}{u} = p \text{ in } \Omega$$
$$\|\underset{\sim}{u}\|_{H^1(\Omega)^2} \leq \tilde{C} \, \|p\|_{L^2(\Omega)} .$$

Let $\underset{\sim}{u}_h \in \underset{\sim}{V}_h$ satisfy

(12.6.4) $$\|\underset{\sim}{u}_h\|_{H^1(\Omega)^2} + \left[\sum_{t \in \mathcal{T}^h} |T|^{-1} \|\underset{\sim}{u} - \underset{\sim}{u}_h\|_{L^2(T)^2}^2\right]^{1/2} \leq C^* \|\underset{\sim}{u}\|_{H^1(\Omega)^2}$$

(see Sect. 4.8 for a construction). Then

$$b(\underset{\sim}{u}_h,p) = \|p\|_{L^2(\Omega)}^2 - b(\underset{\sim}{u} - \underset{\sim}{u}_h, p) \qquad \text{(since } - \text{div}\,\underset{\sim}{u} = p)$$

$$\geq \|p\|_{L^2(\Omega)}^2 - \left[\sum_{T \in \mathcal{T}^h} |T|^{-1} \|\underset{\sim}{u} - \underset{\sim}{u}_h\|_{L^2(T)^2}^2\right]^{1/2} \left[\sum_{T \in \mathcal{T}^h} |T| \, |p|_{H^1(T)}^2\right]^{1/2}$$
$$\text{(by 2.1.5)}$$

$$\geq \|p\|_{L^2(\Omega)}^2 - \tilde{C}^* C \|p\|_{L_2(\Omega)} \left[\sum_{T \in \mathcal{T}^h} |T| \, |p|_{H^1(T)}^2\right]^{1/2}$$
$$\text{(by 12.6.3 \& 12.6.4)}$$

$$\geq \|p\|_{L^2(\Omega)}^2 \left(1 - \frac{\tilde{C}C^*}{K}\right) \qquad \text{(since } p \notin \Pi_h^K)$$

$$\geq \|p\|_{L^2(\Omega)} \left(\frac{1}{\tilde{C}C^*} - \frac{1}{K}\right) \|\underset{\sim}{u}_h\|_{H^1(\Omega)^2} . \qquad \text{(by 12.6.3 \& 12.6.4)}$$

Thus, for K sufficiently large, we have proved (12.5.3) for $p \notin \Pi_h^K$.

Combining the two cases, we have proved the following.

(12.6.6) Theorem. *Suppose $\mathcal{T}^h$ is non-degenerate and has no triangle with two edges on $\partial\Omega$. Let $\underset{\sim}{V}_h$ be as in (12.4.1) and let Π_h be as in (12.4.4). Then condition (12.5.3), namely*

$$\beta\,\|q\|_{L^2(\Omega)} \leq \sup_{\underset{\sim}{v}\in\underset{\sim}{V}_h} \frac{|b(\underset{\sim}{v},q)|}{\|\underset{\sim}{v}\|_{H^1(\Omega)^2}} \quad \forall q \in \Pi_h,$$

(or the equivalent version 12.5.2) holds with $\beta > 0$ independent of h for the case $k = 2$.

As a consequence of this and Corollary 12.5.18, we have the following result.

(12.6.7) Theorem. *Suppose $\mathcal{T}^h$ is non-degenerate and has no triangle with two edges on $\partial\Omega$. Let $\underset{\sim}{V}_h$ be as in (12.4.1) and let Π_h be as in (12.4.4) for $k = 2$. Let $(\underset{\sim}{u},p)$ be the solution to (12.1.6). Let $(\underset{\sim}{u}_h,p_h)$ solve (12.3.1) with the forms given in (12.1.4) and (12.1.5). Then for $0 \leq s \leq 2$*

$$\|\underset{\sim}{u} - \underset{\sim}{u}_h\|_{H^1(\Omega)^2} + \|p - p_h\|_{L^2(\Omega)} \leq Ch^s \left(\|\underset{\sim}{u}\|_{H^{s+1}(\Omega)^2} + \|p\|_{H^s(\Omega)} \right)$$

provided $(\underset{\sim}{u},p) \in H^{s+1}(\Omega)^2 \times H^s(\Omega)$.

Theorems 12.6.6 and 12.6.7 hold for all $k \geq 2$ (cf. Brezzi & Falk 1991). The case $k \geq 4$ is a consequence of Theorem 12.6.10 below under some mild restrictions on the mesh. Similar techniques work for other choices of spaces $\underset{\sim}{V}_h$ and Π_h. The basic philosophy is to pick div $\underset{\sim}{v}$ to match p for "local" $p \in \Pi_h^K$ and then use the general argument (12.6.5) for $p \notin \Pi_h^K$ (Scott & Vogelius 1985a).

Because of the natural pairing of div $\underset{\sim}{V}_h$ with Π_h, it is interesting to ask whether the choice $\Pi_h = $ div $\underset{\sim}{V}_h$ would always satisfy (12.5.2). In Sect. 13.1, we show how (12.5.14) can be solved efficiently without the need for an explicit basis for $\Pi_h = $ div $\underset{\sim}{V}_h$, greatly simplifying the solution process. Since any $p \in \Pi_h$ is of the form $p = $ div $\underset{\sim}{w}$ for some $\underset{\sim}{w} \in \underset{\sim}{V}_h$, choosing $\underset{\sim}{v} = \underset{\sim}{w}$ in (12.5.2) leads to the conclusion that $\beta > 0$ for $\underset{\sim}{V}_h$ finite dimensional, by a compactness argument. However, what is more critical to ask is whether β may be chosen independently of h. This turns out to depend strongly on the degree k of piecewise polynomials.

Let $\underset{\sim}{V}_h$ denote the spaces defined in (12.4.1) for $k \geq 1$ and let $\Pi_h = $ div $\underset{\sim}{V}_h$. For verification of the following results, we refer to (Scott & Vogelius 1985b). For $k = 1$ on quite general meshes, the corresponding Z_h consists only of the function identically zero (cf. (11.3.8)). Comparison with Theorem 12.5.17 shows that β must tend to zero at least as fast as h, since the results of Chapter 4 yield

$$\|\underset{\sim}{u}\|_{H^1(\Omega)^2} = \inf_{\underset{\sim}{v}\in Z_h} \|\underset{\sim}{u} - \underset{\sim}{v}\|_{H^1(\Omega)^2} \leq \left(1 + \frac{1}{\beta}\right) Ch\|\underset{\sim}{u}\|_{H^2(\Omega)^2}.$$

Other examples where $\beta \to 0$ for $k = 2$ and 3 are given in (Scott & Vogelius 1985b). On the other hand, (Scott & Vogelius 1985a) showed that β may be chosen independent of h as soon as $k \geq 4$ under very mild restrictions on the mesh which we now recall.

An interior vertex in a triangulation at which four triangles meet is called *singular* if the corresponding edges that meet there lie on two straight lines. If we label the angles θ_j formed by consecutive edges at such a vertex, we can rephrase the condition by saying that $\theta_i + \theta_{i+1} = \pi$ for $i = 1, 2, 3$. Similarly, we say a vertex on the boundary is singular if $r \leq 4$ triangles meet there and $\theta_i + \theta_{i+1} = \pi$ for $i = 1, \ldots, r - 1$. (Four triangles can only meet at the vertex of a slit.) The case $r = 1$ is somewhat special and the condition must be modified; it occurs when one triangle has two edges on the boundary. In such a case, any piecewise polynomial vanishing on the boundary will have its gradient vanish at such a vertex. For this and other reasons, we will ban such vertices from our triangulations. With the exception of triangles with two edges on the boundary, singularity of vertices poses no problem. However, if a family of triangulations has a sequence of vertices which are nonsingular but tend toward being singular, there is a theoretical possibility of a deterioration of the constant β.

(12.6.8) Definition. *We say a family of triangulations $\mathcal{T}^h$ has no nearly singular vertices if there is a $\sigma > 0$ independent of h such that*

$$(12.6.9) \qquad \sum_{i=1}^{r-1} |\theta_i + \theta_{i+1} - \pi| \geq \sigma$$

holds for all nonsingular vertices in the interior where four triangles meet ($r = 4$ in this case) and all nonsingular vertices on the boundary where $2 \leq r \leq 4$ triangles meet.

The following is a consequence of the results in (Scott & Vogelius 1985a). In the case $V_h = V_h^k \times V_h^k$, no restriction is needed regarding nearly singular vertices on the boundary, nor is any difficulty caused by having triangles with two edges on the boundary, but we ignore this distinction from the case of Dirichlet conditions.

(12.6.10) Theorem. *Suppose $\mathcal{T}^h$ is non-degenerate, has no triangle with two edges on the boundary, and has no nearly singular vertices. Let V_h be either $V_h^k \times V_h^k$ or as in (12.4.1), and let $\Pi_h := \operatorname{div} V_h$. Then condition (12.5.3), namely*

$$\beta \, \|q\|_{L^2(\Omega)} \leq \sup_{v \in V_h} \frac{|b(v,q)|}{\|v\|_{H^1(\Omega)^2}} \qquad \forall q \in \Pi_h,$$

(or the equivalent version (12.5.2)) holds with $\beta > 0$ independent of h for $k \geq 4$.

In the next chapter we will study algorithms that allow one to compute using $\Pi_h = \mathcal{D}V_h$ without having explicit information about the structure of Π_h. The only obstacle to this choice of Π_h is to know that a condition such as (12.4.2) holds. When no Dirichlet boundary conditions are imposed on the functions in $\underset{\sim}{V}_h$, (12.4.2) is a simple consequence of approximability results for $\underset{\sim}{V}_h$, as we now show.

(12.6.11) Lemma. *Condition* (12.4.2) *holds for* $\Pi_h := \operatorname{div}\left(V_h^k\right)^n$.

Proof. Let B be a ball containing Ω, and let Ep be an extension of p (i.e. $Ep|_\Omega = p$) to B so that (see Theorem 1.4.5)

$$\|Ep\|_{H^s(B)} \le C\|p\|_{H^s(\Omega)}.$$

Define ϕ by solving $\Delta\phi = Ep$ in B with Dirichlet boundary conditions, $\phi = 0$ on ∂B. From elliptic regularity (Sect. 5.5),

$$\|\phi\|_{H^{s+2}(B)} \le C\|Ep\|_{H^s(B)} \le C'\|p\|_{H^s(\Omega)}.$$

Write $\underset{\sim}{w} = \operatorname{grad}\phi$ and note that $\|\underset{\sim}{w}\|_{H^{s+1}(\Omega)^n} \le C\|p\|_{H^s(\Omega)}$. Then $p = \operatorname{div}\underset{\sim}{w}$ in Ω and

$$
\begin{aligned}
\inf_{q\in\Pi_h}\|p - q\|_{L^2(\Omega)} &= \inf_{\underset{\sim}{v}\in V_h}\left\|\operatorname{div}\underset{\sim}{w} - \operatorname{div}\underset{\sim}{v}\right\|_{L^2(\Omega)} &&(\Pi_h = \operatorname{div}\underset{\sim}{V}_h)\\
&\le \inf_{\underset{\sim}{v}\in V_h}\sqrt{n}\|\underset{\sim}{w} - \underset{\sim}{v}\|_{H^1(\Omega)^n}\\
&\le Ch^s\|\underset{\sim}{w}\|_{H^{s+1}(\Omega)^n} &&\text{(by 4.4.25)}\\
&\le Ch^s\|p\|_{H^s(\Omega)}.
\end{aligned}
$$

$\square$

Condition (12.4.2) for $\Pi_h = \operatorname{div}\underset{\sim}{V}_h$ does not appear to follow directly from approximability results for $\underset{\sim}{V}_h$ when Dirichlet boundary conditions are imposed on functions in $\underset{\sim}{V}_h$. However, for spaces like (12.4.1), (Scott & Vogelius 1985a) identified $\operatorname{div}\underset{\sim}{V}_h$ and these results allow one to prove (12.4.2) for $\Pi_h = \operatorname{div}\underset{\sim}{V}_h$ for $k \ge 4$. The following is a simple corollary of those results.

(12.6.12) Lemma. *Suppose that* $\mathcal{T}^h$ *is a triangulation of a polygonal domain* $\Omega \subset \mathbb{R}^2$ *having no triangles with two edges on the boundary. Let* $k \ge 4$. *Then*

$$V_h^{k-1} \subset \operatorname{div}\left(V_h^k \times V_h^k\right)$$

and

$$\left\{q \in V_h^{k-1} : \int_\Omega q(x)\,dx = 0\right\} \subset \operatorname{div}\underset{\sim}{V}_h$$

where $\underset{\sim}{V}_h$ *is given by* (12.4.1).

In particular, Lemma 12.6.12 implies that condition (12.4.2) holds for $\Pi_h := \operatorname{div} \underset{\sim}{V}_h$ in both of these cases. Applying this result yields the following in view of Corollary 12.5.13 and the velocity approximation results in Sect. 12.4.

(12.6.13) Theorem. *Suppose $\mathcal{T}^h$ is non-degenerate, has no triangle with two edges on the boundary, and has no nearly singular vertices. For $\underset{\sim}{V}_h$ as in (12.4.1) with $k \geq 4$ and $\Pi_h := \operatorname{div} \underset{\sim}{V}_h$, the error in the pressure approximation to the solution $(\underset{\sim}{u}, p)$ of (12.1.6) satisfies*

$$\|p - p_h\|_{L^2(\Omega)} \leq Ch^s \left(\|\underset{\sim}{u}\|_{H^{s+1}(\Omega)^2} + \|p\|_{H^s(\Omega)} \right), \ 0 \leq s \leq k,$$

provided $(\underset{\sim}{u}, p) \in H^{s+1}(\Omega) \times H^s(\Omega)$. For $\underset{\sim}{V}_h = V_h^k \times V_h^k$ with $k \geq 4$ and $\Pi_h := \operatorname{div} \underset{\sim}{V}_h$, the error in the pressure approximation to the scalar elliptic problem (12.1.7) satisfies

$$\|p - p_h\|_{L^2(\Omega)} \leq Ch^{s+1} \left\| \underset{\approx}{a} \operatorname{grad} p \right\|_{H^{s+1}(\Omega)^2} + Ch^s \|p\|_{H^s(\Omega)}, \ 0 \leq s \leq k,$$

provided $\underset{\approx}{a} \operatorname{grad} p \in H^{s+1}(\Omega)$ and $p \in H^s(\Omega)$.

Recall that corresponding results for the velocity error were proved in Theorems 12.4.10 and 12.4.9, respectively. Thus, for high degree approximations, convergence criteria become simplified and provide few restrictions, whereas low degree approximation can yield widely differing results.

We finish the section by considering the nonconforming method introduced prior to Theorem 12.4.12. We note that $\mathcal{D} = -\operatorname{div}_h$ in this case satisfies (12.5.20) in view of (11.4.13), (11.4.14) and the techniques of Sect. 4.8. Therefore, the following can be proved (see exercise 12.x.14).

(12.6.14) Theorem. *The solution to (12.4.11) satisfies*

$$\|p - p_h\|_{L^2(\Omega)} \leq Ch \left(\|\underset{\sim}{u}\|_{H^2(\Omega)^2} + \|p\|_{H^1(\Omega)} \right)$$

provided $\underset{\sim}{u} \in H^2(\Omega)^2$ and $p \in H^1(\Omega)$.

12.x Exercises

12.x.1 Prove that the form defined in (12.1.4) is coercive on $\overset{\circ}{H}{}^1(\Omega)$. (Hint: apply (5.3.3).)

12.x.2 Prove that $H(\operatorname{div})$ is a Hilbert space with inner-product given by

$$(\underset{\sim}{u}, \underset{\sim}{v})_{H(\operatorname{div})} = (\underset{\sim}{u}, \underset{\sim}{v})_{L^2(\Omega)^n} + (\operatorname{div} \underset{\sim}{u}, \operatorname{div} \underset{\sim}{v})_{L^2(\Omega)}.$$

(Hint: see the proof of Theorem 1.3.2.)

12.x.3 Define a variational form for the Stokes equations by

$$a(\underset{\sim}{u}, \underset{\sim}{v}) := 2 \int_\Omega \sum_{i,j=1}^n e_{ij}(\underset{\sim}{u}) e_{ij}(\underset{\sim}{v}) \, dx,$$

where $e_{ij}(\underset{\sim}{u}) := \frac{1}{2}(u_{i,j} + u_{j,i})$. Prove that this is equal to (12.1.4) for $\underset{\sim}{u}, \underset{\sim}{v} \in Z$. (Hint: $\sum_{i=1}^n v_{i,i} = 0$ for $\underset{\sim}{v} \in Z$.)

12.x.4 Prove that the variational form defined in exercise 12.x.3 is coercive on Z. (Hint: see Korn's inequality in Chapter 11.)

12.x.5 Show that the nonconforming version of the variational form defined in exercise 12.x.3 is not coercive (independent of h) on the space $\underset{\sim}{V}_h^*$ defined in (11.4.5) (see Theorem 12.4.12). (Hint: consider a regular mesh on $\mathbb{R}^2$ and look for $\underset{\sim}{u}$ having a repetitive pattern such that $a(\underset{\sim}{u}, \underset{\sim}{v}) = 0$ for all $\underset{\sim}{v} \in \underset{\sim}{V}_h^*$ having compact support.)

12.x.6 Show that the results of Sect. 12.2 can be applied using the variational form defined in exercise 12.x.3. (Hint: see exercise 12.x.4.)

12.x.7 Suppose Π_h is finite dimensional. Prove that $\beta = 0$ in (12.5.2) implies that there is a $q \in \Pi_h$ such that $b(v, q) = 0$ for all $v \in V_h$. (Hint: use the fact that $\{q \in \Pi_h : \|q\|_\Pi \le 1\}$ is compact.)

12.x.8 Prove that the operator M defined in (12.5.6) is bounded, with norm bounded by $1/\beta$, where β is defined in (12.5.2). (Hint: apply (12.5.3) with $q = Mu$ and use (12.5.6).)

12.x.9 Prove (12.5.9). (Hint: show that any $w \in Z_h^p$ satisfies $Lp - w \in Z_h$ and hence $(Lp, Lp - w)_V = 0$.)

12.x.10 Prove that if the operator L in (12.5.7) is well defined, then (12.5.2) holds with $\beta = 1/\|L\|_{\Pi \to V}$. (Hint: choose $v = Lq$ in (12.5.2).)

12.x.11 Prove that the curl operator maps the Morley space depicted in Fig. 10.5 into the nonconforming piecewise linear space in Theorem 12.4.12. (Hint: the tangential derivative of the Morley elements is continuous at the edge midpoints because the jump across an edge is a quadratic that vanishes at the two vertices.)

12.x.12 Determine a bound for the constant, c, in Corollary 12.5.18 in terms of C, α and β.

12.x.13 Prove (12.5.19).

12.x.14 Prove Theorem 12.6.14.

12.x.15 Prove (12.6.2).

Chapter 13

Iterative Techniques for Mixed Methods

Equations of the form (12.3.1) or (12.5.14) are indefinite and require special care to solve. We will now consider one class of algorithms which involve a penalty method to enforce the second equation in (12.3.1) or (12.5.14). These algorithms transform the linear algebra to positive-definite problems in many cases. Moreover, the number of unknowns in the algebraic system can also be significantly reduced.

We begin with the case when $\Pi_h = \mathcal{D}V_h$, which naturally arises from the *iterated penalty method*. One benefit of this approach is that the degrees of freedom of Π_h do not enter the solution procedure directly, making the linear-algebraic problem smaller. In fact, it is not necessary even to have a basis of Π_h; the iterated penalty method produces $p_h = \mathcal{D}w_h$ for some $w_h \in V_h$.

Subsequently, we consider the general case $\Pi_h \neq \mathcal{D}V_h$. We show that the *augmented Lagrangian method* (Fortin & Glowinski 1983, Glowinski 1984) can be analyzed in a way analogous to the special case $\Pi_h = \mathcal{D}V_h$. We give some examples of the use of these techniques in the solution of the Navier-Stokes equations.

13.1 Iterated Penalty Method

Consider a general mixed method of the form (12.2.4) studied in the previous chapter, namely,

$$
\begin{aligned}
a(u_h, v) + b(v, p_h) &= F(v) \quad \forall v \in V_h \\
b(u_h, q) &= (g, q)_\Pi \quad \forall q \in \Pi_h,
\end{aligned}
$$
(13.1.1)

where $F \in V'$ and $g \in \Pi$. Here V and Π are two Hilbert spaces with subspaces $V_h \subset V$ and $\Pi_h \subset \Pi$, respectively. We assume that the bilinear forms satisfy the continuity conditions

$$
\begin{aligned}
a(u, v) &\leq C_a \|u\|_V \|v\|_V \quad \forall u, v \in V \\
b(v, p) &\leq C_b \|v\|_V \|p\|_\Pi \quad \forall v \in V, \, p \in \Pi
\end{aligned}
$$
(13.1.2)

and the coercivity conditions

(13.1.3)
$$\alpha\|v\|_V^2 \le a(v,v) \quad \forall v \in Z \cup Z_h$$
$$\beta\|p\|_\Pi \le \sup_{v \in V_h} \frac{b(v,p)}{\|v\|_V} \quad \forall p \in \Pi_h.$$

Here Z and Z_h are defined by (12.2.5) and (12.3.2), respectively. Also recall that we are assuming (12.2.3), namely $b(v,p) = (\mathcal{D}v, p)_\Pi$.

Let $r \in \mathbb{R}$ and $\rho > 0$. The iterated penalty method defines $u^n \in V_h$ and p^n by

(13.1.4)
$$a(u^n, v) + r(\mathcal{D}u^n - g, \mathcal{D}v)_\Pi = F(v) - b(v, p^n) \quad \forall v \in V_h$$
$$p^{n+1} = p^n + \rho(\mathcal{D}u^n - P_{\Pi_h} g)$$

where $P_{\Pi_h} g$ denotes the Π-projection of g onto Π_h. Note that the second equation in (13.1.1) says that

(13.1.5)
$$P_{\Pi_h} \mathcal{D}u_h = P_{\Pi_h} g.$$

The algorithm does not require $P_{\Pi_h} g$ to be computed, only $b(v, P_{\Pi_h} g) = (\mathcal{D}v, P_{\Pi_h} g)_\Pi = (P_{\Pi_h} \mathcal{D}v, g)_\Pi$ for $v \in V_h$.

The key point of the iterated penalty method is that the system of equations represented by the first equation in (13.1.4) for u^n, namely

$$a(u^n, v) + r(\mathcal{D}u^n, \mathcal{D}v)_\Pi = F(v) - b(v, p^n) + r(g, \mathcal{D}v)_\Pi \quad \forall v \in V_h,$$

will be symmetric if $a(\cdot, \cdot)$ is symmetric, and it will be positive definite if $a(\cdot, \cdot)$ is coercive and $r > 0$.

If $g = 0$ and we begin with, say, $p^0 = 0$, then $p^n \in \mathcal{D}V_h$ for all n. In particular $p^n = \mathcal{D}w^n$ where $w^n \in V_h$ satisfies

$$a(u^n, v) + r(\mathcal{D}u^n, \mathcal{D}v)_\Pi = F(v) - (\mathcal{D}v, \mathcal{D}w^n)_\Pi \quad \forall v \in V_h$$
$$w^{n+1} = w^n + \rho u^n.$$

Thus, the iterated penalty method implicitly produces an approximation closely connected with the choice of $\Pi_h = \mathcal{D}V_h$. If we can show that $u^n \to u_h \in V_h$ and $p^n \to p_h \in \Pi_h$, then it follows that $\mathcal{D}u^n \to 0$ and that (u_h, p_h) solves (13.1.1). Note that $w^n = \rho \sum_{i=0}^{n-1} u^i$ will not in general converge to anything. See (Scott, Ilin, Metcalfe & Bagheri 1996) for the case of $g \ne 0$.

To study the convergence properties of (13.1.4), let us introduce $e^n := u^n - u_h$ and $\epsilon^n := p^n - p_h$ where (u_h, p_h) solves (13.1.1). We assume that $\Pi_h = \mathcal{D}V_h$, in which case (13.1.5) simplifies to $\mathcal{D}u_h = P_{\Pi_h} g$. Then

(13.1.6)
$$a(e^n, v) + r(\mathcal{D}e^n, \mathcal{D}v)_\Pi = -b(v, \epsilon^n) \quad \forall v \in V_h$$

by subtracting (13.1.1) from (13.1.4), and

(13.1.7)
$$\epsilon^{n+1} = \epsilon^n + \rho(\mathcal{D}u^n - P_{\Pi_h} g) = \epsilon^n + \rho\mathcal{D}e^n.$$

Note that $e^n \in \tilde{Z}_h^\perp$ where we use the tilde to distinguish the space

(13.1.8) $$\tilde{Z}_h^\perp = \{v \in V_h \; : \; a(v,w) = 0 \quad \forall w \in Z_h\}$$

from the space $Z_h^\perp$ defined in (12.5.5) based on the inner-product $(\cdot,\cdot)_V$. Observe that $Z_h = \{v \in V_h \; : \; \mathcal{D}v = 0\}$ since $\Pi_h = \mathcal{D}V_h$. Thus, we can characterize $e^n \in \tilde{Z}_h^\perp$ by

(13.1.9) $$a(e^n, v) + r\,(\mathcal{D}e^n, \mathcal{D}v)_\Pi = -b(v, \epsilon^n) \quad \forall v \in \tilde{Z}_h^\perp.$$

We can then relate the new error to the old by

$$
\begin{aligned}
a(e^{n+1}, v) + r\,(\mathcal{D}e^{n+1}, \mathcal{D}v) &= -b(v, \epsilon^{n+1}) &&\text{(13.1.9 for } n+1) \\
&= -b(v, \epsilon^n) - \rho\,b(v, \mathcal{D}e^n) &&\text{(from 13.1.7)} \\
\text{(13.1.10)} \qquad &= -b(v, \epsilon^n) - \rho\,(\mathcal{D}v, \mathcal{D}e^n)_\Pi &&\text{(from 12.2.3)} \\
&= a(e^n, v) + r\,(\mathcal{D}e^n, \mathcal{D}v)_\Pi - \rho\,(\mathcal{D}v, \mathcal{D}e^n)_\Pi &&\text{(13.1.9 for } n) \\
&= a(e^n, v) + (r - \rho)(\mathcal{D}e^n, \mathcal{D}v)_\Pi \, .
\end{aligned}
$$

Dividing by r, choosing $v = e^{n+1}$ and applying Schwarz' inequality (2.1.5) and the assumptions (13.1.2) we find

(13.1.11)
$$
\begin{aligned}
\left\|\mathcal{D}e^{n+1}\right\|_\Pi^2 + \frac{1}{r}a(e^{n+1}, e^{n+1}) &\leq \frac{C_a}{r}\left\|e^{n+1}\right\|_V \left\|e^n\right\|_V \\
&\quad + \left|1 - \frac{\rho}{r}\right| \left\|\mathcal{D}e^{n+1}\right\|_\Pi \left\|\mathcal{D}e^n\right\|_\Pi \\
&\leq \left(\frac{C_a}{r} + C_b^2\left|1 - \frac{\rho}{r}\right|\right) \left\|e^{n+1}\right\|_V \left\|e^n\right\|_V \, .
\end{aligned}
$$

The key point (which we will make precise shortly) is that the bilinear form $(\mathcal{D}v, \mathcal{D}v)_\Pi$ is coercive on $\tilde{Z}_h^\perp$, provided (12.5.2) holds. Thus, for any ρ in the interval $0 < \rho < 2r$, we find that $e^n \to 0$ as $n \to \infty$. From (13.1.6) it follows that $\epsilon^n \to 0$ as $n \to \infty$ as well, provided (12.5.2) holds.

For $\Pi_h = \mathcal{D}V_h$ (as we are assuming), the second condition in (13.1.3) is equivalent (see Lemma 12.5.10 and Scott & Vogelius 1985a) to the existence of a right-inverse, $L : \Pi_h \to V_h$, for the operator $\mathcal{D}$, that is

(13.1.12) $$\mathcal{D}(Lq) = q \quad \forall q \in \Pi_h,$$

which satisfies

(13.1.13) $$\|Lq\|_V \leq \frac{1}{\beta}\|q\|_\Pi \quad \forall q \in \Pi_h.$$

We now consider the coerciveness of $(\mathcal{D}w, \mathcal{D}w)_\Pi$ on $w \in \tilde{Z}_h^\perp$. Note that since we have defined $\tilde{Z}_h^\perp$ using the bilinear form $a(\cdot,\cdot)$ instead of $(\cdot,\cdot)_V$, we cannot use Hilbert-space properties. Thus, we must re-examine our definitions. Firstly, $\tilde{Z}_h^\perp$ is a closed linear subset of V_h, since $a(\cdot,\cdot)$ is bilinear and continuous. Secondly, $\tilde{Z}_h^\perp \cap Z_h = \{0\}$, since $a(\cdot,\cdot)$ is coercive on Z_h.

Finally, any $w \in V_h$ can be decomposed into $w = z + z^\perp$, with $z \in Z_h$ and $z^\perp \in \tilde{Z}_h^\perp$, as follows. Let $z \in Z_h$ solve

$$a(z, v) = a(w, v) \quad \forall v \in Z_h,$$

which exists in view of Cea's Theorem 2.8.1. Then $z^\perp := w - z$ satisfies $a(z^\perp, v) = 0$ for all $v \in Z_h$, that is, $z^\perp \in \tilde{Z}_h^\perp$. Therefore, $V_h = Z_h \oplus \tilde{Z}_h^\perp$.

Define $L_a : \Pi_h \to V_h$ as follows. For $q \in \Pi_h$, let $z_q \in Z_h$ solve

$$a(z_q, v) = a(Lq, v) \quad \forall v \in Z_h$$

and set $L_a q = Lq - z_q$. We have from Cea's Theorem 2.8.1 that

$$\|z_q\|_V \leq \frac{C_a}{\alpha} \|Lq\|_V$$

using the coercivity and continuity of $a(\cdot, \cdot)$. Thus,

$$(13.1.14) \qquad \|L_a q\|_V \leq \left(1 + \frac{C_a}{\alpha}\right) \|Lq\|_V \leq \left(1 + \frac{C_a}{\alpha}\right) \frac{1}{\beta} \|q\|_\Pi$$

for all $q \in \Pi_h$, from the triangle inequality and (13.1.13). Observe that

$$(13.1.15) \qquad \mathcal{D} L_a q = \mathcal{D} L q = q \quad \forall q \in \Pi_h$$

from the definition of L_a and (13.1.12).

Now we claim that for $w \in \tilde{Z}_h^\perp$, $w = L_a \mathcal{D} w$. For

$$a(L_a \mathcal{D} w, v) = a(Lq - z_q, v) = 0 \quad \forall v \in Z_h$$

(here $q := \mathcal{D} w$) so that $L_a \mathcal{D} w \in \tilde{Z}_h^\perp$. Therefore, $w - L_a \mathcal{D} w \in \tilde{Z}_h^\perp$. But (13.1.15) implies

$$\mathcal{D}(w - L_a \mathcal{D} w) = 0.$$

This implies $w - L_a \mathcal{D} w \in Z_h$. Since $Z_h \cap \tilde{Z}_h^\perp = \{0\}$, we conclude that $w = L_a \mathcal{D} w$. From (13.1.14), it follows that

$$(13.1.16) \qquad \|w\|_V = \|L_a \mathcal{D} w\|_V \leq \left(\frac{1}{\beta} + \frac{C_a}{\alpha\beta}\right) \|\mathcal{D} w\|_\Pi$$

for all $w \in \tilde{Z}_h^\perp$.

Applying this to (13.1.11) and using (13.1.2)–(13.1.3), we find

$$(13.1.17) \qquad \left(1 + \frac{c_1}{r}\right) \|\mathcal{D} e^{n+1}\|_\Pi \leq \left(\left|1 - \frac{\rho}{r}\right| + \frac{c_2}{r}\right) \|\mathcal{D} e^n\|_\Pi.$$

Thus, for $0 < \rho < 2r$ and for r sufficiently large, $\mathcal{D} e^n \to 0$ geometrically as $n \to \infty$. From (13.1.10), it follows that $e^n \to 0$ as well (see exercise 13.x.1). On the other hand, using the coercivity of $a(\cdot, \cdot)$, we find

$$(13.1.18) \quad \left(\left(\frac{1}{\beta} + \frac{C_a}{\alpha\beta}\right)^{-2} + \frac{\alpha^2}{r}\right) \|e^{n+1}\|_V \leq \left(\frac{C_a}{r} + C_b^2 \left|1 - \frac{\rho}{r}\right|\right) \|e^n\|_V.$$

Collecting the above results, we have

(13.1.19) Theorem. *Suppose that the form (13.1.1) satisfies (13.1.2) and (13.1.3). Suppose that V_h and $\Pi_h = \mathcal{D}V_h$ satisfy (13.1.3). Then the algorithm (13.1.4) converges for any $0 < \rho < 2r$ for r sufficiently large. For the choice $\rho = r$, (13.1.4) converges geometrically with a rate given by*

$$C_a \left(\frac{1}{\beta} + \frac{C_a}{\alpha\beta} \right)^2 \bigg/ r \, .$$

Note that (with $r = \rho$) (13.1.18) implies convergence ($e^1 \to 0$ as $r \to \infty$) of the standard penalty method. However, the expression (13.1.6) for ϵ^1 deteriorates as $r \to \infty$ so we cannot conclude anything about ϵ^1. Thus, the *iterated* penalty method is needed to insure convergence of the p approximation.

13.2 Stopping Criteria

For any iterative method, it is of interest to have a good stopping criterion based on information that is easily computable. For the iterated penalty method, this is extremely simple. We will assume that $P_{\Pi_h} g$ can be easily computed by some means. In many applications, $g = 0$ so that this becomes trivial. However, in others it would require some additional work. We show subsequently how this can be avoided.

The error $e^n = u^n - u_h$ is bounded by

$$\|u^n - u_h\|_V = \|e^n\|_V$$

(13.2.1)
$$\leq \left(\frac{1}{\beta} + \frac{C_a}{\alpha\beta} \right) \|\mathcal{D}e^n\|_\Pi \qquad \text{(by 13.1.16)}$$

$$= \left(\frac{1}{\beta} + \frac{C_a}{\alpha\beta} \right) \|\mathcal{D}u^n - P_{\Pi_h} g\|_\Pi \, . \qquad \text{(by 13.1.5)}$$

The error $\epsilon^n = p^n - p_h$ is bounded by

$$\beta \|p^n - p_h\|_\Pi = \beta \|\epsilon^n\|_\Pi$$

$$\leq \sup_{v \in V_h} \frac{|b(v, \epsilon^n)|}{\|v\|_V} \qquad \text{(by 13.1.3)}$$

$$= \sup_{v \in V_h} \frac{|a(e^n, v) + r\,(\mathcal{D}e^n, \mathcal{D}v)_\Pi|}{\|v\|_V} \qquad \text{(by 13.1.6)}$$

$$\leq C_a \|e^n\|_V + rC_b \|\mathcal{D}e^n\|_\Pi \qquad \text{(by 13.1.2)}$$

$$= C_a \|u^n - u_h\|_V + rC_b \|\mathcal{D}u^n - P_{\Pi_h} g\|_\Pi \, . \qquad \text{(by 13.1.5)}$$

Combining the previous estimates, the following theorem is proved.

(13.2.2) Theorem. *Suppose that the form (13.4.9) satisfies (13.1.2) and (13.1.3). Suppose that V_h and $\Pi_h = \mathcal{D}V_h$ satisfy (13.1.3). Then the errors in algorithm (13.1.4) can be estimated by*

$$(13.2.3) \qquad \|u^n - u_h\|_V \leq \left(\frac{1}{\beta} + \frac{C_a}{\alpha\beta}\right) \|\mathcal{D}u^n - P_{\Pi_h}g\|_\Pi$$

and

$$(13.2.4) \qquad \|p^n - p_h\|_\Pi \leq \left(\frac{C_a}{\beta} + \frac{C_a^2}{\alpha\beta} + rC_b\right) \|\mathcal{D}u^n - P_{\Pi_h}g\|_\Pi \, .$$

By monitoring $\|\mathcal{D}u^n - P_{\Pi_h}g\|_\Pi$ as the iteration proceeds, we can determine the convergence properties of the iterated penalty method. Once $\|\mathcal{D}u^n - P_{\Pi_h}g\|_\Pi$ is sufficiently small, the iteration can be terminated.

The main drawback with this approach is that $P_{\Pi_h}g$ has to be computed to determine the error tolerance. Since $\mathcal{D}u^n \in \Pi_h$,

$$\|\mathcal{D}u^n - P_{\Pi_h}g\|_\Pi = \|P_{\Pi_h}(\mathcal{D}u^n - g)\|_\Pi \leq \|\mathcal{D}u^n - g\|_\Pi \, ,$$

and the latter norm may be easy to compute (or bound) in some cases, avoiding the need to compute $P_{\Pi_h}g$. We formalize this observation in the following result.

(13.2.5) Corollary. *Under the conditions of Theorem 13.2.2 the errors in algorithm (13.1.4) can be estimated by*

$$(13.2.6) \qquad \|u^n - u_h\|_V \leq \left(\frac{1}{\beta} + \frac{C_a}{\alpha\beta}\right) \|\mathcal{D}u^n - g\|_\Pi$$

and

$$(13.2.7) \qquad \|p^n - p_h\|_\Pi \leq \left(\frac{C_a}{\beta} + \frac{C_a^2}{\alpha\beta} + rC_b\right) \|\mathcal{D}u^n - g\|_\Pi \, .$$

(13.2.8) Remark. As the above estimates indicate, it can be expected that taking r arbitrarily large will affect the p approximation adversely, although it may have minimal effect on the u approximation.

The results of this and the previous section can be applied directly to several of the mixed methods in Chapter 12. For example, with V_h as in Theorem 12.6.13 (and $\Pi_h := \operatorname{div} V_h$), the algorithm (13.1.4) converges for both the Stokes problem (12.1.6) (see exercise 13.x.3) and for the scalar elliptic problem (12.1.7) (see exercise 13.x.4).

13.3 Augmented Lagrangian Method

In the case of general V_h and $\Pi_h \neq \mathcal{D}V_h$, the equations (13.1.1) have the matrix representation

$$\underset{\approx}{A}\underset{\sim}{U} + \underset{\approx}{B^t}\underset{\sim}{P} = \underset{\sim}{F}$$

$$\underset{\approx}{B}\underset{\sim}{U} = \underset{\sim}{G},$$

where $\underset{\sim}{U}$ (resp. $\underset{\sim}{P}$) denote the coefficients of u_h (resp. p_h) expanded with respect to a basis for V_h (resp. Π_h). The (iterative) augmented Lagrangian method (Fortin & Glowinski 1983, Glowinski 1984) consists of solving

(13.3.1)
$$\underset{\approx}{A}\underset{\sim}{U^n} + r\underset{\approx}{B^t}\left(\underset{\approx}{B}\underset{\sim}{U^n} - \underset{\sim}{G}\right) = \underset{\sim}{F} - \underset{\approx}{B^t}\underset{\sim}{P^n}$$

$$\underset{\sim}{P^{n+1}} = \underset{\sim}{P^n} + \rho\left(\underset{\approx}{B}\underset{\sim}{U^n} - \underset{\sim}{G}\right)$$

for r and ρ non-negative parameters. When $\Pi_h = \mathcal{D}V_h$, (13.3.1) reduces to the iterated penalty method (13.1.4).

The augmented Lagrangian method requires an explicit basis for Π_h in order to construct $\underset{\approx}{B}$, unlike the iterated penalty method. However, it can be used more generally, viz., when $\Pi_h \neq \mathcal{D}V_h$. Like the iterated penalty method, the augmented Lagrangian method reduces the linear-algebraic system to be solved by eliminating the degrees of freedom associated with Π_h, potentially leading to considerable savings in terms of time and storage. Moreover, $\underset{\approx}{A} + r\underset{\approx}{B^t}\underset{\approx}{B}$ is symmetric if $a(\cdot,\cdot)$ is symmetric, and it will be positive definite if $a(\cdot,\cdot)$ is coercive and $r > 0$.

To analyze the convergence properties of (13.3.1), we reformulate it variationally. Let P_{Π_h} denote the Π-projection onto Π_h. Then (13.3.1) is equivalent (exercise 13.x.7) to

(13.3.2)
$$a(u^n, v) + rb(u^n - g, P_{\Pi_h}\mathcal{D}v) = F(v) - b(v, p^n) \quad \forall v \in V_h$$

$$p^{n+1} = p^n - \rho P_{\Pi_h}(\mathcal{D}u^n - g),$$

where g is represented by the coefficients $\underset{\sim}{G}$ in the basis for Π_h and similarly F denotes a linear form in (13.3.2) with coefficients $\underset{\sim}{F}$ in the representation (13.3.1). Note that

$$b(w, P_{\Pi_h}\mathcal{D}v) = (\mathcal{D}w, P_{\Pi_h}\mathcal{D}v) = (P_{\Pi_h}\mathcal{D}w, P_{\Pi_h}\mathcal{D}v)$$

is a symmetric bilinear form on V_h. We now show it is coercive on $\tilde{Z}_h^\perp$ (see (13.1.8)), provided (13.1.3) holds.

Recall the operator $L : \Pi_h \to Z_h^\perp$ given by (12.5.7). We can rephrase (12.5.7) as $P_{\Pi_h}(\mathcal{D}Lq) = q$ (exercise 13.x.5). Define L_a as in the previous section satisfying (13.1.14). Note that $P_{\Pi_h}(\mathcal{D}L_a q) = q$ since $z \in Z_h$ is equivalent to $P_{\Pi_h}(\mathcal{D}z) = 0$ (exercise 13.x.6). Similarly, for any $w \in \tilde{Z}_h^\perp$, $w = L_a(P_{\Pi_h}(\mathcal{D}w))$. Therefore,

$$\|w\|_V = \|L_a P_{\Pi_h} \mathcal{D} w\|_V$$

(13.3.3)
$$\leq \left(\frac{1}{\beta} + \frac{C_a}{\alpha\beta}\right) \|P_{\Pi_h} \mathcal{D} w\|_\Pi$$

for all $w \in \tilde{Z}_h^\perp$. Analogous to (13.1.10), we have

(13.3.4)
$$a(e^{n+1}, v) + r\, (P_{\Pi_h} \mathcal{D} e^{n+1}, P_{\Pi_h} \mathcal{D} v)_\Pi$$
$$= a(e^n, v) + (r - \rho)(P_{\Pi_h} \mathcal{D} e^n, P_{\Pi_h} \mathcal{D} v)_\Pi.$$

Using the same techniques in the previous section, the following is proved.

(13.3.5) Theorem. *Suppose that the form (13.4.9) satisfies (13.1.3) and (13.1.2). Suppose that V_h and Π_h satisfy (13.1.3). Then the algorithm (13.3.1) converges for any $0 < \rho < 2r$ for r sufficiently large. For the choice $\rho = r$, (13.3.1) converges geometrically with a rate given by*

$$C_a \left(\frac{1}{\beta} + \frac{C_a}{\alpha\beta}\right)^2 \bigg/ r.$$

(13.3.6) Remark. The matrix $\underset{\approx}{B}^t \underset{\approx}{B}$ has a large null space (all of Z_h), but (13.3.3) shows that it is positive definite when restricted to $\tilde{Z}_h^\perp$. However, a lower bound for the minimal eigenvalue on this subspace depends precisely on the complementary space $\tilde{Z}_h^\perp$ and is not an invariant dependent only on $\underset{\approx}{B}$.

Stopping criteria can also be developed as in the previous section, based on the following result.

(13.3.7) Theorem. *Suppose that the form (13.4.9) satisfies (13.1.3) and (13.1.2). Suppose that V_h and Π_h satisfy (13.1.3). Then the error in algorithm (13.1.4) can be estimated by*

$$\|u^n - u_h\|_V \leq \left(\frac{1}{\beta} + \frac{C_a}{\alpha\beta}\right) \|P_{\Pi_h} (\mathcal{D} u^n - g)\|_\Pi$$
(13.3.8)
$$\leq \left(\frac{1}{\beta} + \frac{C_a}{\alpha\beta}\right) \|\mathcal{D} u^n - g\|_\Pi$$

and

$$\|p^n - p_h\|_\Pi \leq \left(C\left(\frac{1}{\beta} + \frac{C_a}{\alpha\beta}\right) + rC_b\right) \|P_{\Pi_h} (\mathcal{D} u^n - g)\|_\Pi$$
(13.3.9)
$$\leq \left(C\left(\frac{1}{\beta} + \frac{C_a}{\alpha\beta}\right) + rC_b\right) \|\mathcal{D} u^n - g\|_\Pi.$$

Exercise 13.x.7 implies that

$$\|P_{\Pi_h} \left(\mathcal{D}u^n - g\right)\|_{\Pi}^2 = (\underset{\approx}{B}\underset{\sim}{U}^n - \underset{\sim}{G})^t (\underset{\approx}{B}\underset{\sim}{U}^n - \underset{\sim}{G}),$$

where $\underset{\sim}{U}^n$ (resp. $\underset{\sim}{G}$) denotes the coefficients of u^n (resp. $P_{\Pi_h}g$) with respect to the given basis, so the expressions in Theorem 13.3.7 are easily computed.

The results of this section can be applied directly to the Taylor-Hood methods in Chapter 10 for the Stokes problem. With V_h as in (12.4.1) and Π_h as in (12.4.4), the algorithm (13.3.2) (or equivalently (13.3.1)) converges for the Stokes problem (12.1.6) for all $k \geq 2$ (see exercise 13.x.9).

13.4 Application to the Navier-Stokes Equations

We now consider a more complex application of the theory developed in the previous sections, regarding time-stepping schemes for the Navier-Stokes equations. The Navier-Stokes equations for the flow of a viscous, incompressible, Newtonian fluid can be written

$$\begin{aligned}
-\Delta \underset{\sim}{u} + \nabla p &= -R\left(\underset{\sim}{u} \cdot \nabla \underset{\sim}{u} + \underset{\sim}{u}_t\right) \\
\operatorname{div} \underset{\sim}{u} &= 0.
\end{aligned}$$

(13.4.1)

in $\Omega \subset \mathbb{R}^d$, where $\underset{\sim}{u}$ denotes the fluid velocity and p denotes the pressure. The expression $\underset{\sim}{u} \cdot \nabla \underset{\sim}{v}$ is the vector function whose i-th component is $\underset{\sim}{u} \cdot \nabla v_i$. These equations describe both two- and three-dimensional flows ($d = 2$ and 3, respectively); in the case of two dimensions, the flow field is simply independent of the third variable, and the third component of $\underset{\sim}{u}$ is correspondingly zero. These equations must be supplemented by appropriate boundary conditions, such as the Dirichlet boundary conditions, $\underset{\sim}{u} = 0$ on $\partial\Omega$.

The parameter R in (13.4.1) is called the Reynolds number. When this is very small, the equations reduce to the Stokes equations studied earlier. Numerical techniques for solving (13.4.1) often involve different issues relating separately to the solution of the Stokes (or Stokes-like) equations and to the discretization of the advection term that R multiplies. We will focus here on particularly simple time-stepping schemes, putting emphasis on the affect this has on the particular form of the corresponding Stokes-like equations. More complex time-stepping schemes yield similar Stokes-like equations. For more information, we refer to the survey of (Glowinski & Pironneau 1992).

A complete variational formulation of (13.4.1) takes the form

$$\begin{aligned}
a\left(\underset{\sim}{u}, \underset{\sim}{v}\right) + b\left(\underset{\sim}{v}, p\right) + R\big(c\left(\underset{\sim}{u}, \underset{\sim}{u}, \underset{\sim}{v}\right) + \left(\underset{\sim}{u}_t, \underset{\sim}{v}\right)_{L^2}\big) &= 0 \quad \forall \underset{\sim}{v} \in V, \\
b(\underset{\sim}{u}, q) &= 0 \quad \forall q \in \Pi,
\end{aligned}$$

(13.4.2)

where e.g. $a(\cdot,\cdot)$ and $b(\cdot,\cdot)$ are given in (12.1.4) and (12.1.5), respectively, $(\cdot,\cdot)_{L^2}$ denotes the $L^2(\Omega)^d$-inner-product, and

$$(13.4.3) \qquad c(\underset{\sim}{u},\underset{\sim}{v},\underset{\sim}{w}) := \int_\Omega (\underset{\sim}{u} \cdot \nabla \underset{\sim}{v}) \cdot \underset{\sim}{w}\, dx.$$

The spaces V and Π are the same as for the Stokes equations, as defined in the paragraph following (12.1.5).

One of the simplest time-stepping schemes for the Navier-Stokes equations (13.4.1) is implicit with respect to the linear terms and explicit with respect to the nonlinear terms. Expressed in variational form, it is

$$(13.4.4) \qquad
\begin{aligned}
a\left(\underset{\sim}{u}^\ell,\underset{\sim}{v}\right) + b\left(\underset{\sim}{v},p^\ell\right) &+ R\,c\left(\underset{\sim}{u}^{\ell-1},\underset{\sim}{u}^{\ell-1},\underset{\sim}{v}\right) \\
&+ \frac{R}{\Delta t}\left(\underset{\sim}{u}^\ell - \underset{\sim}{u}^{\ell-1},\underset{\sim}{v}\right)_{L^2} = 0, \\
b\left(\underset{\sim}{u}^\ell,q\right) &= 0,
\end{aligned}$$

where, here and below, $\underset{\sim}{v}$ varies over all V (or V_h) and q varies over all Π (or Π_h) and Δt denotes the time-step size. At each time step, one has a problem to solve of the form (13.1.1) for $(\underset{\sim}{u}^\ell,p^\ell)$ but with the form $a(\cdot,\cdot)$ more general, namely

$$(13.4.5) \qquad \tilde{a}(\underset{\sim}{u},\underset{\sim}{v}) := a(\underset{\sim}{u},\underset{\sim}{v}) + \tau\left(\underset{\sim}{u},\underset{\sim}{v}\right)_{L^2}$$

where the constant $\tau = R/\Delta t$. Numerical experiments will be presented in the next section for such a problem. Note that the linear algebraic problem problem to be solved at each time step is the same.

Equation (13.4.4) may now be written as a problem for $(\underset{\sim}{u}^\ell,p^\ell)$ of the form (13.1.1):

$$
\begin{aligned}
\tilde{a}\left(\underset{\sim}{u}^\ell,\underset{\sim}{v}\right) + b\left(\underset{\sim}{v},p^\ell\right) &= -R\,c\left(\underset{\sim}{u}^{\ell-1},\underset{\sim}{u}^{\ell-1},\underset{\sim}{v}\right) + \tau\left(\underset{\sim}{u}^{\ell-1},\underset{\sim}{v}\right)_{L^2}, \\
b\left(\underset{\sim}{u}^\ell,q\right) &= 0.
\end{aligned}
$$

The iterated penalty method (with $r = \rho$) for (13.4.4) thus takes the form

$$(13.4.6) \qquad
\begin{aligned}
\tilde{a}\left(\underset{\sim}{u}^{\ell,n},\underset{\sim}{v}\right) + r\left(\operatorname{div}\underset{\sim}{u}^{\ell,n},\operatorname{div}\underset{\sim}{v}\right)_{L^2} &= -R\,c\left(\underset{\sim}{u}^{\ell-1},\underset{\sim}{u}^{\ell-1},\underset{\sim}{v}\right) \\
&+ \tau\left(\underset{\sim}{u}^{\ell-1},\underset{\sim}{v}\right)_{L^2} - b\left(\underset{\sim}{v},p^{\ell,n}\right) \\
p^{\ell,n+1} &= p^{\ell,n} - r\operatorname{div}\underset{\sim}{u}^{\ell,n}.
\end{aligned}$$

This would be started with, say, $p^{\ell,0} = 0$ and $n = 0$, and continued until a stopping criterion, such as in Sect. 13.2, is met. At this point we set $\underset{\sim}{u}^\ell = \underset{\sim}{u}^{\ell,n}$ and increment ℓ, going on to the next time step. For purposes of definiteness, we can set $p^\ell = p^{\ell,n}$, but this does not figure in the advancement of the time step.

With the choice $p^{\ell,0} = 0$, (13.4.6) can be written

$$\tilde{a}\left(\underset{\sim}{u}^{\ell,n}, \underset{\sim}{v}\right) + r\left(\operatorname{div}\underset{\sim}{u}^{\ell,n}, \operatorname{div}\underset{\sim}{v}\right)_{L^2} = -Rc\left(\underset{\sim}{u}^{\ell-1}, \underset{\sim}{u}^{\ell-1}, \underset{\sim}{v}\right)$$

(13.4.7)
$$+\tau\left(\underset{\sim}{u}^{\ell-1}, \underset{\sim}{v}\right)_{L^2} + \left(\operatorname{div}\underset{\sim}{v}, \operatorname{div}\underset{\sim}{w}^{\ell,n}\right)_{L^2}$$

$$\underset{\sim}{w}^{\ell,n+1} = \underset{\sim}{w}^{\ell,n} - r\underset{\sim}{u}^{\ell,n}$$

where $\underset{\sim}{w}^{\ell,0} = 0$. If for some reason $p^{\ell} = p^{\ell,n} = \operatorname{div}\underset{\sim}{w}^{\ell,n}$ were desired, it could be computed separately.

A more complex time-stepping scheme could be based on the variational equations

$$a\left(\underset{\sim}{u}^{\ell}, \underset{\sim}{v}\right) + b\left(\underset{\sim}{v}, p^{\ell}\right) + Rc\left(\underset{\sim}{u}^{\ell-1}, \underset{\sim}{u}^{\ell}, \underset{\sim}{v}\right)$$

(13.4.8)
$$+ \frac{R}{\Delta t}\left(\underset{\sim}{u}^{\ell} - \underset{\sim}{u}^{\ell-1}, \underset{\sim}{v}\right)_{L^2} = 0,$$

$$b\left(\underset{\sim}{u}^{\ell}, q\right) = 0,$$

in which the nonlinear term has been approximated in such a way that the linear algebraic problem changes at each time step. It takes the form (13.1.1) with a form $\tilde{a}(\cdot, \cdot)$ given by

(13.4.9)
$$\tilde{a}\left(\underset{\sim}{u}, \underset{\sim}{v}; \underset{\sim}{U}\right) = a(\underset{\sim}{u}, \underset{\sim}{v}) + \int_{\Omega} \tau \underset{\sim}{u} \cdot \underset{\sim}{v} + \underset{\sim}{U} \cdot \nabla \underset{\sim}{u} \cdot \underset{\sim}{v}\, dx$$

where $\underset{\sim}{U} = R\underset{\sim}{u}^n$ arises from linearizing the nonlinear term. The iterated penalty method (with $r = \rho$) for (13.4.8) takes the form

$$\tilde{a}\left(\underset{\sim}{u}^{\ell,n}, \underset{\sim}{v}; R\underset{\sim}{u}^{\ell-1}\right) + r\left(\operatorname{div}\underset{\sim}{u}^{\ell,n}, \operatorname{div}\underset{\sim}{v}\right)_{L^2} = \tau\left(\underset{\sim}{u}^{\ell-1}, \underset{\sim}{v}\right)_{L^2}$$

(13.4.10)
$$- b\left(\underset{\sim}{v}, p^{\ell,n}\right)$$

$$p^{\ell,n+1} = p^{\ell,n} - r\operatorname{div}\underset{\sim}{u}^{\ell,n},$$

with $p^{\ell,0}$ initialized in some way and $\underset{\sim}{u}^{\ell} = \underset{\sim}{u}^{\ell,n}$ for appropriate n. If $p^{\ell,0} = 0$, then (13.4.10) becomes

$$\tilde{a}\left(\underset{\sim}{u}^{\ell,n}, \underset{\sim}{v}; R\underset{\sim}{u}^{\ell-1}\right) + r\left(\operatorname{div}\underset{\sim}{u}^{\ell,n}, \operatorname{div}\underset{\sim}{v}\right)_{L^2} = \tau\left(\underset{\sim}{u}^{\ell-1}, \underset{\sim}{v}\right)_{L^2}$$

(13.4.11)
$$+ \left(\operatorname{div}\underset{\sim}{v}, \operatorname{div}\underset{\sim}{w}^{\ell,n}\right)_{L^2}$$

$$\underset{\sim}{w}^{\ell,n+1} = \underset{\sim}{w}^{\ell,n} - r\underset{\sim}{u}^{\ell,n}$$

where $\underset{\sim}{w}^{\ell,0} = 0$.

Even though the addition of the $\underset{\sim}{U}$ term makes it non-symmetric, $\tilde{a}(\cdot, \cdot)$ will be coercive for τ sufficiently large (i.e., for Δt sufficiently small, cf. the proof of Gårding's inequality, Theorem 5.6.8), as we now assume:

(13.4.12)
$$\alpha\|\underset{\sim}{v}\|_V^2 \leq \tilde{a}(\underset{\sim}{v}, \underset{\sim}{v}) \quad \forall \underset{\sim}{v} \in V$$

for $\alpha > 0$. Of course, $a(\cdot, \cdot)$ is continuous:

(13.4.13) $$\tilde{a}(\underset{\sim}{v}, \underset{\sim}{w}) \le C_a \|\underset{\sim}{v}\|_V \|\underset{\sim}{w}\|_V \quad \forall \underset{\sim}{v}, \underset{\sim}{w} \in V$$

but now C_a depends on τ and $\underset{\sim}{U}$. Applying the theory in Sect. 13.1, we prove the following.

(13.4.14) Theorem. *Let V_h and Π_h be as in Theorem 12.6.13. Suppose that (13.4.12) and (13.4.13) hold. Then the algorithms (13.4.6) (or 13.4.7) and (13.4.10) (or 13.4.11) converge geometrically for r sufficiently large. The error can be estimated by*

$$\left\| p^{\ell,n} - p^\ell \right\|_{L^2(\Omega)} + \left\| \underset{\sim}{u}^{\ell,n} - \underset{\sim}{u}^\ell \right\|_{H^1(\Omega)^2} \le C \left\| \operatorname{div} \underset{\sim}{u}^{\ell,n} \right\|_{L^2(\Omega)}.$$

(13.4.15) Remark. If the constant τ in (13.4.9) is very large, then the ratio C_a/α will be as well (more precisely, C_a will be comparable to τ for τ large, while α will stay fixed). Computational experience (Bagheri, Scott & Zhang, 1994) indicates that it is necessary to take r comparable to τ for large τ, whereas the estimate in Theorem 13.1.19 suggests that it might be necessary to take r much larger. It is unclear whether the constant in (13.1.16) is sharp with regard to the dependence on C_a/α.

We will leave the formulation of algorithms and convergence results regarding other discretizations as exercises. In particular, we formulate the use of augmented Lagrangian methods for the Taylor-Hood spaces as exercises 13.x.10 and 13.x.11.

13.5 Computational Examples

We give here the results of some computational experiments which indicate the performance of the iterated penalty method. In keeping with the results described at the end of Sect. 12.6, we consider V_h as defined in Theorem 12.4.5 for $k = 4$ and 6. We pick a problem for which the solution is nontrivial yet can be computed by independent means, namely, Jeffrey-Hamel flow in a converging duct (Bagheri, Scott & Zhang, 1994). This similarity solution is of the form

(13.5.1) $$\underset{\sim}{u}(x, y) := \frac{u(\operatorname{atan}(y/x))}{x^2 + y^2} \underset{\sim}{x}, \quad \underset{\sim}{x} = (x, y) \in \Omega$$

where the (scaled) radial velocity u satisfies

(13.5.2) $$u'' + 4u + u^2 = C_{\text{JH}}, \quad u(0) = u(\theta) = 0,$$

where differentiation is with respect to the polar angle ϕ and θ is the angle of convergence of the channel. We have taken Ω to be the segment of the wedge depicted in Fig. 13.1, namely the quadrilateral with vertices

$\{(1,0),(2,0),(1,1),(2,2)\}$, and we have used (13.5.1) to provide Dirichlet boundary conditions on all of $\partial\Omega$. Fig. 13.1 shows the velocity field for Jeffrey-Hamel flow in Ω with $C_{JH} = 10^4$ (normalizing by the maximum velocity yields a Reynolds number of approximately 600). Depicted is the velocity field calculated with two different meshes and with $k = 4$ on the finer mesh and $k = 6$ on the coarser mesh. These yield an error in $L^2(\Omega)$-norm of less than 0.028 (resp. 0.038) times $\|u\|_{L^2(\Omega)}$ for $k = 4$ (resp. $k = 6$).

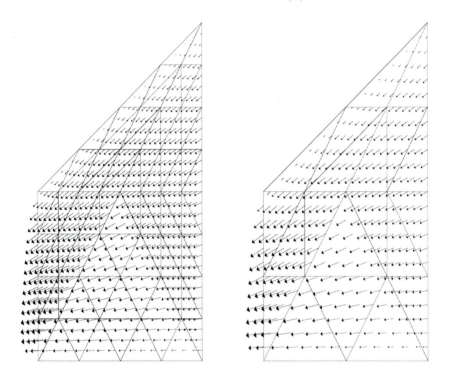

Fig. 13.1. Jeffrey-Hamel flow computed using fourth- and sixth-degree piecewise polynomials

The equations (13.4.1) were approximated by the time-stepping scheme (13.4.4) (Bagheri, Scott & Zhang, 1994) which involves solving, at each time step, a problem of the form (13.1.1) with $\tilde{a}(\cdot,\cdot)$ as in (13.4.5) with $\tau = 10^5$. Note that the maximum velocity $\|\underset{\sim}{u}\|_{L^\infty(\Omega)^2}$ is approximately 600 in these calculations, and this may be taken as a good measure of the Reynolds number. If the equations were scaled so that the maximum velocity were unity, then the corresponding time-step would be increased by a factor of approximately 600, and τ would be decreased to less than 200. However, the coercivity constant α would be correspondingly decreased, with the ratio τ/α being invariant.

We are particularly interested in the number of penalty iterations needed to reduce $\left\|\mathcal{D}\underset{\sim}{u}^n\right\|_a/\left\|\underset{\sim}{u}^n\right\|_a$ to be, say, less than 10^{-10}, for different values of the penalty parameter $r = \rho$ (these are taken equal for simplicity). Typically, the number of penalty iterations was the same for each time step, after the initial step. Table 13.1 indicates the results by giving this "typical" number of penalty iterations for each case.

Table 13.1. Typical number of penalty iterations per solution of (13.1.1), with $r = \rho$, needed to make $\left\|\operatorname{div}\underset{\sim}{u}^n\right\|_a/\left\|\underset{\sim}{u}^n\right\|_a < 10^{-10}$.

	$r = 10^4$	$r = 10^5$	$r = 10^6$	$r = 10^7$
$k = 4$	46	6	5	4
$k = 6$	51	11	6	4

In view of the estimate for the pressure error in Theorem 13.2.2, we are also interested in the number of penalty iterations needed to reduce $\left\|\operatorname{div}\underset{\sim}{u}^n\right\|_a/\left\|\underset{\sim}{u}^n\right\|_a$ to be less than a fixed parameter times r for different values of r. Again, the number of penalty iterations was the same for each time step, after the initial step. Table 13.2 indicates the results by giving this "typical" number of penalty iterations for each case.

Table 13.2. Typical number of penalty iterations per solution of (13.1.1), with $r = \rho$, needed to make $\left\|\operatorname{div}\underset{\sim}{u}^n\right\|_a/\left\|\underset{\sim}{u}^n\right\|_a < 10^{-5}/r$.

	$r = 10^4$	$r = 10^5$	$r = 10^6$	$r = 10^7$	$r = 10^8$
$k = 4$	38	6	6	5	4
$k = 6$	43	11	6	5	4

In both experiments, note that the higher-degree approximation can require a larger number of iterations. This is consistent with the fact that β decreases as k increases (Jensen & Vogelius 1990) since Theorem 13.1.19 predicts a convergence rate proportional to β^2. In view of Theorem 13.2.2, it might be appropriate to require $\left\|\operatorname{div}\underset{\sim}{u}^n\right\|_a/\left\|\underset{\sim}{u}^n\right\|_a$ to be even smaller for larger k (to guarantee the same accuracy), leading possibly to a slightly larger number of iterations.

The main conclusion to be drawn from the experiments is that the iterated penalty method is extremely robust and efficient, for $r = \rho$ sufficiently large.

(13.5.3) *Remark.* We note the strong similarity, for large τ, between the form $\tilde{a}(\cdot, \cdot)$ defined in (13.4.5) and $\tau a(\cdot, \cdot)$ where $a(\cdot, \cdot)$ is the form defined in (12.1.8) for (a_{ij}) (and hence A) being the identity matrix. Thus, we can expect convergence properties for the iterated penalty method applied to the problem (12.1.8–12.1.9) similar to the experiments described here.

13.x Exercises

13.x.1 Prove that $\|\mathcal{D}e^n\|_{\varPi} \leq C\gamma^n$ for some $C < \infty$ and $0 < \gamma < 1$ implies $e^n \to 0$ using (13.1.10). (Hint: pick $v = e^{n+1} - e^n$ in (13.1.10) and show that $\{e^n\}$ is a Cauchy sequence in V.)

13.x.2 Prove that the iterated penalty method (13.1.4), for suitable ρ, converges geometrically even with $r = 0$ in the case $a(\cdot, \cdot)$ is symmetric. (Hint: show that for suitable ρ, there is a $\gamma < 1$ such that

$$|a(u, v) - \rho\,(\mathcal{D}u, \mathcal{D}v)|^2 \leq \gamma a(u, u)a(v, v).$$

See Appendix III of (Glowinski 1984).)

13.x.3 Suppose $\mathcal{T}^h$ satisfies the conditions of Theorem 12.6.13. For $\underset{\sim}{V}_h$ as in (12.4.1) with $k \geq 4$, prove that the iteration (13.1.4) with $r = \rho$ sufficiently large converges to the solution of the mixed formulation of the Stokes equations, cf. (12.1.6). Show that the linear system that needs to be solved in (13.1.4) is symmetric, positive definite. Describe stopping criteria for the iteration and give error estimates.

13.x.4 Suppose $\mathcal{T}^h$ satisfies the conditions of Theorem 12.6.13. For $V_h = V_h^k \times V_h^k$ with $k \geq 4$, prove that the iteration (13.1.4) with $r = \rho$ sufficiently large converges to the solution of the mixed formulation of the scalar elliptic problem, cf. (12.1.9). Show that the linear system that needs to be solved in (13.1.4) is symmetric, positive definite. Describe stopping criteria for the iteration and give error estimates.

13.x.5 Prove that (12.5.7) implies $P_{\varPi_h}\,(\mathcal{D}Lp) = p$ for any $p \in \varPi_h$. (Hint: recall that $b(v, q) = (\mathcal{D}v, q)_{\varPi}$.)

13.x.6 Prove that $z \in Z_h$ is equivalent to $P_{\varPi_h}\,(\mathcal{D}z) = 0$. (Hint: recall that $b(v, q) = (\mathcal{D}v, q)_{\varPi}$.)

13.x.7 Prove that (13.3.1) and (13.3.2) are equivalent. (Hint: let W denote the coefficients of v with respect to the given basis of V_h and show that $W^t B^t B U^n = b(u^n, P_{\varPi_h}\mathcal{D}v)$.)

13.x.8 Prove that the augmented Lagrangian method (13.3.1), for suitable ρ, converges geometrically even with $r = 0$ in the case $a(\cdot, \cdot)$ is symmetric. (Hint: show that for suitable ρ, there is a $\gamma < 1$ such that

$$|a(u, v) - \rho\, (P_{\Pi_h} \mathcal{D}u, P_{\Pi_h} \mathcal{D}v)|^2 \le \gamma a(u, u) a(v, v)$$

as in 13.x.2.)

13.x.9 Let V_h be as in (12.4.1) and Π_h as in (12.4.4), for $k \ge 2$. Prove that the algorithm (13.3.2) (or equivalently (13.3.1)) converges to the solution of the mixed formulation of the Stokes equations, cf. (12.1.6). Show that the linear system that needs to be solved in (13.1.4) is symmetric, positive definite.

13.x.10 Formulate the augmented Lagrangian method for solving the equations associated with the Taylor-Hood discretization (see Theorem 12.6.6) of the time-stepping scheme (13.4.4). State and prove the corresponding convergence result.

13.x.11 Formulate the augmented Lagrangian method for solving the equations associated with the Taylor-Hood discretization (see Theorem 12.6.6) of the time-stepping scheme (13.4.8). State and prove the corresponding convergence result.

Chapter 14

Applications of Operator-Interpolation Theory

Interpolation spaces are useful technical tools. They allow one to bridge between known results, yielding new results that could not be obtained directly. They also provide a concept of fractional-order derivatives, extending the definition of the Sobolev spaces used so far. Such extensions allow one to measure more precisely, for example, the regularity of solutions to elliptic boundary value problems.

There are several methods of defining interpolation spaces, and hence several (not necessarily equivalent) definitions of fractional order Sobolev spaces. Two of these are the "real" and "complex" methods. For the special case of Hilbert spaces, there is also a technique based on fractional powers of an operator. Finally, for Sobolev spaces, there are intrinsic norms that can be defined. For s a real number in the interval $(0, 1)$, for $1 \le p < \infty$ and for k a non-negative integer, we define

$$(14.0.1) \quad \|u\|_{W_p^{k+s}(\Omega)}^p := \|u\|_{W_p^k(\Omega)}^p + \sum_{|\alpha|=k} \int_\Omega \int_\Omega \frac{|u^{(\alpha)}(x) - u^{(\alpha)}(y)|^p}{|x-y|^{n+sp}} \, dx dy.$$

Furthermore, in the case of special domains (e.g. $\mathbb{R}^n$) it is possible to give alternate characterizations of Sobolev norms using the Fourier transform (Adams 1975).

It is known (Adams 1975) that the real and complex methods of interpolation yield different spaces when applied to the Sobolev spaces with $p \ne 2$. Some of the possible relationships are discussed in (Bergh & Löfstrom 1976).

14.1 The Real Method of Interpolation

Given two Banach spaces, B_0 and B_1, we will define Banach spaces that "interpolate" between them. For simplicity, we will make the assumption that $B_1 \subset B_0$. For example, one can think of $B_0 = W_p^k(\Omega)$ and $B_1 = W_p^m(\Omega)$ where $k \le m$. For any $u \in B_0$ and $t > 0$, define

$$K(t, u) := \inf_{v \in B_1} \left(\|u - v\|_{B_0} + t\|v\|_{B_1} \right).$$

In a sense, K measures how well u can be approximated by B_1 with a penalty factor t times the B_1-norm of the approximant.

Some simple behavior for $K(t, u)$ can be deduced immediately. For $u \in B_1$, we conclude that $K(t, u) \leq t \|u\|_{B_1}$ by choosing $v = u$. On the other hand, $K(t, u) \leq \|u\|_{B_0}$ by choosing $v = 0$. For $0 < \theta < 1$ and $1 \leq p < \infty$, define a norm

$$(14.1.1) \qquad \|u\|_{[B_0, B_1]_{\theta, p}} := \left(\int_0^\infty t^{-\theta p} K(t, u)^p \, \frac{dt}{t} \right)^{1/p}.$$

When $p = \infty$, we define

$$\|u\|_{[B_0, B_1]_{\theta, \infty}} := \sup_{0 < t < \infty} t^{-\theta} K(t, u).$$

The set

$$[B_0, B_1]_{\theta, p} = B_{\theta, p} := \left\{ u \in B_0 \; : \; \|u\|_{[B_0, B_1]_{\theta, p}} < \infty \right\}$$

forms a Banach space with norm (14.1.1) (Bergh & Löfstrom 1976).

Some simple properties can be deduced immediately from the definition. First, suppose that A_i and B_i are two pairs of Banach spaces as above, and that $B_i \subset A_i$ ($i = 0, 1$). Then $B_{\theta, p} \subset A_{\theta, p}$ (see exercise 14.x.1). An application of this is the following. Let $\Omega \subset \tilde{\Omega}$. Then

$$(14.1.2) \qquad \left[W_p^k(\tilde{\Omega}), W_p^m(\tilde{\Omega}) \right]_{\theta, p} \subset \left[W_p^k(\Omega), W_p^m(\Omega) \right]_{\theta, p}$$

together with the corresponding norm inequality (see exercise 14.x.2).

The interpolation spaces form a "scale" because of our simplifying assumption that $B_1 \subset B_0$: we always have $B_1 \subset B_{\theta, p} \subset B_0$ (cf. exercise 14.x.4). The second of these inclusions follows in part from the estimate (cf. exercise 14.x.3)

$$(14.1.3) \qquad K(t, u) \leq C_{\theta, p} t^\theta \|u\|_{B_{\theta, p}}.$$

This can be interpreted as saying that $u \in B_{\theta, p}$ implies that u can be approximated to order t^θ in B_0. In fact, if $p < \infty$, then $u \in B_{\theta, p}$ implies (exercise 14.x.5) that

$$(14.1.4) \qquad K(t, u) = o(t^\theta).$$

An application of this more-refined approximation estimate will be given in the section on finite element applications.

One immediate consequence of (14.1.3) is that $B_{\theta, p} \subset B_{\theta, \infty}$ for all $1 \leq p \leq \infty$. Further, it is known (Bergh & Löfstrom 1976) that $B_{\theta, 1} \subset B_{\theta, p}$ for all $1 \leq p \leq \infty$. If $\theta_1 \leq \theta_2$ then $B_{\theta_2, p} \subset B_{\theta_1, p}$ for all $1 \leq p \leq \infty$, and if $p \leq q$ then $B_{\theta, p} \subset B_{\theta, q}$ for all $0 < \theta < 1$ (Bergh & Löfstrom 1976).

The key result of interpolation theory concerns operators on Banach spaces.

(14.1.5) Proposition. *Suppose that A_i and B_i are two pairs of Banach spaces as above, and that T is a linear operator that maps A_i to B_i ($i = 0, 1$). Then T maps $A_{\theta,p}$ to $B_{\theta,p}$. Moreover,*

$$(14.1.6) \qquad \|T\|_{A_{\theta,p} \to B_{\theta,p}} \leq \|T\|_{A_0 \to B_0}^{1-\theta} \|T\|_{A_1 \to B_1}^{\theta}.$$

Proof. Let $M_i := \|T\|_{A_i \to B_i}$. For any $v \in A_1$,

$$K_B(t, Tu) \leq \|Tu - Tv\|_{B_0} + t\|Tv\|_{B_1}$$
$$\leq M_0 \|u - v\|_{A_0} + t M_1 \|v\|_{A_1}.$$

Taking the infimum over $v \in A_1$, we find

$$K_B(t, Tu) \leq M_0 K_A(t M_1/M_0, u).$$

Integrate this and make the change of variables $s = t M_1/M_0$ in the integral on the right-hand side of the inequality.   □

There are two additional results regarding the real method of interpolation that we require but will not prove. For proofs, see (Bergh & Löfstrom 1976). The first, called the "reiteration" theorem, says that

$$(14.1.7) \qquad \left[[B_0, B_1]_{\theta_0, p_0}, [B_0, B_1]_{\theta_1, p_1} \right]_{\lambda, q} = [B_0, B_1]_{(1-\lambda)\theta_0 + \lambda\theta_1, q}$$

for any $0 \leq \theta_0 < \theta_1 \leq 1$, $1 \leq p_0, p_1, q \leq \infty$ and $0 < \lambda < 1$. The second relates to dual spaces. Let us assume that B_1 is dense in B_0. Then

$$(14.1.8) \qquad [B_0, B_1]_{\theta, p}' = [B_1', B_0']_{1-\theta, p'}$$

for any $0 < \theta < 1$ and $1 \leq p < \infty$, where $\frac{1}{p} + \frac{1}{p'} = 1$. Applications of these results will be given in the next section with regard to negative-index Sobolev spaces.

Finally, we note that the restriction $B_1 \subset B_0$ is unnecessary. In general, it is only necessary that sums $v_0 + v_1$ be well defined for $v_i \in B_i$. It then holds that $[B_0, B_1]_{\theta, p} = [B_1, B_0]_{1-\theta, p}$. Thus, the duality theorem can be written $[B_0, B_1]_{\theta, p}' = [B_0', B_1']_{\theta, p'}$

14.2 Real Interpolation of Sobolev Spaces

The most basic interpolation result, the Riesz convexity theorem, relates Lebesgue spaces. In our setting, this is just the result that

$$(14.2.1) \qquad L^p(\Omega) = \left[L^1(\Omega), L^\infty(\Omega) \right]_{1-\frac{1}{p}, p} \qquad \forall 1 < p < \infty$$

(cf. Bergh & Löfstrom 1976 and Stein & Weiss 1971). This also extends to the Sobolev spaces:

$$(14.2.2) \qquad W_p^k(\Omega) = \left[W_1^k(\Omega), W_\infty^k(\Omega) \right]_{1-\frac{1}{p},p} \qquad \forall 1 < p < \infty$$

with equivalent norms (cf. (DeVore & Scherer 1979)). So far, we have only interpolated in the p variable, with k fixed. Now we consider keeping p fixed and varying k.

Another possible definition of fractional-order Sobolev space is given by interpolation. We will prove the following result which shows their equivalence.

(14.2.3) Theorem. *Let* $0 < s < 1$. *If* Ω *has a Lipschitz boundary, then*

$$W_p^{k+s}(\Omega) = \left[W_p^k(\Omega), W_p^{k+1}(\Omega) \right]_{s,p}$$

and the norms are equivalent.

Proof. When $\Omega = \mathbb{R}^n$, it is known that the two definitions yield identical spaces and equivalent norms. For example, this is proved in (Adams 1975) using an interpolation method (method of traces) that is equivalent (Bergh & Löfstrom 1976) to the "K" method described earlier. Also see Theorem 6.4.5, item (4), together with exercise 7 of Chapter 6, in (Bergh & Löfstrom 1976). From the extension result (1.4.5) of Stein mentioned earlier, we know that there is an operator $E_S : W_p^k(\Omega) \longrightarrow W_p^k(\mathbb{R}^n)$ for all k. Interpolating this operator, we find

$$
\begin{aligned}
\|u\|_{W_p^{k+s}(\Omega)} &= \|E_S u\|_{W_p^{k+s}(\Omega)} \\
&\le \|E_S u\|_{W_p^{k+s}(\mathbb{R}^n)} && \text{(by 14.0.1)} \\
&\le C \|E_S u\|_{\left[W_p^k(\mathbb{R}^n), W_p^{k+1}(\mathbb{R}^n) \right]_{s,p}} && \text{(using the } \mathbb{R}^n \text{ case)} \\
&\le C \|u\|_{\left[W_p^k(\Omega), W_p^{k+1}(\Omega) \right]_{s,p}} && \text{(by 14.1.5).}
\end{aligned}
$$

Note that it is important to know that the operator E_S is defined for both $W_p^k(\Omega)$ and $W_p^{k+1}(\Omega)$ in this argument, otherwise operator interpolation would not be applicable.

Conversely, in (Grisvard 1985), it is stated that there exists an extension operator, E_G, that maps $W_p^{k+s}(\Omega)$ to $W_p^{k+s}(\mathbb{R}^n)$ such that $E_G u|_\Omega = u$, provided that Ω is a Lipschitz domain, where, the definition of fractional-order Sobolev norm is as given in (14.0.1). See (DeVore & Sharpley 1993) for a proof. Then

$$
\begin{aligned}
\|u\|_{\left[W_p^k(\Omega), W_p^{k+1}(\Omega) \right]_{s,p}} &= \|E_G u\|_{\left[W_p^k(\Omega), W_p^{k+1}(\Omega) \right]_{s,p}} \\
&\le \|E_G u\|_{\left[W_p^k(\mathbb{R}^n), W_p^{k+1}(\mathbb{R}^n) \right]_{s,p}} && \text{(by 14.1.2 \& 14.x.2)} \\
&\le C \|E_G u\|_{W_p^{k+s}(\mathbb{R}^n)} && \text{(using the } \mathbb{R}^n \text{ case)} \\
&\le C \|u\|_{W_p^{k+s}(\Omega)}.
\end{aligned}
$$

Note that we can allow E_G to depend on s. Thus, we conclude that the intrinsic norm (14.0.1) is equivalent to the one obtained by real interpolation in the case of a Lipschitz domain.   □

For more complex domains, it is not clear that the spaces considered in Theorem 14.2.3 would necessarily be equivalent.

The spaces $W_p^{k+1}(\Omega)$ and $\left[W_p^k(\Omega), W_p^{k+2}(\Omega)\right]_{\frac{1}{2},p}$ are not the same, unless $p = 2$, although they are known (Bergh & Löfstrom 1976) to be very close, in that

$$\left[W_p^k(\Omega), W_p^{k+2}(\Omega)\right]_{\frac{1}{2},1} \subset W_p^{k+1}(\Omega) \subset \left[W_p^k(\Omega), W_p^{k+2}(\Omega)\right]_{\frac{1}{2},\infty}.$$

However, the proof of Theorem 14.2.3 also shows that

$$(14.2.4) \qquad \left[W_p^m(\Omega), W_p^k(\Omega)\right]_{\theta,p} = W_p^{(1-\theta)m+\theta k}(\Omega),$$

provided that $(1 - \theta)m + \theta k$ is *not* an integer. Moreover, the reiteration theorem (14.1.7) allows us to conclude that (14.2.4) holds even if m and k are not integers.

We can define Sobolev spaces for negative, fractional indices by duality as was done in the integer case, that is, $W_p^{-s}(\Omega) := W_{p'}^s(\Omega)'$. The duality theorem (14.1.8) allows us to conclude that (14.2.4) holds as well for negative indices.

We emphasize that the relation (14.2.4) does hold without restriction when $p = 2$, that is

$$(14.2.5) \qquad \left[H^m(\Omega), H^k(\Omega)\right]_{\theta,2} = H^{(1-\theta)m+\theta k}(\Omega).$$

The proof of Theorem 14.2.3 demonstrates this, once we know the result is true for the case $\Omega = \mathbb{R}^n$. The latter can be shown by using another, equivalent definition of the $H^s(\mathbb{R}^n)$ norm, namely

$$(14.2.6) \qquad \|u\|_{H_F^s(\mathbb{R}^n)} := \left(\int_{\mathbb{R}^n} \left(1 + |\xi|^2\right)^s |\widehat{u}|^2 \, d\xi\right)^{1/2},$$

where $\widehat{u}$ denotes the Fourier transform of u. Using this as an intermediary (cf. exercises 14.x.6 and 14.x.7), one completes the equivalence.

Finally, we observe that (14.2.5) actually holds without the assumption that m and k have the same sign. To begin with

$$\left[H^{-k}(\Omega), H^k(\Omega)\right]_{1/2,2} = L^2(\Omega) = H^0(\Omega)$$

(cf. Bergh & Löfstrom 1976). Using the reiteration theorem (14.1.7), the following is obtained.

(14.2.7) Theorem. *If Ω has a Lipschitz boundary, then (14.2.5) holds for arbitrary real numbers m and k, and the norms are equivalent.*

14.3 Finite Element Convergence Estimates

Consider a situation in which the solutions $u \in H^m(\Omega)$ and $u_h \in V_h$ to a variational problem and its approximation satisfy (cf. Cea's Theorem 2.8.1)

$$(14.3.1) \qquad \|u - u_h\|_{H^m(\Omega)} \leq c_0 \inf_{v \in V_h} \|u - v\|_{H^m(\Omega)}.$$

Provided there is an interpolant $\mathcal{I}^h u \in V_h$ (as in Theorem 4.4.20) such that

$$\left\|u - \mathcal{I}^h u\right\|_{H^m(\Omega)} \leq Ch^{k-m}\|u\|_{H^k(\Omega)}$$

and u is sufficiently smooth, this implies

$$(14.3.2) \qquad \|u - u_h\|_{H^m(\Omega)} \leq c_1 h^{k-m}\|u\|_{H^k(\Omega)}.$$

However, if u is less smooth, say such that $\mathcal{I}^h u$ is not well defined, we might not be able to conclude anything. For example, the standard Hermite cubic interpolant is not defined on functions in $H^2(\Omega)$ in two dimensions, yet the above holds for $m = 1$ and $k = 4$. The Argyris element gives rise to an example with $m = 2$ and $k \geq 6$, yet its standard nodal interpolant is undefined on $H^3(\Omega)$. Since the data of the problem may not guarantee a smooth solution, these restrictions can be a serious concern. However, we can use Banach-space interpolation to extend (14.3.2) to apply for values of k unrelated to a given interpolant.

(14.3.3) Theorem. *Suppose that (14.3.1) and (14.3.2) hold. Let $m < s < k$. Then for any $u \in H^s(\Omega)$,*

$$(14.3.4) \qquad \|u - u_h\|_{H^m(\Omega)} \leq Ch^{s-m}\|u\|_{H^s(\Omega)}.$$

Proof. Let us define an operator, T, that maps u to the error $u - u_h$, that is,

$$Tu := u - u_h.$$

Estimate (14.3.2) implies that T maps $H^k(\Omega)$ to $H^m(\Omega)$, with

$$\|T\|_{H^k(\Omega) \to H^m(\Omega)} \leq c_1 h^{k-m}.$$

In addition, (14.3.1) implies (take $v = 0$)

$$\|T\|_{H^m(\Omega) \to H^m(\Omega)} \leq c_0.$$

Thus, we have the setting of Banach-space interpolation: $A_0 = H^m(\Omega)$, $A_1 = H^k(\Omega)$, and $B_0 = B_1 = H^m(\Omega)$. We conclude that

$$\|T\|_{H^s(\Omega) \to H^m(\Omega)} \leq Ch^{s-m}$$

for any $m < s < k$. This is equivalent to (14.3.4). $\square$

It is important to know that the constant, C, in (14.3.4) does not depend on h or V_h in any way. A key to this is the fact that the constant in inequality (14.1.6) is equal to one. However, there is an additional technicality regarding the constants involved. Proposition 14.1.5 implies that

$$\|T\|_{[H^m(\Omega),H^k(\Omega)]_{\theta,2}\to H^m(\Omega)} \le c_0^{1-\theta}c_1^\theta h^{s-m}, \quad s = m + \theta(k-m).$$

Of course, $[H^m(\Omega), H^k(\Omega)]_{\theta,2} = H^s(\Omega) = W_2^s(\Omega)$, but if the norm of the latter is given by (14.0.1), the norm of $[H^m(\Omega), H^k(\Omega)]_{\theta,2}$ will not be identical, only equivalent. Thus, the constant, C, in (14.3.4) depends on the constant in the equivalence of norms in Theorem 14.2.3. How this constant depends on k, s, p and Ω is beyond the scope of this book. However, it is easy to see that C in (14.3.4) can be taken to be a continuous function of s in the open interval $m < s < k$.

Estimates such as (14.3.2), which relate to the order of approximation of piecewise polynomial spaces, are sharp, that is $\|u - \mathcal{I}^h u\|_{H^m(\Omega)}$ would not in general go to zero faster than $\mathcal{O}(h^{k-m})$, as can be seen by applying the interpolant to a polynomial of higher order. However, when the function u being approximated is less smooth, the degree of smoothness no longer determines the order of approximation so precisely.

(14.3.5) Theorem. *Suppose that (14.3.1) and (14.3.2) hold. Let $m < s < k$. Then for any $u \in H^s(\Omega)$,*

$$(14.3.6) \qquad \|u - u_h\|_{H^m(\Omega)} = o(h^{s-m}).$$

Proof. Recall the definition of T in the proof of Theorem 14.3.3. For any $v \in H^k(\Omega)$,

$$
\begin{aligned}
\|u - u_h\|_{H^m(\Omega)} &= \|Tu\|_{H^m(\Omega)} \\
&\le \|Tu - Tv\|_{H^m(\Omega)} + \|Tv\|_{H^m(\Omega)} \\
&\le c_0 \|u - v\|_{H^m(\Omega)} + c_1 h^{k-m} \|v\|_{H^k(\Omega)}.
\end{aligned}
$$

Taking the infimum over all $v \in H^k(\Omega)$, we find

$$\|u - u_h\|_{H^m(\Omega)} \le c_0 K((c_1/c_0)h^{k-m}, u).$$

Applying (14.1.4), we conclude that (14.3.6) holds. □

Such techniques are not restricted to the Hilbert spaces. In Chapter 7, estimates of the form

$$(14.3.7) \qquad \|u_h\|_{W_p^1(\Omega)} \le C \|u\|_{W_p^1(\Omega)}$$

are proved for $p \ne 2$. From this it follows that

$$(14.3.8) \qquad \|u - u_h\|_{W_p^1(\Omega)} \le C h^{k-1} \|u\|_{W_p^k(\Omega)}$$

provided u is sufficiently smooth, again using an interpolant. Via Banach-space interpolation as above, this result can be extended to hold as follows:

$$(14.3.9) \qquad \|u - u_h\|_{W_p^1(\Omega)} \le Ch^{s-1}\|u\|_{W_p^s(\Omega)}, \quad 1 < s < k.$$

Moreover, for $u \in W_p^s(\Omega)$, we also conclude that

$$(14.3.10) \qquad \|u - u_h\|_{W_p^1(\Omega)} = o(h^{s-1}).$$

So far we have considered estimates of $u - u_h$ in the case that u lies in an intermediate (interpolation) space. Banach-space interpolation can also be applied to obtain estimates of $u - u_h$ in intermediate spaces. Recall that in deriving estimates for the biharmonic problem in Sect. 5.9, an estimate in $H^1(\Omega)$ was not obtained, only estimates in $H^2(\Omega)$ and $H^{-s}(\Omega)$ for $s \ge 0$. Moreover, the techniques of proof, say for Theorem 5.9.9, require an approximation result of the sort discussed above for functions with less than maximal regularity. Thus, we may only know that

$$(14.3.11) \qquad \|u - u_h\|_{H^{-r}(\Omega)} \le Ch^{r+k}\|u\|_{H^k(\Omega)}.$$

The following captures this situation.

(14.3.12) Theorem. *Suppose that (14.3.2) and (14.3.11) hold. Let $-r < s < m$. Then*

$$\|u - u_h\|_{H^s(\Omega)} \le Ch^{k-s}\|u\|_{H^k(\Omega)}.$$

Proof. We apply (14.1.6) to the operator, T, defined in the proof of Theorem 14.3.3 that maps u to the error $u - u_h$, except that we now take $A_0 = A_1 = H^k(\Omega)$, and $B_0 = H^m(\Omega)$ and $B_1 = H^{-r}(\Omega)$.   □

14.4 The Simultaneous Approximation Theorem

There have been several instances in the book, and numerous more in the literature, where it is essential to know that a particular approximant (such as an interpolant) has appropriate orders of approximation in two different norms simultaneously. Assumption (8.1.7) is just one such example. It turns out that in many cases, such an assumption is redundant, with the simultaneous approximability being a consequence of approximability in a single norm. We give a simple example of this and refer to (Bramble & Scott 1978) for the general theory.

Let us suppose that we have a family of spaces $V_h \subset H^m(\Omega)$ with the property that, for all $u \in H^k(\Omega)$ and $0 < h \le 1$,

$$(14.4.1) \qquad \inf_{v \in V_h} \|u - v\|_{H^m(\Omega)} \le Ch^{k-m}\|u\|_{H^k(\Omega)}.$$

The following result can be found in (Bramble & Scott 1978).

(14.4.2) Theorem. *Suppose that condition (14.4.1) holds, and let $s < m$ and $m \leq r \leq k$. Then there is a constant C such that*

$$\inf_{v \in V_h} \left(h^s \|u - v\|_{H^s(\Omega)} + h^m \|u - v\|_{H^m(\Omega)} \right) \leq C h^r \|u\|_{H^r(\Omega)}$$

provided $u \in H^r(\Omega)$. If $r < k$ and $u \in H^r(\Omega)$, then

$$\inf_{v \in V_h} \left(h^s \|u - v\|_{H^s(\Omega)} + h^m \|u - v\|_{H^m(\Omega)} \right) / h^r \to 0 \quad as \quad h \to 0.$$

Note that s can be negative in the theorem, and the result can easily be extended to any finite sum of norms instead of just two. Corresponding results for other Sobolev spaces ($p \neq 2$) can be derived from (Bramble & Scott 1978). However, a condition such as (8.1.7) would not follow so easily, as it involves a family of weighted norms where the weights depend on the approximation parameter h.

14.5 Precise Characterizations of Regularity

So far, the second interpolation index has not been used in a significant way. However, it allows precise characterizations of various regularity properties. For example, functions that are piecewise smooth (Scott 1979) lie in the space $\left[L^2(\Omega), H^1(\Omega) \right]_{1/2, \infty}$ but do not lie in any smaller space (e.g., functions with discontinuities across an internal boundary do not lie in $H^{1/2}$). If such a function, f, is the data for a variational problem as discussed in Chapter 5, then the solution u will lie in a corresponding interpolation space. For example, suppose that a regularity estimate

$$(14.5.1) \qquad \|u\|_{H^{k+2m}(\Omega)} \leq C \|f\|_{H^k(\Omega)}$$

holds for some $k > 1/2$ (cf. Dauge 1988) for the solution to

$$a(u, v) = (f, v) \quad \forall v \in V$$

where $a(\cdot, \cdot)$ is continuous and coercive on $V \subset H^m(\Omega)$. Recall that (cf. exercise 2.x.9) we also have $\|u\|_{H^m(\Omega)} \leq C\|f\|_{H^m(\Omega)'} \leq C\|f\|_{H^{-m}(\Omega)}$. Interpolating the solution operator $f \to u$ we find

$$(14.5.2) \qquad \|u\|_{[H^{2m}(\Omega), H^{2m+1}(\Omega)]_{1/2, \infty}} \leq C\|f\|_{[H^0(\Omega), H^1(\Omega)]_{1/2, \infty}}.$$

Combining this regularity result with the techniques used to prove Theorem 14.3.3, we obtain the following.

(14.5.3) Theorem. *Suppose that f is piecewise smooth (Scott 1979), that (14.5.1) holds for some $k > 1/2$ and that (14.3.2) holds for $k \geq 2m + 1$. Then*

$$\|u - u_h\|_{H^m(\Omega)} \leq Ch^{m+\frac{1}{2}}.$$

14.x Exercises

14.x.1 Suppose that A_i and B_i are two pairs of Banach spaces as above, and that $B_i \subset A_i$ where the embeddings are continuous ($i = 0, 1$). Prove that $B_{\theta,p} \subset A_{\theta,p}$. (Hint: let C be such that $\|v\|_{A_i} \leq C\|v\|_{B_i}$ and prove that $K_A(t, u) \leq CK_B(t, u)$.)

14.x.2 Prove that the norm of the inclusion operator (14.1.2) is 1.

14.x.3 Prove (14.1.3). (Hint: observe that $\min\{1, s/t\}K(t, u) \leq K(s, u)$ and integrate this inequality.)

14.x.4 Given that $B_1 \subset B_0$, prove that $B_1 \subset B_{\theta,p} \subset B_0$, where all inclusions are continuous. (Hint: suppose that $\|v\|_{B_0} \leq C\|v\|_{B_1}$ and show that $\|u\|_{B_0} \leq K(C, u)$, then use exercise 14.x.3 to verify the second inclusion. To prove the first, show that $K(t, u) \leq \min\{C, t\}\|u\|_{B_1}$ using the choices for v given just prior to (14.1.1).)

14.x.5 Prove (14.1.4). (Hint: observe that $t^{-\theta}K(t, u)$ is in $L^p(0, \infty; dt/t)$ and hence must go to zero as $t \to 0$.)

14.x.6 Prove (14.2.6) is an equivalent norm for $H^s(\Omega)$ in the case s is an integer. (Hint: recall that $\widehat{D^\alpha u} = (i\xi)^\alpha \hat{u}$ and that the Fourier transform is an isomorphism in L^2. Note that $|\xi|^r \leq 1 + |\xi|^s$ if $0 < r < s$.)

14.x.7 Prove (14.2.6) is equivalent to (14.0.1) in the case s is fractional. (Hint: use Plancherel's and Fubini's Theorems to write

$$\int_{\mathbb{R}^n} \int_{\mathbb{R}^n} \frac{|u(x + r) - u(x)|^2}{|r|^{d+2s}} \, dx\,dr$$

$$= \int_{\mathbb{R}^n} \int_{\mathbb{R}^n} |\hat{u}(\xi)| \left(e^{ir\cdot\xi} - 1\right)|^2 \, d\xi \frac{dr}{|r|^{d+2s}}$$

and prove that

$$\int_{\mathbb{R}^n} \frac{|e^{ir\cdot\xi} - 1|^2}{|r|^{d+2s}} \, dr \leq C \left(1 + |\xi|^2\right)^s$$

for $\xi \in \mathbb{R}^n$ and $0 < s < 1$.)

References

In addition to the references appearing in the text, we have included a number of textbooks and monographs for general reference. One category is books on finite element methods and includes (Aubin 1980), (Axelsson & Barker 1984), (Becker, Carey & Oden 1981), (Braess 2001), (Ciarlet 1976), (Ciarlet 1991), (Fairweather 1978), (Fried 1979), (Hughes 2000), (Oden & Reddy 1976), (Schwab 1998), (Szabó & Babuška 1991), and (Schultz 1972). Another category is books on partial differential equations and includes (Ciarlet 1988), (Gilbarg & Trudinger 1983), (Lions & Magenes 1972), (Showalter 1977), (Sobolev 1964), (Taylor 1996), (Wendland 1979) and (Wloka 1987).

Adams, R. A. (1975) *Sobolev Spaces.* Academic Press, New York, 1975

Agmon, S. (1965) *Lectures on Elliptic Boundary Value Problems.* Van Nostrand, Princeton, NJ, 1965

Ainsworth, M., Oden, J.T. (2000) *A Posteriori Error Estimation in Finite Element Analysis.* John Wiley & Sons, New York, 2000

Arnold, D.N., Boffi, D., Falk, R.S. (2000) Approximation by quadrilateral finite elements. Math. Comp. (submitted)

Arnold, D.N., Brezzi, Douglas, J. Jr. (1984) PEERS: A new mixed finite element for plane elasticity. Japan J. Appl. Math. **1** (1984) 347-367

Arnold, D.N., Douglas, J. Jr., Gupta, C.P. (1984) A family of higher order mixed finite element methods for plane elasticity. Numer. Math. **45** (1984) 1-22

Arnold, D.N., Scott, L.R., Vogelius, M. (1988) Regular inversion of the divergence operator with Dirichlet boundary conditions on a polygon. Ann. Scuola Norm. Sup. Pisa Cl. Sci.-Serie IV **XV** (1988) 169-192

Aubin, J.P. (1980) *Approximation of Elliptic Boundary-value Problems.* R. E. Krieger Pub. Co., Huntington, NY (and now Melbourne, FL), 1980

Axelsson, O., Barker, V.A. (1984) *Finite Element Solution of Boundary Value Problems.* Academic Press, Orlando, 1984

Babuška, I., Aziz, A.K. (1972) Survey lectures on the mathematical foundations of the finite element method. in: *The Mathematical Foundations of the Finite Element Method with Applications to Partial Differential Equations.* A.K. Aziz, ed. Academic Press, New York and London, 1972

Babuška, I., Chandra, J., Flaherty, J.E., eds. (1983) *Adaptive Computational Methods for Partial Differential Equations.* Society for Industrial and Applied Mathematics, Philadelphia, 1983

Babuška, I., Strouboulis, T. (2001) *The Finite Element Method and Its Reliability.* Oxford University Press, Oxford, 2001

Babuška, I., Suri, M. (1992a) On locking and robustness in the finite element method. SIAM J. Num. Anal. **29** (1992) 1261-1293

Babuška, I., Suri, M. (1992b) Locking effects in the finite element approximation of elasticity problems. Numer. Math. **62** (1992) 439-463

Babuška, I. (1973) The finite element method with Lagrangian multipliers. Numer. Math. **20** (1973) 179-192

Babuška, I., Zienkiewicz, O. C., Gago, J., de A. Oliveira, E.R, eds. (1986) *Accuracy Estimates and Adaptive Refinements in Finite Element Computation.* John Wiley and Sons, New York, 1986

Bagheri, B., Scott, L.R., Zhang, S. (1992) Implementation issues for high-order finite element methods. Finite Elem. Anal. Des. **16** (1994), 175-189

Bank, R.E., Dupont, T. (1981) An optimal order process for solving finite element equations. Math. Comp. **36** (1981) 35-51

Bank, R.E., Scott, L.R. (1989) On the conditioning of finite element equations with highly refined meshes. SIAM J. Num. Anal. **26** (1989) 1383-1394

Becker, E.B., Carey, G.F., Oden, J.T. (1981) *Finite Elements.* Prentice-Hall, Englewood Cliffs, NJ, 1981

Becker, R., Rannacher, R. (2001) An optimal control approach to a posteriori error estimation in finite element methods. Acta Numerica **10** (2001) 1–102

Bergh, J., Löfstrom, J. (1976) *Interpolation Spaces, an Introduction.* Springer-Verlag, Berlin, 1976

Bergman, S., Schiffer, M. (1953) *Kernel Functions and Elliptic Differential Equations in Mathematical Physics.* Academic Press, New York, 1953

Bjørstad, P., Mandel, J. (1991) On the spectra of sums of orthogonal projections with applications to parallel computing. BIT **49** (1991) 76-88

Blum, H., Rannacher, R. (1980) On the boundary value problem of the biharmonic operator on domains with angular corners. Math. Meth. Appl. Sci. **2** (1980) 556-581

Braess, D. (2001) Finite Elements. Second Edition. Cambridge University Press, 2001

Braess, D., Hackbusch, W. (1983) A new convergence proof for the multigrid method including the V-cycle. SIAM J. Numer. Anal. **20** (1983) 967-975

Bramble, J.H. (1966) A second order finite difference analog of the first biharmonic boundary value problem. Numer. Math. **9** (1966) 236-249

Bramble, J.H. (1995) Interpolation between Sobolev spaces in Lipschitz domains with an application to multigrid theory. Math. Comp. **64** (1995) 1359-1365

Bramble, J.H. (1993) *Multigrid Methods.* Pitman Research Notes in Mathematics, v. 294, John Wiley and Sons, 1993

Bramble, J.H., Pasciak, J.E. (1987) New convergence estimates for multigrid algorithms. Math. Comp. **49** (1987) 311-329

Bramble, J.H., Pasciak, J.E. (1993) New estimates for multilevel algorithms including the V-cycle. Math. Comp. **60** (1993) 447-471

Bramble, J.H., Pasciak, J.E., Schatz, A.H. (1986) The construction of preconditioners for elliptic problems by substructuring I. Math. Comp. **47** (1986) 103-134

Bramble, J.H., Pasciak, J.E., Schatz, A.H. (1989) The construction of preconditioners for elliptic problems by substructuring, IV. Math. Comp. **53** (1989) 1-24

Bramble, J.H., Pasciak, J.E., Wang, J., Xu, J. (1991) Convergence estimates for multigrid algorithms without regularity assumptions. Math. Comp. **57** (1991) 23-45

Bramble, J.H., Pasciak, J.E., Xu, J. (1990) Parallel multilevel preconditioners. Math. Comp. **55** (1990) 1-22

Bramble, J.H., Scott, L.R. (1978) Simultaneous approximation in scales of Banach spaces. Math. Comp. **32** (1978) 947-954

Bramble, J.H., Zhang, X. (2000) The Analysis of Multigrid Methods. Handbook of Numerical Analysis, vol. VII, P.G. Ciarlet and J.L. Lions, eds., Elsevier Science, (2000) 173-415

Brenner, S.C. (1996) Preconditioning complicated finite elements by simple finite elements. SIAM J. Sci. Comput. **17** (1996) 1269-1274

Brenner, S.C. (2002) Convergence of the multigrid V-cycle algorithm for second order boundary value problems without full elliptic regularity. Math. Comp. **71** (2002) 507-525

Brenner, S.C. (2000) Lower bounds for two-level additive Schwarz preconditioners with small overlap. SIAM J. Sci. Comput. **21** (2000) 1657-1669

Brenner, S.C., Sung, L.-Y. (1992) Linear finite element methods for planar linear elasticity. Math. Comp. **59** (1992) 321-338

Brenner, S.C., Sung, L.-Y. (1999) Balancing domain decomposition for nonconforming plate elements. Numer. Math. **83** (1999) 25-52

Brenner, S.C., Sung, L.-Y. (2000) Discrete Sobolev and Poincaré inequalities via Fourier series, East-West J. Numer. Math. **8** (2000) 83-92

Brenner, S.C., Sung, L.-Y. (2000) Lower bounds for nonoverlapping domain decomposition preconditioners in two dimensions. Math. Comp. **69** (2000) 1319-1339

Brezzi, F. (1974) On the existence, uniqueness and approximation of saddle-point problems arising from Lagrangian multipliers. RAIRO Anal. Numér. **8** (1974) 129-151

Brezzi, F., Falk, R. (1991) Stability of higher-order Hood-Taylor methods. SIAM J. Numer. Anal. **28** (1991) 581-590

Brezzi, F., Fortin, M. (1991) *Mixed and Hybrid Finite Element Methods*. Springer-Verlag, New York, 1991

Chan, T.F., Mathew, T.P. (1994) Domain decomposition algorithms. Acta Numerica (1994) 61-143

Ciarlet, P.G. (1976) *Numerical Analysis of the Finite Element Method*. Les Presses de l'Université de Montreal, Quebec, 1976

Ciarlet, P.G. (1978) *The Finite Element Method for Elliptic Problems*. North-Holland, Amsterdam, New York, Oxford, 1978

Ciarlet, P.G. (1988) *Mathematical Elasticity, Vol. I: Three-dimensional Elasticity*. North-Holland, Amsterdam, New York, Oxford, 1988

Ciarlet, P.G. (1991) Basic Error Estimates for Elliptic Problems. Handbook of Numerical Analysis, vol. *II*, P.G. Ciarlet and J.L. Lions, eds., Elsevier Science, (1991) 17-351

Ciarlet, P.G., Lions, J.L. (1991) *Handbook of Numerical Analysis, vol. II. Finite Element Methods (Part 1)*. North-Holland, Amsterdam, New York, Oxford, 1991

Ciarlet, P.G., Raviart, P.-A. (1972a) General Lagrange and Hermite interpolation in $\mathbb{R}^n$ with applications to finite element methods. Arch. Rational Mech. Anal. **46** (1972) 177-199

Ciarlet, P.G., Raviart, P.-A. (1972b) Interpolation theory over curved elements, with applications to finite element methods. Comp. Methods Appl. Mech. Engrg. **1** (1972) 217-249

Clément, P. (1975) Approximation by finite element functions using local regularization. RAIRO Anal. Numér. **9** (1975) 77–84

Crouzeix, M., Raviart, P.-A. (1973) Conforming and nonconforming finite element methods for solving the stationary Stokes equations. RAIRO Anal. Numér. **7** (1973) 33-75

Dauge, M. (1988) Elliptic Boundary Value Problems on Corner Domains. Lecture Notes in Math. v. 1341, Springer-Verlag, Berlin, 1988

Dechevski, L.T., Quak, E. (1990) On the Bramble-Hilbert lemma. Numer. Funct. Anal. Optimiz. **11** (1990) 485-495

DeVore, R.A., Howard, R., Micchelli, C. (1989) Optimal nonlinear approximation. Manuscripta Math. 63 (1989), 469–478

DeVore, R., Scherer, K. (1979) Interpolation of linear operators on Sobolev spaces. Annals Math. **109** (1979) 583-599

DeVore, R., Sharpley, R.C. (1993) Besov spaces on domains in $\mathbb{R}^d$. Trans. AMS. **335** (1993) 843-864

Dörfler, W., Nochetto, R.H. (2001) Small data oscillation implies the saturation assumption. Numer. Math. (2001) DOI 10.10007/s002110100321 (to appear in print)

Douglas, J., Dupont, T. (1975) A Galerkin method for a nonlinear Dirichlet problem. Math. Comp. **29** (1975) 689-696

Draganescu, A., Dupont, T., Scott, L.R. (2002) On the positivity of the discrete Green's function (to appear)

Dryja, M., Smith, B.F., Widlund, O.B. (1994) Schwarz analysis of iterative substructuring algorithms for elliptic problems in three dimensions. SIAM J. Numer. Anal. **31** (1994) 1662-1694

Dryja, M., Widlund, O.B. (1987) An additive variant of the Schwarz alternating method in the case of many subregions. Technical Report 339, Department of Computer Science, Courant Institute, 1987

Dryja, M., Widlund, O.B. (1989) Some domain decomposition algorithms for elliptic problems. *Iterative Methods for Large Linear Systems*, L. Hayes and D. Kincaid, eds., Academic Press, San Diego, CA, 1989, 273-291

Dryja, M., Widlund, O.B. (1990) Towards a unified theory of domain decomposition algorithms for elliptic problems. *Proceedings of the Third International Symposium on Domain Decomposition Methods for Partial Differential Equations*, T. Chan, R. Glowinski, J. Périaux and O.B. Widlund, eds., SIAM, Philadelphia, 1990, 3-21

Dryja, M., Widlund, O.B. (1992) Additive Schwarz methods for elliptic finite element problems in three dimensions. *Proceedings of the Fifth International Symposium on Domain Decomposition Methods for Partial Differential Equations*, D.E. Keyes, T.F. Chan, G. Meurant, J.S. Scroggs and R.G. Voigt, eds., SIAM, Philadelphia, 1992, 3-18

Dryja, M., Widlund, O.B. (1994) Domain decomposition algorithms with small overlap. SIAM J. Sci. Comput. **15** (1994) 604-620

Dryja, M., Widlund, O.B. (1995) Schwarz methods of Neumann-Neumann type for three dimensional elliptic finite element problems. Comm. Pure Appl. Math. **48** (1995) 121-155

Dugundji, J. (1966) *Topology*. Allyn and Bacon, Boston, 1966

Dupont, T., Scott, R. (1980) Polynomial approximation of functions in Sobolev spaces. Math. Comp. **34** (1980) 441-463

Duvaut, G., Lions, J.L. (1972) *Les Inéquations en Mécanique et en Physique*. Dunod, Paris, 1972

Eriksson, L., Estep, D., Hansbo, P. Johnson, C. (1995) Introduction to adaptive methods for differential equations. Acta Numerica **4** (1995) pp. 105-158

Fairweather, G. (1978) *Finite Element Galerkin Methods for Differential Equations*. M. Dekker, New York, 1978

Falk, R.S. (1991) Nonconforming finite element methods for the equations of linear elasticity. Math. Comp. **57** (1991) 529-550

Fortin, M., Glowinski, R. (1983) *Augmented Lagrangian Methods: Applications to the Numerical Solution of Boundary-value Problems*. North-Holland, Amsterdam, New York, Oxford, 1983

Fried, I. (1979) *Numerical Solution of Differential Equations*. Academic Press, New York, 1979

Friedman, A. (1976) *Partial Differential Equations*. R. E. Krieger Pub. Co., Huntington, NY (and now Melbourne, FL), 1976

Gilbarg, D., Trudinger, N.S. (1983) *Elliptic Partial Differential Equations of Second Order*. 2nd ed. Springer-Verlag, Berlin, 1983

Girault, V., Raviart, P.-A. (1979) *Finite Element Approximation of the Navier-Stokes Equations*. Lecture Notes in Math. v. 749, Springer-Verlag, Berlin, 1979

Girault, V., Raviart, P.-A. (1986) Finite Element *Methods for Navier-Stokes Equations, Theory and Algorithms*. Springer-Verlag , Berlin, New York, 1986

Girault, V., Scott, L.R. (2002) Hermite interpolation of non-smooth functions preserving boundary conditions. (to appear in Math. Comp.)

Glowinski, R. (1984) *Numerical Methods for Nonlinear Variational Problems*. Springer-Verlag, New York, 1984

Glowinski, R., Pironneau, O. (1992) Finite element methods for Navier-Stokes equations. Annu. Rev. Fluid Mech. **24** (1992) 167-204

Golub, G.H., Van Loan, C.F. (1989) *Matrix Computations*. The Johns Hopkins University Press, Baltimore, 1989

Griebel, M., Oswald, P. (1995) On the abstract theory of additive and multiplicative Schwarz algorithms. Numer. Math. **70** (1995) 163-180

Grisvard, P. (1985) *Elliptic Problems in Nonsmooth Domains*. Pitman Advanced Publishing Program, Boston, 1985

Grisvard, P. (1986) Problémes aux limites dans les polygones. Mode d'emploi. Bulletin de la Direction des Etudes et Recherches Series C Mathematiques, Informatique **1** (1986) 21-59

Grisvard, P. (1989) Singularités en Elasticité. Arch. Rat. Mech. Anal. **107** (1989) 157-180

Hackbusch, W. (1985) *Multi-grid Methods and Applications*. Springer-Verlag, Berlin, 1985

Halmos, P.R. (1957) *Introduction to Hilbert Space and the Theory of Spectral Multiplicity*. 2nd ed. Chelsea Pub. Co., New York, 1957

Halmos, P.R. (1991) *Measure Theory*. Springer-Verlag, New York, 1991

Hartman, S., Mikusinski, J. (1961) *The Theory of Lebesgue Measure and Integration*. Pergamon Press, New York, 1961

Haverkamp, R. (1984) Eine Aussage zur L_∞-Stabilität und zur genauen Konvergenzordnung der H_0^1-Projektionen. Numer. Math. **44** (1984) 393-405

Hughes, T.J.R. (2000) *The Finite Element Method*. Dover, New York, 2000

Isaacson, E., Keller, H.B. (1966) *Analysis of Numerical Methods*. John Wiley and Sons, New York, 1966

Jensen, S., Vogelius, M. (1990) Divergence stability in connection with the *p*-version of the finite element method. RAIRO Math. Modeling & Numer. Anal. **24** (1990) 737-764

Johnson, C. (1987) *Numerical Solutions of Partial Differential Equations by the Finite Element Method*. Cambridge University Press, Cambridge, 1987

Kreiss, H.-O., Manteuffel, T.A., Swartz, B., Wendroff, B., White, A.B., Jr. (1986) Supra-convergent schemes on irregular grids. Math. Comp.**47** (1986) 537-554

Lenoir, M. (1986) Optimal isoparametric finite elements and error estimates for domains involving curved boundaries. SIAM J. Numer. Anal. **23** (1986) 562-580

LeTallec, P. (1994) Domain decomposition methods in computational mechanics. Computational Mechanics Advances **1** (1994) 121-220

LeTallec, P., Mandel, J., Vidrascu, M. (1998) A Neumann-Neumann domain decomposition algorithm for solving plate and shell problems. SIAM J. Numer. Anal. **35** (1998) 836-867

Liao, X., Nochetto, R.H. (2002) Local a posteriori error estimates and adaptive control of pollution effects, preprint

Lions, J.L., Magenes, E. (1972) *Non-homogeneous Boundary Value Problems and Applications*. Springer-Verlag, New York, 1972

Mandel, J. (1993) Balancing Domain Decomposition. Comm. Numer. Methods Engrg. **9** (1993) 233-241

Mandel, J., Brezina, M. (1996) Balancing domain decomposition for problems with large jumps in coefficients. Math. Comp. **65** (1996) 1387-1401

McCormick, S.F., ed. (1987) Multigrid Methods. SIAM Frontiers in Applied Mathematics 3. Society for Industrial and Applied Mathematics, Philadelphia, 1987

Meyers, N.G. (1963) An L^p-estimate for the gradient of solutions of second order elliptic divergence equations. Annali della Scuola Normale Superiore di Pisa. Ser. III. **XVII** (1963) 189-206

Meyers, N.G., Serrin, J. (1964) H = W. Proc. Nat. Acad. Sci. USA **51** (1964) 1055-1056

Morgan, J., Scott, R. (1975) A nodal basis for C^1 piecewise polynomials of degree $n \geq 5$. Math. Comp. **29** (1975) 736-740

Nečas, J. (1967) *Les Méthodes Directes en Théorie des Équations Elliptiques*. Masson, Paris, 1967

Nepomnyaschikh, S.V. (1989) On the application of the bordering method to the mixed boundary value problem for elliptic equations and on mesh norms in $W_2^{1/2}(S)$. Sov. J. Numer. Anal. Math. Modelling **4** (1989) 493-506

Nitsche, J. (1981) On Korn's second inequality. RAIRO Anal. Numér. **15** (1981) 237-248

Nitsche, J., Schatz, A. (1974) Interior estimates for Ritz-Galerkin methods. Math. Comp. **28** (1974) 937-958

Nochetto, R.H. (1995) Pointwise a posteriori error estimates for elliptic problems on highly graded meshes. Math. Comp. **64** (1995) 1–22

Oden, J.T., Reddy, J.N. (1976) *An Introduction to the Mathematical Theory of Finite Elements*. Wiley-Interscience, New York, 1976

Oswald, P. (1994) *Multilevel Finite Element Approximation: Theory and Applications*. B.G. Teubner, Stuttgart, 1994

Quarteroni, A., Valli, A. (1999) *Domain Decomposition Methods for Partial Differential Equations*. Clarendon Press, Oxford, 1999

Rannacher, R., Scott, R. (1982) Some optimal error estimates for piecewise linear finite element approximations. Math. Comp. **38** (1982) 437-445

Rauch, J. (1991) *Partial Differential Equations*. Springer-Verlag, New York, 1991

Royden, H. L. (1988) *Real Analysis*. 3rd ed. Macmillan, New York, 1988

Rudin, W. (1987) *Real and Complex Analysis*. 3rd ed. McGraw-Hill, New York, 1987

Rudin, W. (1991) *Functional Analysis*. 2nd ed. McGraw-Hill, 1991

Saad, Y. (1996) *Iterative Methods for Sparse Linear Systems*. PWS Publishing Company, Boston, 1996

Saavedra, P., Scott, L.R. (1991) A variational formulation of free-boundary problems. Math. Comp. **57** (1991) 451-475

Schwarb, Ch. (1998) *p- and hp- Finite Element Methods*. Clarendon Press, 1998

Schwartz, L. (1957) *Theorie des Distributions*. 2nd ed. Hermann, Paris, 1957

Schultz, M.H. (1972) *Spline Analysis*. Prentice-Hall, Englewood Cliffs, NJ, 1972

Scott, R. (1975) Interpolated boundary conditions in the finite element method. SIAM J. Numer. Anal. **12** (1975) 404-427

Scott, R. (1976) Optimal L^∞ estimates for the finite element method on irregular meshes. Math. Comp. **30** (1976) 681-697

Scott, R. (1979) *Applications of Banach space interpolation to finite element theory. Functional Analysis Methods in Numerical Analysis*, M. Z. Nashed, ed. Lecture Notes in Math. v. 701, Springer-Verlag, Berlin, 1979, 298-313

Scott, L.R., Ilin, A., Metcalfe, R.W., Bagheri, B. Fast algorithms for solvin high-order finite element equations for incompressible flow. ICOSAHOM'95 (Proceedings of the Third International Conference on Spectral and High Order Methods, Houston, TX, 1995) (1996) pp. 221-232

Scott, L.R., Vogelius, M. (1985a) Norm estimates for a maximal right inverse of the divergence operator in spaces of piecewise polynomials. M^2AN **19** (1985) 111-143

Scott, L.R., Vogelius, M. (1985b) Conforming finite element methods for incompressible and nearly incompressible continua. Lectures in Applied Math. **22** (Part 2) (1985) B. E. Engquist, *et al.*, eds., AMS, Providence 221-244

Scott, L.R., Zhang, S. (1990) Finite element interpolation of non-smooth functions satisfying boundary conditions. Math. Comp. **54** (1990) 483-493

Scott, L.R., Zhang, S. (1992) Higher dimensional non-nested multigrid methods. Math. Comp. **58** (1992) 457-466

Shi, Z. (1990) Error estimates of Morley element. Chinese J. Num. Math. & Appl. **12** (1990) 102-108

Showalter, R.E. (1977) *Hilbert Space Methods for Partial Differential Equations*. Pitman, London, San Francisco, 1977

Simader, C. G. (1972) *On Dirichlet's Boundary Value Problem*. Lecture Notes in Math. v. 268, Springer-Verlag, Berlin, 1972

Smith, B.F. (1991) A domain decomposition algorithm for elliptic problems in three dimensions. Numer. Math. **60** (1991) 219-234

Smith, B., Bjørstad, P., Gropp, W. (1996) *Domain Decomposition*, Cambridge University Press, Cambridge, 1996.

Sobolev, S. L. (1963) *Applications of Functional Analysis in Mathematical Physics.* American Mathematical Society, Providence, 1963 (reprinted 1987)

Sobolev, S. L. (1991) *Some Applications of Functional Analysis in Mathematical Physics.* 3rd ed. American Mathematical Society, Providence, 1991

Sobolev, S. L. (1964) *Partial Differential Equations of Mathematical Physics.* Pergamon Press, Oxford, New York, 1964

Spijker, M.N. (1971) On the structure of error estimates for finite difference methods. Numer. Math. **18** (1971) 73-100

Stein, E.M. (1970) *Singular Integrals and Differentiability Properties of Functions.* Princeton University Press, Princeton, 1970

Stein, E.M., Weiss, G. (1971) *Introduction to Fourier Analysis on Euclidean Spaces.* Princeton University Press, Princeton, 1971

Stenberg, R. (1988) A family of mixed finite elements for the elasticity problem. Numer. Math. **53** (1988) 513-538

Strang, G., Fix, G.J. (1973) *An Analysis of the Finite Element Method.* Prentice-Hall, Englewood Cliffs, NJ, 1973 (now published by Wellesley-Cambridge Press, Wellesley, MA; Cambridge, MA)

Szabó, B., Babuška, I. (1991) *Finite Element Analysis.* John Wiley and Sons, New York, 1991

Taylor, M.E. (1996) *Partial Differential Equations. Basic Theory.* Texts in Applied Mathematics 23, Springer-Verlag, New York, 1996

Temam, R. (1984) *Navier-Stokes Equations: Theory and Numerical Analysis.* 3rd revised ed. North-Holland, Amsterdam, New York, Oxford, 1984

Thomée, V. (1984) *Galerkin Finite Element Methods for Parabolic Problems.* Lecture Notes in Math. v. 1054, Springer-Verlag, Berlin, 1984

Thomasset, F. (1981) *Implementation of Finite Element Methods for Navier-Stokes Equations.* Springer-Verlag, Berline, New York, Heidelberg, 1981

Trèves, F. (1967) *Topological Vector Spaces, Distributions and Kernels.* Academic Press, New York, 1967

Valent, T. (1988) *Boundary Value Problems of Finite Elasticity.* Springer-Verlag, New York 1988

Verfürth, R. (1996) *A Review of A Posteriori Error Estimation and Adaptive Mesh-Refinement Techniques.* Wiley-Teubner, Chichester, 1996

Vogelius, M. (1983) An analysis of the p-version of the finite element method for nearly incompressible materials. Uniformly valid, optimal error estimates. Numer. Math. **41** (1983) 39-53

Wachspress, E.L. (1975) *A Rational Finite Element Basis.* Academic Press, New York, 1975

Wendland, W. L. (1979) *Elliptic Systems in the Plane.* Pitman, London, San Francisco, 1979

Widlund, O.B. (1986) An extension theorem for finite element spaces with three applications. *Numerical Techniques in Continuum Mechanics* (Notes on Numerical Fluid Mechanics V. 16), W. Hackbusch and K. Witsch, eds., Friedr. Vieweg und Sohn, 1986, 110-122

Widlund, O.B. (1999) Domain decomposition methods for elliptic partial differential equations. NATO Sci. Ser. C Math. Phys. Sci. 536, Kluwer Acad. Publ., Dordrecht, 1999, 325-354

Wloka, J. (1987) *Partial Differential Equations.* Cambridge University Press, Cambridge, 1987

Xu, J. (1992) Iterative methods by space decomposition and subspace correction. SIAM Review **34** (1992) 581-613

Xu, J., Zou, J. (1998) Some nonoverlapping domain decomposition methods (1998) SIAM Review **40** 857-914

Yserentant, H. (1986) On the multi-level splitting of finite element spaces. Numer. Math. **49** (1986) 379-412

Zhang, S. (1995) Successive subdivisions of tetrahedra and multigrid methods on tetrahedral meshes. Houston J. Math. **21** (1995) 541-555

Zhang, X. (1991) Studies in Domain Decomposition: Multilevel Methods and the
    Biharmonic Dirichlet Problem. Dissertation, Courant Institute, 1991
Zhang, X. (1992) Multilevel Schwarz methods. Numer. Math. **63** (1992) 521-539

Index

We adopt the conventions used in this book for the references. In other words, for integers i, j and k,

"$i.j$" refers to a section,

"$i.\text{x}.j$" refers to an exercise, and

"$(i.j.k)$" refers to a display number, Definition, Theorem, Remark, etc. If a particular item has no numbered reference, we give the nearest prior reference or the section number.

Printed in the United Kingdom
by Lightning Source UK Ltd.
121014UK00006B/142

9 780387 954516

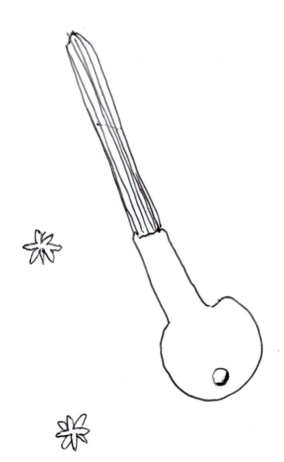